The IMA Volumes
in Mathematics
and its Applications

Volume 136

Series Editors
Douglas N. Arnold   Fadil Santosa

Springer Science+Business Media, LLC

Institute for Mathematics and its Applications (IMA)

The **Institute for Mathematics and its Applications** was established by a grant from the National Science Foundation to the University of Minnesota in 1982. The primary mission of the IMA is to foster research of a truly interdisciplinary nature, establishing links between mathematics of the highest caliber and important scientific and technological problems from other disciplines and industry. To this end, the IMA organizes a wide variety of programs, ranging from short intense workshops in areas of exceptional interest and opportunity to extensive thematic programs lasting a year. IMA Volumes are used to communicate results of these programs that we believe are of particular value to the broader scientific community.

The full list of IMA books can be found at the Web site of the Institute for Mathematics and its Applications:

http://www.ima.umn.edu/springer/full-list-volumes.html.

Douglas N. Arnold, Director of the IMA

* * * * * * * * * *

IMA ANNUAL PROGRAMS

1982–1983	Statistical and Continuum Approaches to Phase Transition
1983–1984	Mathematical Models for the Economics of Decentralized Resource Allocation
1984–1985	Continuum Physics and Partial Differential Equations
1985–1986	Stochastic Differential Equations and Their Applications
1986–1987	Scientific Computation
1987–1988	Applied Combinatorics
1988–1989	Nonlinear Waves
1989–1990	Dynamical Systems and Their Applications
1990–1991	Phase Transitions and Free Boundaries
1991–1992	Applied Linear Algebra
1992–1993	Control Theory and its Applications
1993–1994	Emerging Applications of Probability
1994–1995	Waves and Scattering
1995–1996	Mathematical Methods in Material Science
1996–1997	Mathematics of High Performance Computing
1997–1998	Emerging Applications of Dynamical Systems
1998–1999	Mathematics in Biology

Continued at the back

Naoufel Ben Abdallah   Anton Arnold
Pierre Degond   Irene M. Gamba
Robert T. Glassey   C. David Levermore
Christian Ringhofer
Editors

Dispersive Transport Equations and Multiscale Models

With 113 Illustrations

Springer

Naoufel Ben Abdallah
Laboratoire MIP
Université Paul Sabatier
Toulouse Cedex 4, 31062
France
nauofel@mip.ups-tlse.fr

Anton Arnold
Angewandte Mathematik
Universität des Saarlandes
Saarbrucken, D-66041
Germany
Arnold@num.uni-sb.de

Pierre Degond
Laboratoire MIP
Université Paul Sabatier
Toulouse Cedex 4, 31062
France
degond@mip.ups-tlse.fr

Irene M. Gamba
Department of Mathematics
University of Texas at Austin
Austin, TX 78712
USA
gamba@math.utexas.edu

Robert T. Glassey
Department of Mathematics
Indiana University
Bloomington, IN 47405-5701
USA
glassey@indiana.edu

C. David Levermore
CSCAMM
University of Maryland
College Park, MD 20742-4015
USA
lvrmr@math.umd.edu

Christian Ringhofer
Department of Mathematics
Arizona State University
Tempe, AZ 85287
USA
ringhofer@asu.edu

Series Editors:
Douglas N. Arnold
Fadil Santosa
Institute for Mathematics and
    its Applications
University of Minnesota
Minneapolis, MN 55455
USA
http://www.ima.umn.edu

Mathematics Subject Classification (2000): 35Qxx, 65Mxx, 65Nxx, 65Z05, 76-xx, 78-xx, 80-xx, 81-xx, 82-xx, 85-xx

Library of Congress Cataloging-in-Publication Data
Dispersive transport equations and multiscale models / Naoufel Abdallah . . . [et al.]
      p. cm. — (The IMA volumes in mathematics and its applications ; v. 136)
      ISBN 0-387-40496-1
      1. Transport theory—Congresses. 2. Semiconductors—Mathematical models—Congresses.
   I. Ben-Abdallah, Naoufel. II. Series.
   QC175.25.A1D57 2003

Printed on acid-free paper.   2003054315

ISBN 978-1-4612-6473-6   ISBN 978-1-4419-8935-2 (eBook)
DOI 10.1007/978-1-4419-8935-2

9 8 7 6 5 4 3 2 1   SPIN 10939246

Camera-ready copy provided by the IMA.

www.springer-ny.com

FOREWORD

This IMA Volume in Mathematics and its Applications

DISPERSIVE TRANSPORT EQUATIONS
AND MULTISCALE MODELS

along with the accompanying volume, "Transport in Transition Regimes" which will be published as IMA Volume 135 contains papers presented at three one-week workshops. The first workshop "Dispersive Corrections to Transport Equations" which took place on May 1–5, 2000 was organized by Anton Arnold (Universitaet Muenster), Naoufel Ben Abdallah (Université Paul Sabatier), C. David Levermore (University of Maryland), and Ken T.-R. McLaughlin (University of Arizona). The second workshop "Simulation of Transport in Transition Regimes" was held on May 22–26, 2000. The organizers were Pierre Degond (Université Paul Sabatier), Irene M. Gamba (University of Texas at Austin), and Philip Roe (University of Michigan). Leonard J. Borucki (Motorola, Inc.) and Christian Ringhofer (Arizona State University) were the organizers of the third workshop "Multiscale Models for Surface Evolution and Reacting Flows" which took place on June 5–9, 2000. The three workshops were integral parts of the 1999-2000 IMA program on "REACTIVE FLOW AND TRANSPORT PHENOM-ENA."

We would like to thank the organizers and all the participants for making the events successful. We also appreciate the organizers for their vital role as editors of the two proceedings.

We take this opportunity to thank the National Science Foundation, whose financial support of the IMA made the annual program possible.

Series Editors

Douglas N. Arnold, Director of the IMA

Fadil Santosa, Deputy Director of the IMA

PREFACE

IMA Volumes 135: Transport in Transition Regimes and 136: Dispersive Transport Equations and Multiscale Models are the compilation of papers presented in 3 related workshops held at the IMA in the spring of 2000. The focus of the program was the modeling of processes for which transport is one of the most complicated components. This includes processes that involve a wide range of length scales over different spatio-temporal regions of the problem, ranging from the order of mean-free paths to many times this scale. Consequently, effective modeling techniques require different transport models in each region.

In some cases the underlying kinetic description is relatively well understood, such as is the case for the Boltzmann equation for rarified gases, or the transport equation for radiation. In such cases the main issue is one of economy, a fully resolved kinetic simulation being impractical. One therefore develops homogenization, stochastic, or moment based subgrid models. This was the focus of two of the workshops: "Model Hierarchies for the Evolution of Surfaces under Chemically Reacting Flows" and "Transport Phenomena in Transition Regimes."

In other cases there is considerable disagreement about the underlying kinetic description, especially when dispersive effects become macroscopic, for example due to quantum effects in semiconductors and superfluids. These disagreements are the focus of the workshop: "Dispersive Corrections to Transport Equations."

Workshop on "Dispersive Corrections to Transport Equations," May 1–5, 2000 (Organized by D. Levermore, A. Arnold, N. Ben Abdallah, K. McLaughlin)

Dispersive corrections to classical and semiclassical transport equations arise from the rudimentary incorporation of quantum effects into macroscopic flow descriptions. These models play an increasing role in the study of nanometer scale electronic devices and of fluids at extremely low temperatures. One of the advantages of dispersively corrected transport equations is that they allow for a more classical coupling of the quantum system to the environment than the fully quantum mechanical descriptions. The main topics of this workshop were, on one hand, the mathematical derivation of dispersive correction terms, and, on the other hand, the computational issues raised by the interplay between nonlinear and dispersive effects in quatum dots and wires, superfluids and dispersive phenomena in nonlinear optics.

Workshop on "Simulation of Transport in Transition Regimes," May 22–26. 2000 (Organized by P. Degond, I. Gamba, P. Roe, R. Glassey)

Technology is increasingly advancing into regimes in which particle mean-free paths are comparable to the length scales of interest, whereby traditional transport models breakdown. For example, drift-diffusion models of electron-hole transport break down for submicron semiconductors, while Navier-Stokes approximations of fluid dynamics break down in outer planetary atmospheres or hypersonic flight. The cost of particle simulations is usually much larger than that of fluid simulations. This makes the simulation of problems in which transition regimes coexist with fluid regimes particularly difficult. This difficulty is compounded when the geometry of the problem is complex or even random. This workshop explored advanced moment based models, both deterministic and stochastic in origin, in the context of the simulation of high-altitude flight, charged particles in outer planetary atmospheres, electron and holes in submicon semiconductor devices, and radiation through inhomogenous media, together with hybrid numerical schemes that properly match transition regimes.

Workshop on "Multiscale Models for Surface Evolution and Reacting Flows," June 5–9, 2000 (Organized by L. Borucki and C. Ringhofer)

Multilayered compound materials with microscopically structured surfaces play a key role in semiconductor manufacturing. These structures are produced by a variety of processes, such as the deposition of thin films, etching techniques and controlled crystal growth. The topic of this workshop was the integration of different models describing these processes on different spatial and temporal scales. Well-developed models exist for each stage of the above processes on the microscopic-atomistic and macroscopic-fluid scale. However, in order to describe completely the whole process, it is necessary to link these models via an appropriate mathematical description of the transition regimes. This involves a mixture of boundary layer and homogenization techniques as well as a mathematical analysis of the transition process from the atomistic description of the early stages of thin film growth to the evolution of continuous films. Computational issues covered by this workshop were the deterministic and probabilistic representation of film surfaces and numerical methods for the transitional models.

Anton Arnold (Institut fuer Numerische Mathematik, Universitaet Muenster)

Naoufel Ben Abdallah (Laboratoire MIP, Universit Paul Sabatier)

Pierre Degond (Mathématiques pour l'Industrie et la Physique, CNRS, Universite Paul Sabatier)

Irene Gamba (Department of Mathematics, University of Texas at Austin)

Robert Glassey (Department of Mathematics, Indiana University)

C. David Levermore (Applied Mathematical and Scientific Computation Program, University of Maryland)

Christian Ringhofer (Department of Mathematics, Arizona State University)

CONTENTS

ON THE DERIVATION OF NONLINEAR SCHRÖDINGER AND VLASOV EQUATIONS

CLAUDE BARDOS[*], FRANÇOIS GOLSE[†], ALEX GOTTLIEB[‡], AND NORBERT J. MAUSER[§]

Abstract. We present and discuss derivations of nonlinear 1-particle equations from linear N-particle Schrödinger equations with pair interaction in the time dependent case.

We regard both the "classical" limit of vanishing Planck constant $\hbar \to 0$ which leads to Vlasov type equations and the "weak coupling" limit $1/N \to 0$ which leads to nonlinear 1 particle equations.

We use an approach to weak coupling limits where the so-called "finite Schrödinger hierarchy" and the limiting "(infinite) Schrödinger hierarchy" play a central role. Convergence of solutions of the first to solutions of the second is established using "physically relevant" estimates (L^2 and energy conservation) under very general assumptions on the interaction potential, including in particular the Coulomb potential.

The goal of this work is to give an overview of the existing results, including some minor improvements, and clearly state the open problems.

1. Introduction. In this work we give a survey of the derivation of nonlinear 1-particle Schrödinger and Vlasov equations starting from the linear N-particle Schrödinger equation. We regard both the "classical" limit of vanishing Planck constant $\hbar \to 0$ which leads to Vlasov type equations and the "weak-coupling" limit $1/N \to 0$ which leads to nonlinear 1-particle equations. The relevant particle systems and limits are illustrated in the following (presumably commutative) diagram:

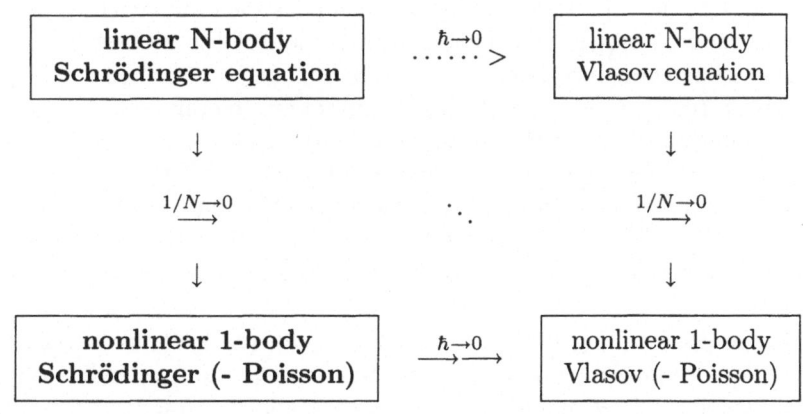

[*]CMLA, ENS-Cachan and LAN (Univ. Paris 6), France (bardos@math.jussieu.fr).

[†]ENS-Ulm and LAN (Univ. Paris 6), France (Francois.Golse@ens.fr).

[‡]Inst. f. Mathematik, Univ Wien, Strudlhofg. 4, A–1090 Wien, Austria (gottlieb @math.berkeley.edu).

[§]Inst. f. Mathematik, Univ Wien, Strudlhofg. 4, A–1090 Wien, Austria (mauser @courant.nyu.edu).

The vertical arrows represent the weak-coupling limits and the horizontal arrows represent the classical limits. The diagonal limit, i.e., the classical + weak-coupling limit corresponding to letting $\hbar \to 0^+$ and $N \to \infty$ (simultaneously) shall also be considered in the present contribution.

The vertical limit from the linear N-particle Schrödinger equation to a nonlinear 1-particle Schrödinger equation has been given by Spohn in [Sp1], with recent minor improvements in [BGM1]; both contributions assuming bounded potentials.

The vertical limit from the linear N-particle Vlasov equation to a nonlinear 1-particle Vlasov equation has been given by Braun and Hepp in [BH], assuming bounded and regular potentials.

The diagonal limit from the linear N-particle Schrödinger equation to a nonlinear 1-particle Vlasov equation is given by Narnhofer and Sewell in [NS] for the case of a bounded, real analytic interaction potential.

The lower horizontal limit has been given by Lions and Paul in [LP], and Markowich and Mauser in [MM] thus deriving the Vlasov-Poisson from the Schrödinger-Poisson system.

Validating the upper horizontal limit, from the linear N-particle Schrödinger equation to the linear N-particle Vlasov equation, is still an open problem for the case of Coulomb interaction.

The three succeeding sections of this article investigate the three non-horizontal limits just mentioned. Our ultimate goal is to prove the existence and uniqueness of solutions to an *infinite hierarchy* of equations associated to the relevant N-particle dynamics and scaling. Existence and uniqueness of solutions to the infinite hierarchy easily implies the limits with which we are concerned. Uniqueness can be shown to hold if the 2-body potential satisfies strong enough conditions, like boundedness, but existence is easier to establish (cp. [BGM1]. Solutions to the infinite hierarchy are obtained as accumulation points of the finite-particle dynamics, in a suitable topology. This convergence does not require boundedness of the interparticle potential. Since we have in mind the case of the Coulomb potential, which leads to the Schrödinger-Poisson system, we value the existence theorems announced below because their hypotheses accommodate unbounded potentials.

The weak-coupling limit of N-particle quantum systems is the tool for the derivation of a time-dependent Hartree equation, as indicated by the left vertical "$1/N$ arrow" in the diagram above. However, the same technique seems not to work for deriving a (local approximation of) time dependent Hartree-Fock equation based on Pauli's exclusion principle. As shown in Section 5, the weak limit of the density matrix is vanishing in this "fermion case".

How to derive a nonlinear 1-body time-dependent Schrödinger equation for fermions in the infinite-particle limit (but not *classical* limit) is not known. The following diagram outlines an approach to this problem in the stationary case which is outside the scope of this article. For the stationary case indeed rigorous results are available (see e.g. [BM] and references therein), for the time dependent case a heuristic model is given in [M6].

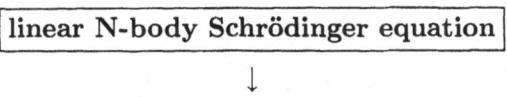

$$\downarrow$$

Hartree Fock ansatz
minimization of total energy

$$\downarrow$$

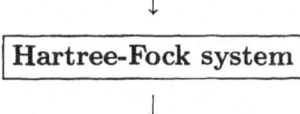

$$\downarrow$$

local approximation of exchange term
$$N \to \infty$$

$$\downarrow$$

| Schrödinger - Poisson-$X\alpha$ equation |

2. The weak coupling limit of the linear N Schrödinger equation. The starting point is the Schrödinger equation for the wave function $\Psi_N = \Psi_N(x_1, x_2, ..., x_N, t)$ of N interacting particles, which reads

$$(1) \quad i\hbar\partial_t\Psi_N = -\frac{\hbar^2}{2}\sum_{1\leq j\leq N}\Delta_{x_j}\Psi_N + \frac{1}{N}\sum_{1\leq j<k\leq N}V(|x_j - x_k|)\Psi_N =: H_N\Psi_N$$

$$(2) \quad \Psi_N(t = 0) = \Psi_N^I(x_1, x_2, ..., x_N).$$

The factor $1/N$ in front of the potential V is the standard weak-coupling scaling, as discussed e.g. by Spohn in [Sp2]. It corresponds to assuming that collective effects of order 1 can be observed over a unit length of the macroscopic time scale.

The potential V is assumed to be real-valued and bounded from below, but no assumption is made as to its sign. In other words, attractive as well as repulsive interactions are amenable to the methods presented in this paper.

The following notations will be used constantly in the sequel

$$(3) \quad \begin{aligned} X_N &:= (x_1, x_2, ..., x_N), & X_n &:= (x_1, x_2, ..., x_n), & X_N^n &:= (x_{n+1}, ..., x_N) \\ Y_N &:= (y_1, y_2, ..., y_N), & Y_n &:= (y_1, y_2, ..., y_n), & Y_N^n &:= (y_{n+1}, ..., y_N) \\ Z_N &:= (z_1, z_2, ..., z_N), & Z_n &:= (z_1, z_2, ..., z_n), & Z_N^n &:= (z_{n+1}, ..., z_N) \\ V_N &:= (v_1, v_2, ..., v_N), & V^n &:= (v_1, v_2, ..., v_n), & V_N^n &:= (v_{n+1}, ..., v_N). \end{aligned}$$

The state of the N-particle system can also be described (see for example [LL3]) by the density operator $\rho_N(t)$ acting on $L^2(\mathbb{R}^3)^N$ or equivalently by its integral kernel, known as the density matrix $\rho_N(X_N, Y_N, t)$. For a general "mixed state" we have

$$(4) \qquad \rho_N(X_N, Y_N, t) = \sum_{k \in \mathbb{N}} \lambda_k \Psi_{N,k}(X_N, t) \overline{\Psi_{N,k}(Y_N, t)},$$

where $\lambda_k > 0$ are the "occupation probabilities" satisfying $\sum_k \lambda_k = 1$. However, the N-particle Schrödinger equation is linear, so that we can assume without loss of generality that the density matrix is that of a "pure state":

$$(5) \qquad \begin{aligned} \rho_N(X_N, Y_N, t) &= \rho_N(x_1, x_2, ..., x_N, y_1, y_2, ..., y_N, t) \\ &= \Psi_N(X_N, t) \overline{\Psi_N(Y_N, t)}. \end{aligned}$$

This density matrix $\rho_N(X_N, Y_N, t)$ is the integral kernel of the density operator $\rho_N(t)$, the time evolution of which is given by

$$(6) \qquad \rho_N(t) = e^{-i \frac{t H_N}{\hbar}} \rho_N(0) e^{i \frac{t H_N}{\hbar}}$$

i.e. the density operator $\rho_N(t)$ satisfies the "von Neumann equation":

$$(7) \qquad i\hbar \partial_t \rho_N = H_N \rho_N - \rho_N H_N.$$

Equivalently, the density matrix must satisfy

$$(8) \qquad \begin{aligned} i\hbar \partial_t \rho_N(X_N, Y_N, t) = &-\frac{\hbar^2}{2}[\Delta_{X_N} - \Delta_{Y_N}]\rho_N(X_N, Y_N, t) \\ &+\frac{1}{N} \sum_{1 \le j < k \le N} [V(|x_j - x_k|)dis - V(|y_j - y_k|)]\rho_N(X_N, Y_N, t). \end{aligned}$$

The operator ρ_N is of trace class, its trace being given by:

$$(9) \qquad \begin{aligned} \mathrm{Tr}\rho_N(t) &= \int \rho_N(X_N, X_N, t) dX_N = \int |\Psi_N(X_N, t)|^2 dX_N \\ &= \int |\Psi_N^I(X_N)|^2 dX_N = 1 \end{aligned}$$

after normalization.

The "marginal distributions" or "partial traces" are introduced according to the formula:

$$(10) \qquad \rho_{N,n}(t) := \mathrm{Tr}_{[n+1,N]}\rho_N(t) = \int \rho_N(X_n, Z_N^n, Y_n, Z_N^n, t) dZ_N^n.$$

We further assume that the initial data satisfy the relation

$$(11) \qquad \begin{aligned} \rho_N(x_1, x_2, ..., x_n, y_1, y_2, ..., y_n, 0) = \\ \rho_N(x_{\sigma(1)}, x_{\sigma(2)}, ..., x_{\sigma(n)}, y_{\sigma(1)}, y_{\sigma(2)}, ..., y_{\sigma(n)}, 0) \end{aligned}$$

for any permutation σ of the set $\{1, 2, 3, ..., N\}$. This encodes the fact that we are considering the statistics of undistinguishable particles. This property is preserved by the time evolution of the von Neumann equation, so that (11) implies that

$$
(12) \quad
\begin{aligned}
&\rho_N(x_1, x_2, ..., x_n, y_1, y_2, ..., y_n, t) = \\
&\qquad \rho_N(x_{\sigma(1)}, x_{\sigma(2)}, ..., x_{\sigma(n)}, y_{\sigma(1)}, y_{\sigma(2)}, ..., y_{\sigma(n)}, t)
\end{aligned}
$$

holds for all $t \in \mathbb{R}$.

Assuming that the initial N-particle distribution satisfies (11), we obtain from a rather straightforward computation that the marginal distributions $\rho_{N,n}(t)$ solve the system

$$
(13) \quad
\begin{aligned}
i\hbar \partial_t \rho_{N,n}(X_n, Y_n, t) = &-\frac{\hbar^2}{2}[\Delta_{X_n} - \Delta_{Y_n}]\rho_{N,n}(X_n, Y_n, t) \\
&+ \frac{1}{N} \sum_{1 \le j < k \le n} [V(|x_j - x_k|) - V(|y_j - y_k|)]\rho_{N,n}(X_n, Y_n, t) \\
&+ \frac{N-n}{N} \sum_{1 \le j \le n} \int [V(|x_j - z|) - V(|y_j - z|)]\rho_{N,n+1}(X_n, z, Y_n, z, t)dz.
\end{aligned}
$$

Observe indeed that the missing term in (13) is the one corresponding to applying the partial trace $\mathrm{Tr}_{[n+1,N]}$ to the summation that appears as the last term in the right hand side of (8) restricted to the subset of indices $\{(j, k) \mid n+1 \le j, k \le N\}$. Since this restricted sum involves only terms that obviously vanish on the set $\{(X_N, Y_N) \mid X_N^n = Y_N^n\}$, applying the partial trace $\mathrm{Tr}_{[n+1,N]}$ does not contribute any additional term in (13).

The system (13) is called the "N-particle (finite) Schrödinger hierarchy". Observe in particular that for $n = N$ one recovers the equation (8) for $\rho_{N,N} = \rho_N$.

Introducing the operators $\mathcal{C}_{n,n+1}$ mapping $n + 1$-particle densities to n-particle functions formally defined by

$$
(14) \quad
\begin{aligned}
&\mathcal{C}_{n,n+1}(\rho_{N,n+1})(X_n, Y_n) = \\
&\qquad \sum_{1 \le j \le n} \int [V(|x_j - z|) - V(|y_j - z|)]\rho_{N,n+1}(X_n, z, Y_n, z)dz \,,
\end{aligned}
$$

the N-particle Schrödinger hierarchy is rewritten as:

$$
(15) \quad
\begin{aligned}
i\hbar \partial_t \rho_{N,n}(X_N, Y_N, t) = &-\frac{\hbar^2}{2}[\Delta_{X_n} - \Delta_{Y_n}]\rho_{N,n}(X_n, Y_n, t) \\
&+ \frac{1}{N} \sum_{1 \le j < k \le n} [V(|x_j - x_k|) - V(|y_j - y_k|)]\rho_{N,n}(X_n, Y_n, t) \\
&+ \frac{N-n}{N}(\mathcal{C}_{n,n+1}\rho_{N,n+1})(X_n, Y_n, t)\,, \quad \forall n = 1, \ldots, N\,,
\end{aligned}
$$

(16) $$\rho_{N,n}(X_n, Y_n, t) = 0 , \quad \forall n > N .$$

The "infinite Schrödinger hierarchy" is obtained from the N-particle (finite) Schrödinger hierarchy by letting $N \to +\infty$ while keeping $\hbar$ fixed and giving up the constraint (16). We denote by ρ_n the n-particle marginal distribution involved in the infinite Schrödinger hierarchy which of course differs from $\rho_{N,n}$, the n-th marginal distribution involved in the N-particle (finite) hierarchy. Letting formally $N \to +\infty$ in (13) leads to:

(17)
$$i\hbar\partial_t\rho_n(X_n, Y_n, t) = -\frac{\hbar^2}{2}[\Delta_{X_n} - \Delta_{Y_n}]\rho_n(X_n, Y_n, t)$$
$$+ \sum_{1 \le j \le n} \int [V(|x_j - z|) - V(|y_j - z|)]\rho_{n+1}(X_n, z, Y_n, z, t)dz .$$

A function ρ_n of the variables (X_n, Y_n) is henceforth said to be factorized if it is the n-th fold tensor power of a function $\rho \equiv \rho(x_1, y_1)$, i.e.

(18) $$\rho_n(X_n, Y_n) = \prod_{1 \le k \le n} \rho(x_k, y_k) .$$

Observe that if $\psi(x, t)$ is a solution of the (nonlinear) "self-consistent, 1-particle Schrödinger equation"

(19) $$i\hbar\partial_t\psi(x, t) = -\frac{\hbar^2}{2}\Delta_x\psi(x, t) + \psi(x, t) \int V(|x - z|)|\psi(z, t)|^2 dz$$

then

(20) $$\rho = \psi(x, t)\overline{\psi(y, t)}$$

is a solution of the "self-consistent von Neumann equation"

(21)
$$i\hbar\partial_t\rho(x, y, t) = -\frac{\hbar^2}{2}[\Delta_x - \Delta_y]\rho(x, y, t)$$
$$+\rho(x, y, t) \int [V(|x - z|) - V(|y - z|)]\rho(z, z, t)dz ,$$

while the (sequence of) factorized n-particle densities

(22) $$\rho_n(X_n, Y_n, t) = \prod_{1 \le k \le n} \rho(x_k, y_k, t)$$

is a solution of the (infinite) Schrödinger hierarchy. On the other hand, at $t = 0$ (cf (9)):

(23)
$$\rho_{N,n+1}(X_n, Y_n, 0) = \prod_{1 \le k \le n} \psi(x_k, 0)\overline{\psi(y_k, 0)} \prod_{n+1 \le k \le N} \int |\psi(z_k)|^2 dz_k$$
$$= \prod_{1 \le k \le n} \psi(x_k, 0)\overline{\psi(y_k, 0)} = \prod_{1 \le k \le n} \rho(x_k, y_k, 0).$$

As a consequence a uniqueness result for the hierarchy (Corollary 2.1 below) implies that, with initial data factorized as in (18), the solution of the hierarchy is given by

$$(24) \qquad \rho_n(x_n, y_n, t) = \prod_{1 \le k \le n} \psi(x_k, t)\overline{\psi(y_k, t)}$$

with $\psi(x_k, t)$ solution of the self-consistent Schrödinger equation (19). The factorization, assumed at $t = 0$ for the finite hierarchy, will in general get lost at later times due to the presence of the interaction potential V; however it is recovered in the limit as $N \to +\infty$.

2.1. A priori estimates for the N-particle Schrödinger hierarchy. The starting point is a variant of the Cauchy-Schwarz inequality applied to the marginal distributions. While straightforward, it provides useful estimates.

PROPOSITION 2.1. *The marginal distributions satisfy the inequalities*

$$(25) \qquad \iint |\rho_{N,n}(X_n, Y_n, t)|^2 dX_n dY_n \le 1$$

and

$$(26) \qquad \begin{aligned} |\rho_{N,n+1}(X_n, z, Y_n, z, t)| &\le \\ \rho_{N,n+1}(X_n, z, X_n, z, t)^{\frac{1}{2}} &\rho_{N,n+1}(Y_n, z, Y_n, z, t)^{\frac{1}{2}} \end{aligned}$$

for all $t \in \mathbb{R}$.

Another basic result is a $\hbar$-dependent estimate on the kinetic energy of the N-particle system.

PROPOSITION 2.2. *Assume that the interacting potential is of the form*

$$(27) \qquad \begin{aligned} V(|x|) &= V_+(|x|) + V_-(|x|) \\ with \quad V_+(|x|) &\ge 0, \, V_+ \in L^2(\mathbb{R}^3), \, V_-(|x|)) \ge -C_{\text{pot}} > -\infty. \end{aligned}$$

Assume further that the initial data $\Psi_N^I(x_1, \ldots, x_N)$ satisfies the assumption of undistinguishable particles (11) and has energy

$$(28) \qquad \begin{aligned} \mathcal{E}_{N,\hbar} &= \frac{1}{2}\hbar^2 \sum_{1 \le j \le N} \int |\nabla_{x_j} \Psi_N^I(X_N)|^2 dX_N \\ &+ \frac{1}{N} \sum_{1 \le j < k \le m} \int V(|x_j - x_k|)|\Psi_N^I(X_N)|^2 dX_N = O(N) \end{aligned}$$

as $N \to +\infty$.

Then, for any j such that $1 \le j \le n$, the solution Ψ_N of the N-particle Schrödinger equation satisfies

$$(29) \qquad \sup_{1 \le j \le N} \int |\nabla_{x_j} \Psi_N(X_N, t)|^2 dX_N \le C_{\text{pot}} \frac{N(N-1)}{N^2 \hbar^2} + 2 \frac{\mathcal{E}_{N,\hbar}}{N\hbar^2}.$$

2.2. Results. In [BGM1] two types of results were established:

1) under very general assumptions (containing in particular the physically relevant case of the Coulomb potential) the solution of the finite hierarchy is shown to converge to the solution of the infinite hierarchy;

2) for bounded potentials, the infinite hierarchy is shown to possess a single solution determined by its initial data; in particular, the limit as $N \to +\infty$ of the sequence $\rho_{N,n+1}$ is factorized and coincides with the functions given by the self-consistent, 1-particle, nonlinear Schrödinger equation.

However, the method for the proof of uniqueness does not encompass the case of the Coulomb potential, and fails to provide a derivation of the Schrödinger-Poisson equation which remains an open problem. The precise results are as follows:

THEOREM 2.1. *Assume that the potential* $x \mapsto V(|x|)$ *is bounded from below, belongs to* $C^0(\mathbb{R}^3 \setminus \{0\}) \cap L^2_{loc}(\mathbb{R}^3)$ *and vanishes at infinity:*

$$(30) \qquad \lim_{r \to +\infty} V(r) = 0.$$

Let $\Psi_N \in C^0(\mathbb{R}_+; L^2(\mathbb{R}^3))$ *be a weak solution of (1) with initial data* Ψ_N^I *satisfying the assumption of undistinguishable particles (11), the normalization (9) and the finite energy condition (28). It is assumed that* [1], *for all* $n \geq 1$,

$$\rho_{N,n}^I \equiv \rho_{N,n}^I(X_n, Y_n)$$

$$= \int \Psi_N^I(X_n, Z_N^n) \overline{\Psi_N^I}(Y_n, Z_N^n) dZ_N^n \longrightarrow \rho_n^I \equiv \rho_n^I(X_n, Y_n)$$

in $\mathcal{L}^1(L^2(\mathbb{R}^{3n}))$ *weak-* *as* $N \to +\infty$.

Let ρ_N *be defined by (6) with its marginal distributions* $\rho_{N,n}$ *defined by (10). Then, any limit point as* $N \to +\infty$ *of the family of partial traces* $(\rho_{N,n})_{n \geq 1}$ *solves the infinite Schrödinger hierarchy (17) in the sense of distributions* [2] *and satisfies the initial condition*

$$(31) \qquad \rho_{n|t=0} = \rho_n^I.$$

Limit points for the sequence $(\rho_{N,n})_{n \geq 1}$ *as* $N \to +\infty$ *are to be understood in the sense of the product topology on* $\Pi_{n \geq 1} L^\infty(\mathbb{R}_+; \mathcal{L}^1(L^2(\mathbb{R}^{3n})))$, *each factor being equipped with the weak-* *topology.*

REMARK 2.1. *The assumption on* $\mathcal{E}_{N,\hbar}$ *is satisfied if initial data is factorized:*

$$(32) \quad \Psi_N^I(X_n) = \prod_{1 \leq j \leq n} \psi(x_j), \quad \int |\psi(x)|^2 dx = 1, \quad \int |\nabla \psi(x)|^2 dx < +\infty.$$

---

[1]Since the problem is linear, the result can trivially be extended to initial data that are finite convex combinations of pure states satisfying the hypothesis.

[2]With the assumptions above, the interaction integrands $[V(|x_j - z|) - V(|y_j - z|)]\rho_{n+1}(X_n, z, Y_n, z, t)$ may fail to belong to $L^1(\mathbb{R}^3; dz)$ for each (t, X_n, Y_n); yet the integral $\int [V(|x_j - z|) - V(|y_j - z|)]\rho_{n+1}(X_n, z, Y_n, z, t) dz$ is defined as a Radon measure (distribution of order 0) in the variables X_n and Y_n.

REMARK 2.2. *The notion of limit points in the product topology described above can be given a somewhat more concrete interpretation by the diagonal extraction procedure. Let $(\rho_n)_{n\geq 1}$ be a weak-* limit point of $(\rho_{N,n})_{n\geq 1}$ as $N \to +\infty$. For each $n \geq 1$, there exists a subsequence of $\rho_{N,n}$ converging to ρ_n in $L^\infty(\mathbb{R}_+; \mathcal{L}^1(L^2(\mathbb{R}^{3n})))$ weak-*. In other words, there exists an increasing function $\phi_n : \mathbb{N}^* \to \mathbb{N}^*$ such that $\rho_{\phi_n(N),n} \to \rho_n$ in $L^\infty(\mathbb{R}_+; \mathcal{L}^1(L^2(\mathbb{R}^{3n})))$ weak-*. Define $\phi(N) = \phi_N \circ \ldots \circ \phi_1(N)$; clearly $\phi(N)$ is an increasing sequence of integers. By the construction of ϕ_n and the relation (16), one sees that $\rho_{\phi(N),n} \to \rho_n$ in $L^\infty(\mathbb{R}_+; \mathcal{L}^1(L^2(\mathbb{R}^{3n})))$ weak-* for all $n \geq 1$. In other words, the same subsequence of N leads to weak-* convergence for all $n \geq 1$.*

REMARK 2.3. *In the particular case of the Coulomb interaction, the same result holds with the trace norm replaced by the $\mathcal{L}^2(L^2(\mathbb{R}^{3n})$ norm. The proof relies on Leray's 3D variant of the Hardy inequality (formula (1.13) of [Le]), called the "uncertainty principle" in [RS2] and [LY].*

Now we make the additional assumption that $V \in L^\infty(\mathbb{R}_+)$. The infinite hierarchy can be recast in the abstract form

$$(33) \qquad \hbar\partial_t\rho_n = A_n\rho_n + C_{n,n+1}\rho_{n+1}, \qquad n \geq 1,$$

with $C_{n,n+1}$ defined in (14) and A_n denoting the skew-adjoint operator

$$(34) \qquad A_n = i\frac{\hbar^2}{2}(\Delta_{X_n} - \Delta_{Y_n}).$$

The trace norm is a "good" norm for both the operators A_n and $C_{n,n+1}$; this, together with its analogy with the L^1 norm on functions, makes it a natural tool in studying these Schrödinger hierarchies, as noticed by Spohn [Sp1].

More precisely we have

THEOREM 2.2. *Assume that $V \in L^\infty(\mathbb{R}_+)$, and denote by H_n the Banach space $\mathcal{L}^1(L^2(\mathbb{R}^{3n}))$. Let $(\rho_n)_{n\geq 1} \in \prod_{n\geq 1} L^\infty(\mathbb{R}_+; H_n)$ be a weak solution of the infinite Schrödinger hierarchy (33) with initial data $(\rho_n^I)_{n\geq 1}$ satisfying*

$$\|\rho_n^I\|_{H_n} \leq \epsilon_n, \quad n \geq 1$$

where $(\epsilon_n)_{n\geq 1}$ is a given sequence of positive numbers. In addition, assume the existence of a positive constant D such that

$$(35) \qquad \sup_{t\geq 0}\|\rho_n(t)\|_{H_n} \leq D, \quad \text{for all} \quad n \geq 1.$$

Then, for all $0 < t < \frac{\hbar}{2\|V\|_{L^\infty}}$,

$$\|\rho_n(t)\|_{H_n} \leq \sum_{m\geq 0} \binom{n+m}{m} \left(\frac{2\|V\|_{L^\infty} t}{\hbar}\right)^m \epsilon_{n+m}.$$

This stability statement, although local and valid over time intervals of the order of $\hbar$, obviously implies a global uniqueness result for the infinite Schrödinger hierarchy.

COROLLARY 2.1. *Let* $(\rho_n^1)_{n\geq 1}$ *and* $(\rho_n^2)_{n\geq 1}$ *be two solutions of the infinite Schrödinger hierarchy, obtained by the limiting procedure of the previous section and which coincide at time* $t = 0$. *Then they are equal for all* $t \in \mathbb{R}_+$.

This uniqueness result proves convergence to the self-consistent von Neumann equation as described in the introduction:

THEOREM 2.3. *Assume that* $V \in L^\infty(\mathbb{R}_+)$ *and that* $V(r) \to 0$ *as* $r \to +\infty$. *For any* $\psi^I \in H^1(\mathbb{R}^3)$ *and* $\|\psi^I\|_{L^2} = 1$, *define*

$$\rho_n^I(X_n, Y_n) = \prod_{1 \leq j \leq n} \psi^I(x_j)\overline{\psi^I(y_j)}.$$

Let $\rho_{N,n}$ *be the weak solution of the finite Schrödinger hierarchy defined by*

$$\rho_{N,n}(X_n, Y_n, t) = e^{-i\frac{t}{\hbar}H_N}\rho_n^I(X_n, Y_n)e^{+i\frac{t}{\hbar}H_N},$$

where H_N *is the* N*-body Schrödinger operator defined in (1). As* $N \to +\infty$, $\rho_{N,n} \to \rho_n$ *in* $L^\infty(\mathbb{R}_+; \mathcal{L}^1(L^2(\mathbb{R}^{3n})))$ *weak-*, with* ρ_n *given by the formula*

$$(36) \qquad \rho_n(X_n, Y_n, t) = \prod_{1 \leq j \leq n} \psi(x_j, t)\overline{\psi(y_j, t)}$$

where ψ *is the solution of the self-consistent, 1-particle Schrödinger equation*

$$(37) \qquad i\hbar\partial_t\psi(x, t) = -\frac{\hbar^2}{2}\Delta\psi(x, t) + \psi(x, t)\int V(|x - y|)|\psi(y, t)|^2 dy,$$

with initial data

$$\psi(x, 0) = \psi^I(x).$$

3. Convergence of the N-particle Liouville equation to the Vlasov equation.

In this section it is shown that the infinite Vlasov hierarchy, defined below, can be obtained as the limit of the Liouville hierarchy under some reasonably weak assumptions on the potential. However there is at least one important step missing before one can derive from this infinite hierarchy the genuine Vlasov-Poisson equation.

What is missing is a uniqueness theorem for solutions of the infinite hierarchy. Up to now such a theorem has been proven [NS] only under very stringent conditions on the potential V. In [NS] it is assumed that the

Fourier transform of V has compact support (implying in particular that V is analytic). Indeed, the "Vlasov" operators appearing in the hierarchy

$$\nabla_{v_i} \iint \nabla_{x_i} V(|x_i - x^*|) f_N^{n+1}(X_n, V_n, x^*, v^*, t) dx^* dv^*$$

"lose one derivative" in the v_i variable.

By comparison with this approach, the proof of [BH], valid for C^2 potentials, circumvents this difficulty by avoiding the infinite Vlasov hierarchy as a step in the derivation of the self-consistent Vlasov equation.

Consider the N-particle Liouville equation in the phase-space variables

$$(38) \qquad (X_N, V_N) = (x_1, x_2, ...x_N) \times (v_1, v_2, ...v_N)$$

with phase flow defined by

$$(39) \qquad \partial_t f_N(X_N, V_N, t) + \{H_N, f_N(X_N, V_N, t)\} = 0$$

where

$$(40) \quad H_N(X_N, V_N) = \sum_{1 \leq i \leq N} \frac{1}{2} |v_i|^2 + \frac{1}{N} \sum_{1 \leq i < j \leq N} V(|x_i - x_j|) = 0$$

is the "Hamilton function" corresponding to the "Hamilton operator" of (1).

Then introduce the n-th order marginal defined by the formula (cf (10)):

$$(41) \qquad f_N^n(X_n, V_n, t) = \int f_N(X_N, V_N, t) dX_N^n dV_N^n .$$

Assuming that the particles are indistinguishable (cf (11)) for $t = 0$ (and therefore for any $t \in \mathbb{R}$):

$$(42) \qquad f_N^n(x_{\sigma(1)}, v_{\sigma(1)}, \ldots, x_{\sigma(N)}, v_{\sigma(N)}) = f_N^n(x_1, v_1, \ldots, x_N, v_N),$$

for all permutations $\forall \sigma \in \Sigma_N$, one deduces from (39) the formula:

$$
\begin{aligned}
(43) \qquad & \partial_t f_N^n(X_n, V_n, t) + \sum_{1 \leq i \leq n} v_n \nabla_{x_i} f_N^n(X_n, V_n, t) \\
& -\frac{1}{N} \sum_{1 \leq i < j \leq n} \nabla_{v_i} (\nabla_{x_i} V(|x_i - x_j|) f_N^n(X_n, V_n, t)) \\
& -\frac{N-n}{N} \sum_{1 \leq i \leq n} \nabla_{v_i} \iint \nabla_{x_i} V(|x_i - x^*|) \\
& \qquad \times f_N^{n+1}(X_n, V_n, x^*, v^*, t) dx^* dv^* = 0.
\end{aligned}
$$

Letting N go to infinity in the equation (43) one obtains formally that the density

(44) $$f^n = \lim_{N \to \infty} f_N^n$$

satisfies the Vlasov hierarchy:

(45)
$$\partial_t f^n + V_n \nabla_{X_n} f^n(X_n, V_n, t)$$
$$- \sum_{1 \le i \le n} \nabla_{v_i} \iint \nabla_{x_i} V(|x_i - x^*|) f^{n+1}(X_n, V_n, x^*, v^*, t) dx^* dv^* = 0.$$

Observe that any density such that

$$f(x, v, t) \ge 0, \quad \iint f(x, v, t) dx dv = 1$$

that solves the Vlasov equation

(46)
$$\partial_t f(x, v, t) + v \nabla_x f(x, v, t)$$
$$- \left(\iint \nabla_x V(|x - x^*|) f(x^*, v^*) dx^* dv^* \right) \nabla_v f(x, v, t) = 0$$

produces a solution of the Vlasov hierarchy by the factorisation formula

(47) $$f^n(X_n, V, n, t) = \prod_{1 \le i \le n} f(x_i, v_i, t).$$

Thus the Vlasov hierarchy can be viewed as a generalized form of the Vlasov equation.

The formal computation above can be made precise, as follows.

PROPOSITION 3.1. *Assume that $V(x) = V(|x|)$ is continuous and bounded from below while ∇V is continuous and bounded.*

Assume that the initial data $f_N(X_N, V_N, t)$ are positive probability densities with energy of order N, i.e., satisfy

(48)
$$f_N \ge 0, \quad \iint f_N(X_N, V_N, 0) dX_N dV_N = 1,$$
$$\iint H_N(X_N, V_N) f_N(X_N, V_N, 0) dX_N dV_N \le CN,$$

and are invariant under permutation of the phase variables. Assume further that

$$\iint \frac{|X_N|^2}{2} f_N(X_N, V_N, 0) dX_N dV_N < CN.$$

Then (after passing to a subsequence N') the marginals

(49) $$f_{N'}^n(X_n, V_n, t) = \iint f_{N'}(X_{N'}, V_{N'}, t) dX_{N'}^n dV_{N'}^n$$

converge weakly (in the dual of the bounded continuous functions) to prob-
ability measures μ_n which are solutions in the sense of distributions of the
Vlasov hierarchy (46).*

Proof. Observe that the positivity, the total density, the energy estimate (48) and the invariance under permutation are preserved under the
Hamiltonian flow H_N. In particular one has:

$$
(50) \qquad
\begin{aligned}
\iint \frac{|v_i|^2}{2} f_N(X_N, V_N, t) dX_N dV_N = & \\
\frac{1}{N} \iint \frac{|V_N|^2}{2} f_N(X_N, V_N, t) dX_N dV_N & \leq C.
\end{aligned}
$$

Furthermore

$$
(51) \qquad
\begin{aligned}
\frac{d}{dt} \iint \frac{|x_i|^2}{2} f_N(X_N, V_N, t) dX_N dV_N & \\
= - \iint \frac{|x_i|^2}{2} \{H_N, f_N(X_N, V_N, t)\} dX_N dV_N & \\
= \iint x_i v_i f_N(X_N, V_N, t) dX_N dV_N & \\
\leq \iint \frac{|x_i|^2 + |v_i|^2}{2} f_N(X_N, V_N, t) dX_N dV_N & \\
\leq C + \iint \frac{|x_i|^2}{2} f_N(X_N, V_N, t) dX_N dV_N. &
\end{aligned}
$$

With the Gronwall lemma one deduces from (51) the existence, for all time
$t \leq T$, of a constant C_T independent of N such that

$$
(52) \qquad \iint \frac{|x_i|^2 + |v_i|^2}{2} f_N(X_N, V_N, t) dX_N dV_N \leq C_T.
$$

This bound shows that, for each fixed n, the sequence $\{f_N^n\}$ is "tight" in
the terminology of probability theory; it follows that a subsequence $\{f_{N'}^n\}$
converges to a *probability* measure μ_n. By diagonalization one extracts a
subsequence on which this convergence occurs for all n at once. "Recycle"
the symbol N and call this subsequence $\{f_N^n\}$.

From the indistinguishability hypothesis (42) the equation for the
marginal f_N^n is

$$
(53) \qquad
\begin{aligned}
\partial_t f_N^n + \sum_{1 \leq i \leq n} v_i \cdot \nabla_{x_i} f_N^n - \frac{1}{N} \sum_{1 \leq i < j \leq n} \nabla_{v_i} (\nabla_{x_i} V(x_i - x_j) f_N^n) \\
= \frac{N-n}{N} \sum_{1 \leq i \leq n} \nabla_{v_i} \left(\iint \nabla V(x_i - x^*) f_N^{n+1}(X_n, x^*, V_n v^*) dx^* dv^* \right).
\end{aligned}
$$

The measures μ_n solve the hierarchy of equations

(54)
$$\partial_t \mu_n + \sum_{1 \leq i \leq n} v_i \cdot \nabla_{x_i} \mu_n$$
$$= \lim_{N \to \infty} \sum_{1 \leq i < j \leq n} \nabla_{v_i} \left(\iint \nabla V(x_i - x^*) f_N^{n+1}(X_n, x^*, V_n v^*) dx^* dv^* \right).$$

To conclude the proof one has to analyze the limit in the sense of distributions of the right hand side of (54). Introduce a smooth test function of compact support $\Theta(X_n, V_n)$ and write:

(55)
$$\iint \Theta(X_n, V_n) \nabla_{v_i} \left(\iint \nabla V(x_i - x^*) f_N^{n+1}(X_n, x^*, V_n v^*) dx^* dv^* \right) dX_n dV_n$$
$$= -\iiiint \nabla_{v_i} \Theta(V_n, X_n) \nabla V(x_i - x^*)$$
$$\times f_N^{n+1}(X_n, x^*, V_n, v^*) dX_n dV_n dx^* dv^*$$

Since $\nabla_{v_i} \Theta(X_n, V_n) \nabla V(x_i - x^*)$ is continuous and bounded, the weak* convergence of f_N^n to μ_n and the tightness property (52) imply the convergence of (55) to

$$-\iiiint \nabla_{v_i}(\Theta(V_n, X_n)) \nabla V(x_i - x^*) \mu^{n+1}(X_n, x^*, V_n, v^*) dX_n dV_n dx^* dv^*.$$

This shows that

$$\lim_{N \to \infty} \sum_{1 \leq i < j \leq n} \nabla_{v_i} \left(\iint \nabla V(x_i - x^*) f_N^{n+1}(X_n, x^*, V_n v^*) dx^* dv^* \right)$$
$$= \sum_{1 \leq i < j \leq n} \nabla_{v_i} \iint \nabla V(x_i - x^*) \mu_{n+1}(X_n, dx^*, V_n, dv^*)$$

in the distribution sense, concluding the proof.   □

4. Convergence of the N-particle Schrödinger equation to the Vlasov equation. With the introduction of the Wigner transforms f_N^n (see e.g. [LP], [GMMP]) of the density operators $\rho_{N,n}$ the convergence of the solution of the N-particle Schrödinger hierarchy to the infinite hierarchy can be done under

1) the assumption that $N \to +\infty$, $\hbar \to 0$ and $N\hbar \to +\infty$,

2) and an extra assumption of uniform integrability (hypothesis **H1** below, see (69) below) which, in the present state of our knowledge, remains an essential ingredient removed for the treatment of rather general potentials.

Consider, as in section 2, the N-particle Schrödinger equation:

(56)
$$i\hbar_N \partial_t \Psi_N = -\frac{1}{2} \hbar_N^2 \Delta_x \Psi_N + \frac{1}{N} \sum_{1 \leq j < k \leq N} V(|x_j - x_k|) \Psi_N,$$

and the density matrix $\rho_N(X_N, Y_N)$. Introduce its Wigner transform (see eg. [LP], [GMMP]) at the scale $\hbar$:

$$
\begin{aligned}
&f_N(X_N, V_N, t) = \\
&\quad (2\pi)^{-N} \int_{\mathbb{R}^N} \rho_N(X_N + \hbar\frac{Y_N}{2}, X_N - \hbar\frac{Y_N}{2}, t) e^{-iY_N \cdot V_N} dY_N .
\end{aligned}
\tag{57}
$$

Observe that the Wigner transforms of the partial trace density matrices coincide with the marginals $f_{N,n}$ of $f_N(X_N, V_N, t)$:

$$
\begin{aligned}
&(\frac{1}{2\pi})^n \int_{\mathbb{R}^{3n}} \rho_{N,n}(X_n + \hbar\frac{Y_n}{2}, X_n - \hbar\frac{Y_n}{2}, t) e^{-iY_n V_n} dY_n = \\
&\quad \iint f_N(X_n, V_N, V_N^n, X_N^n, t) dV_N^n dX_N^n = f_{N,n}(X_n, V_n, t) .
\end{aligned}
\tag{58}
$$

Therefore dividing

$$
\begin{aligned}
i\hbar\partial_t \rho_{N,n}(X_n, Y_n, t) = \\
-\frac{\hbar^2}{2}[\Delta_{X_n} - \Delta_{Y_n}]\rho_{N,n}(X_n, Y_n, t) \\
+\frac{1}{N} \sum_{1\le j<k\le n} [V(|x_j - x_k|) - V(|y_j - y_k|)]\rho_{N,n}(X_n, Y_n, t) \\
+\frac{N-n}{N} \sum_{1\le j\le n} \int [V(|x_j - z|) - V(|y_j - z|)]\rho_{N,n+1}(X_n, z, Y_n, z, t)dz
\end{aligned}
\tag{59}
$$

by $i\hbar$ one obtains for the finite hierarchy of Wigner transforms the equations:

$$
\begin{aligned}
&\partial_t f_{N,n}(X_n, V_n, t) + V_n \cdot \nabla_{X_n} f_{N,n} = \\
&\frac{1}{i\hbar N} \sum_{1\le j<k\le n} \int_{\mathbb{R}^{3n}} \frac{e^{-iY_n \cdot V_n}}{(2\pi)^n} \left[V(|x_j + \frac{\hbar y_j}{2} - x_k + \frac{\hbar y_k}{2}|) \right. \\
&\left. -V(|x_j - \frac{\hbar y_j}{2} - x_k - \frac{\hbar y_k}{2}|) \right] \rho_{N,n}(X_n + \frac{\hbar Y_n}{2}, X_n - \frac{\hbar Y_n}{2}, t)dY_n \\
&+\frac{N-n}{(2\pi)^n N} \sum_{1\le j\le n} \int_{\mathbb{R}^{3n}} e^{-iY_n \cdot V_n} \int \frac{[V(|x_j + \frac{\hbar y_j}{2} - z|) - V(|x_j - \frac{\hbar y_j}{2} - z|]}{i\hbar} \\
&\times \rho_{N,n+1}(X_n + \frac{\hbar Y_n}{2}, z, X_n - \frac{\hbar Y_n}{2}, z, t)dzdY_n
\end{aligned}
\tag{60}
$$

With $N \to \infty$ $\hbar \to 0$ and $N\hbar \to \infty$ one obtains, at least formally, the equation:

$$
\begin{aligned}
&\partial_t f_N^n(X_n, V_n, t) + V_n \cdot \nabla_{X_n} f_N^n = \\
&\sum_{1\le j\le n} \lim_{N\to\infty, \hbar\to 0} \int_{\mathbb{R}^{3n}} \frac{e^{-iY_n \cdot V_n}}{(2\pi)^n} \int \frac{[V(|x_j + \frac{\hbar y_j}{2} - z|) - V(|x_j - \frac{\hbar y_j}{2} - z|]}{i\hbar} \\
&\times \rho_{N,n+1}(X_n + \frac{\hbar Y_n}{2}, z, X_n - \frac{\hbar Y_n}{2}, z, t)dzdY_n .
\end{aligned}
\tag{61}
$$

Furthermore, the condition $\hbar \to 0$ implies that

$$\int_{\mathbb{R}^{3n}} \frac{e^{-iY_n \cdot V_n}}{(2\pi)^n} \int \frac{[V(|x_j + \frac{\hbar y_j}{2} - z|) - V(|x_j - \frac{\hbar y_j}{2} - z|)]}{i\hbar}$$
$$\times \rho_{N,n+1}\left(X_n + \frac{\hbar Y_n}{2}, z, X_n - \frac{\hbar Y_n}{2}, z, t\right)dz\,dY_n$$

$$= \int dz \int_{\mathbb{R}^{3n}} \frac{e^{-iY_n \cdot V_n}}{(2\pi)^n} \frac{[V(|x_j + \frac{\hbar y_j}{2} - z|) - V(|x_j - \frac{\hbar y_j}{2} - z|)]}{i\hbar}$$
$$\times \rho_{N,n+1}\left(X_n + \frac{\hbar Y_n}{2}, z, X_n - \frac{\hbar Y_n}{2}, z, t\right)dY_n$$

$$(62) \qquad = -\int dz \int_{\mathbb{R}^{3n}} \frac{e^{-iY_n \cdot V_n}}{(2\pi)^n} \nabla_{x_j} V(|x_j - z|) y_j \rho_{N,n+1}\left(X_n\right.$$
$$\left. + \frac{\hbar Y_n}{2}, z, X_n - \frac{\hbar Y_n}{2}, z, t\right)dY_n + O(\hbar^2)$$

$$= \nabla_{v_j} \int dz \int_{\mathbb{R}^{3n}} \frac{e^{-iY_n \cdot V_n}}{(2\pi)^n} \nabla_{x_j} V(|x_j - z|) \rho_{N,n+1}\left(X_n\right.$$
$$\left. + \frac{\hbar Y_n}{2}, z, X_n - \frac{\hbar Y_n}{2}, z, t\right)dY_n + O(\hbar^2)$$

$$= \nabla_{v_j} \iint \nabla_{x_j} V(|x_j - z|) f_{N,n+1}(X_n, z, V_n, v^*)dz\,dv^* + O(\hbar^2)$$

which shows that the limit of the function $f_{N,n}$ are solution of the Vlasov hierarchy.

The next proposition attempts to provide a rigorous proof of the above statement based on the tools developed in [LP] and the hypothesis **H1**. To this end, we recall the definition of the convenient functional framework for Wigner transforms. Consider the space of functions

$$(63) \quad \mathcal{A}_n = \{f(x,v) \in C_0(\mathbb{R}_x^n \times \mathbb{R}_v^n) : (\mathcal{F}_{v \to y} f)(x,y) \in L^1(\mathbb{R}_y^n; C_0(\mathbb{R}_x^n))\}$$

equipped with the norm

$$(64) \qquad \|\mathcal{F}f\|_{L^1(C_0)} = \int_{\mathbb{R}^{3n}} \sup_x |\mathcal{F}f|(x,y)dy.$$

PROPOSITION 4.1. *Assume that the potential $V(|x|)$ satisfies the following conditions:*

$$(65) \qquad \nabla_x V(|x|) \text{ is uniformly bounded and vanishes as } |x| \to \infty,$$

and

$$(66) \qquad \frac{V(|x + \frac{\hbar}{2}y|) - V(|x - \frac{\hbar}{2}y|)}{\hbar} = \nabla_x V(|x|) \cdot y + R(x, \hbar, y)$$

where the remainder $R(x, h, y)$ satisfies

$$(67) \qquad |R(x, h, y)| \le a(h) \text{ with } \lim_{h \to 0} a(h) = 0.$$

Assume that the initial densities are pure states invariant under permutations

(68) $\rho_N(X_N, X_N) = \Psi(X_N)\overline{\Psi(Y_N)}$ *with* $\int |\Psi_N(X_N)|^2 dX_N = 1$.

Assume finally hypothesis **H1**, *that the marginals are uniformly integrable in the sense that*

(69) $\forall n \geq 1 :$ $\displaystyle\sup \int_{|v^*|\geq M} |f_{N,n+1}(X_n, z, V_n, v^*)| dv^* \leq A_n(M)$

where $A_n(M)$ *denotes a positive function independent of* N *that vanishes as* $M \to \infty$.

Then in the limit

(70) $N \to \infty,\ \hbar \to 0,\ \ and\ N\hbar \to \infty$

and modulo the extraction of a subsequence, the marginals

$$
(71) \quad
\begin{aligned}
f_N^n(X_n, V_n, t) &= \int_{I\!R^{3n}} \frac{e^{-iY_n \cdot V_n}}{(2\pi)^n} \rho_{N,n}\left(X_n + \hbar\frac{Y_n}{2}, X_n - \hbar\frac{Y_n}{2}, t\right) dY_n \\
&= \iint f_N(X_n, V_N, V_N^n, X_N^n, t) dV_N^n dX_N^n
\end{aligned}
$$

converge in the dual of $C_0(I\!R_t; \mathcal{A}_n)$ *to positive measures that are invariant under permutation of the phase variables and solve the Vlasov hierarchy.*

Proof. From the conservation of density (cf (68)) one deduces with the proposition III.1 of ([LP]) that up to the extraction of a subsequence the functions f_N^n converge in the dual of $C_0(I\!R_t; \mathcal{A}_n)$ positive distributions $f^n(X_n, V_n, t)$ which satisfy the relation:

(72) $\displaystyle\iint f^n(X_n, V_n, t) dX_n dV_n \leq 1$.

The positivity is obtained as in [LP] via the introduction of the Husimi transform. At this point, we are unable to establish the equality in (72).

Then the next (and only) step is to prove, for any function $\phi(X_n, V_n, t)$ such that $\nabla_{V_n}\phi(X_n, V_n, t) \in C_0(I\!R_t; \mathcal{A}_n)$, the following convergence (omitting the integration in the t variable):

$$
(73) \quad
\begin{aligned}
\lim_{\hbar \to 0, N \to \infty} &< \int dz \int_{I\!R^{3n}} \frac{e^{-iY_n \cdot V_n}}{2\pi^n} \frac{[V(|x_j + \frac{\hbar y_j}{2} - z|) - V(|x_j - \frac{\hbar y_j}{2} - z|]}{i\hbar} \\
&\times \rho_{N,n+1}\left(X_n + \frac{\hbar Y_n}{2}, z, X_n - \frac{\hbar Y_n}{2}, z, t\right) dY_n, \phi > \\
&= - < \iint \nabla_{x_j} V(|x_j - z|) \mu_{n+1}(X_n, z, V_n, v^*) dz dv^*, \nabla_{v_j}\phi(X_n, V_n) >
\end{aligned}
$$

Following [LP] write

$$
(74)
\begin{aligned}
&< \int dz \int_{\mathbb{R}^{3n}} e^{-iY_n \cdot V_n} \frac{[V(|x_j + \frac{\hbar y_j}{2} - z|) - V(|x_j - \frac{\hbar y_j}{2} - z|]}{i\hbar} \\
&\qquad\qquad \times \rho_{N,n+1}(X_n + \frac{\hbar Y_n}{2}, z, X_n - \frac{\hbar Y_n}{2}, z, t)dY_n, \phi > \\
&= -i < \int dz \int_{\mathbb{R}^{3n}} e^{-iY_n \cdot V_n} [\nabla_{x_j} V(|x_j - z|) \cdot y_j + R(x_j - z, \hbar, y)] \\
&\qquad\qquad \times \rho_{N,n+1}(X_n + \frac{\hbar Y_n}{2}, z, X_n - \frac{\hbar Y_n}{2}, z, t)dY_n, \phi > \\
&= < \nabla_{v_j} \int dz \int_{\mathbb{R}^{3n}} e^{-iY_n \cdot V_n} [\nabla_{x_j} V(|x_j - z|) \\
&\qquad\qquad \times \rho_{N,n+1}(X_n + \frac{\hbar Y_n}{2}, z, X_n - \frac{\hbar Y_n}{2}, z, t)dY_n, \phi > \\
&\quad + < \int dz \int_{\mathbb{R}^{3n}} e^{-iY_n \cdot V_n} R(x_j - z, \hbar, y) \\
&\qquad\qquad \times \rho_{N,n+1}(X_n + \frac{\hbar Y_n}{2}, z, X_n - \frac{\hbar Y_n}{2}, z, t)dY_n, \phi > .
\end{aligned}
$$

The condition (66) guarantees that the last term in (74) vanishes with $\hbar$. In fact,

$$
(75)
\begin{aligned}
&\left| < \int dz \int_{\mathbb{R}^{3n}} e^{-iY_n \cdot V_n} R(x_j - z, \hbar, y) \right. \\
&\qquad\qquad \left. \times \rho_{N,n+1}(X_n + \frac{\hbar Y_n}{2}, z, X_n - \frac{\hbar Y_n}{2}, z, t)dY_n, \phi > \right| \\
&\leq a(h) \int dz \int_{\mathbb{R}^{3n}} \rho_{N,n+1}(X_n + \frac{\hbar Y_n}{2}, z, X_n - \frac{\hbar Y_n}{2}, z, t) \\
&\qquad\qquad \times |(\mathcal{F}_{v \to y}\phi)(X_n, Y_n)|dX_n dY_n \\
&\leq a(h) \int dz \int_{\mathbb{R}^{3n}} \rho_{N,n+1}^{\frac{1}{2}}(X_n + \frac{\hbar Y_n}{2}, z, X_n + \frac{\hbar Y_n}{2}, z, t) \\
&\qquad\qquad \times \rho_{N,n+1}^{\frac{1}{2}}(X_n - \frac{\hbar Y_n}{2}, z, X_n - \frac{\hbar Y_n}{2}, z, t) \\
&\qquad\qquad \times |(\mathcal{F}_{v \to y}\phi)(X_n, Y_n)|dX_n dY_n \\
&\leq a(h) \Big(\iint_{\mathbb{R}^{3n}} \rho_{N,n+1}(X_n + \frac{\hbar Y_n}{2}, z, X_n + \frac{\hbar Y_n}{2}, z, t) \\
&\qquad\qquad \times |(\mathcal{F}_{v \to y}\phi)(X_n, Y_n)|dX_n dY_n dz \Big)^{\frac{1}{2}} \\
&\qquad\qquad \times \Big(\iint_{\mathbb{R}^{3n}} \rho_{N,n+1}(X_n + \frac{\hbar Y_n}{2}, z, X_n + \frac{\hbar Y_n}{2}, z, t) \\
&\qquad\qquad \times |(\mathcal{F}_{v \to y}\phi)(X_n, Y_n)|dX_n dY_n dz \Big)^{\frac{1}{2}}
\end{aligned}
$$

Eventually observe that one has

$$\iint_{\mathbb{R}^{3n}} \rho_{N,n+1}(X_n \pm \frac{\hbar Y_n}{2}, z, X_n \pm \frac{\hbar Y_n}{2}, z, t)|(\mathcal{F}_{v \to y}\phi)(X_n, Y_n)|dX_n dY_n dz$$

(76) $$= \iint_{\mathbb{R}^{3n}} \rho_{N,n+1}(X_n, z, X_n, z, t)|(\mathcal{F}_{v \to y}\phi)(X_n \mp \frac{\hbar Y_n}{2}, Y_n)|dX_n dY_n dz$$

$$\leq C \iint_{\mathbb{R}^{3n}} \rho_{N,n+1}(X_n, z, X_n, z, t)dX_n dY_n dz$$

which shows that the right hand side of (75) vanishes with $\hbar$.

Then write:

$$< \nabla_{v_j} \int dz \int_{\mathbb{R}^{3n}} e^{-iY_n \cdot V_n}[\nabla_{x_j} V(|x_j - z|)\rho_{N,n+1}$$
$$\times (X_n + \frac{\hbar Y_n}{2}, z, X_n - \frac{\hbar Y_n}{2}, z, t)dY_n, \phi >$$

(77) $$= - < \int dz \int_{\mathbb{R}^{3n}} e^{-iY_n \cdot V_n} \nabla_{x_j} V(|x_j - z|)\rho_{N,n+1}$$
$$\times (X_n + \frac{\hbar Y_n}{2}, z, X_n - \frac{\hbar Y_n}{2}, z, t)dY_n, \nabla_{v_j}\phi >$$

$$= - \iiint \nabla_{x_j} V(|x_j - z|)\nabla_{v_j}\phi(X_n, V_n)$$
$$\times f_{N,n+1}(X_n, z, V_n, v^*)dX_n dz dV_n dv^* .$$

With (65) and the hypothesis **H1** (see(69)) it is enough to consider the limit of

$$\iiiint \Theta_1(z)\Theta_2(v^*)\nabla_{x_j} V(|x_j - z|)\nabla_{v_j}\phi(X_n, V_n)$$
$$\times f_{N,n+1}(X_n, z, V_n, v^*)dX_n dz dV_n dv^*$$

where Θ_1 and Θ_2 denote continuous functions with compact support. Then, the result follows from Theorem III.1 of [LP].   □

5. Remarks on (the scaling for) fermions. The infinite-particle *cum* classical limit treated in the preceeding section,

(78) $$N \to \infty, \hbar \to 0 \text{ and } \lim_{N \to \infty} \lim_{\hbar \to 0} N\hbar \to \infty$$

is applicable to the "Jellium" model of plasmas, in which N electron occupy a cube of volume L^3 with the density

(79) $$\bar{n} = \frac{N}{L^3}$$

held fixed as $N \to \infty$.

Introducing the new variable $x = X/L$ the equation (1) for the Coulomb potential

$$V(|x|) = \frac{1}{|x|}$$

is exactly transformed into the equation:

$$(80) \qquad i\hbar\partial_t\Psi = -\frac{1}{2}\frac{\hbar^2}{N^{\frac{2}{3}}}\Delta_x\Psi + \frac{1}{N^{\frac{1}{3}}}\sum_{1\le j<k\le N}V(|x_j - x_k|)\Psi$$

Multiplying both side of the equation by $N^{\frac{2}{3}}$ changes this equation into a standard Schrödinger equation with a new Planck constant:

$$(81) \qquad \hbar_N = \hbar N^{-\frac{2}{3}}$$

with a weak coupling potential:

$$(82) \qquad i\hbar_N\partial_t\Psi_N = -\frac{1}{2}\hbar_N^2\Delta_x\Psi_N + \frac{1}{N}\sum_{1\le j<k\le N}V(|x_j - x_k|)\Psi_N.$$

Since $\hbar_N N \to \infty$ we are in the situation treated in the preceding section. Sewell was the first to derive the classical Vlasov equation for a plasma in this manner from the quantum dynamics of electrons [Sew].

In contrast to this, the weak-coupling limit

$$N \longrightarrow \infty \qquad \hbar = \text{constant}$$

is problematic for systems of electrons, indeed for fermions in general. The approach of Section 2 produces only trivial limits:

THEOREM 5.1. *Let ρ_N^I be an N-particle fermionic density for each N, in the notation of Theorem 2.1. Then we have*

$$(83) \qquad \lim_{N\to\infty}\rho_{N,1}^I = 0$$

and the infinite hierarchy becomes trivial.

REMARK 5.1. *The failure to obtain a nontrivial limit can be understood by considering the energy: the Pauli exclusion principle forces the energy per particle of the N-particle system (1) to diverge to infinity, despite the weak coupling. It may still be possible to derive a (fermion version of the) Schrödinger-Poisson equation (like the "Schrödinger-Poisson-Xα" model [M6]) directly from the quantum mechanics of electron systems, but not through the weak-coupling limit of [Sp1].* The proof of (83) follows from Theorem 5.2 below.

Let $\mathcal{H}$ be a Hilbert space with orthonormal basis $\{\phi_b\}_{b\in B}$, where B is a possibly uncountable index set. Let $\mathcal{B}_n$ be the set of all subsets of B of size n. For $\mathbf{s} = \{s(1), s(2), \ldots, s(n)\}$ in $\mathcal{B}_n$ define the n-particle Fock vector

$$(84) \qquad \Phi_{\mathbf{s}} = \frac{1}{\sqrt{n!}}\sum_{\sigma\in\Sigma_n}\text{sign}(\sigma)\phi_{s(\sigma(1))}\otimes\phi_{s(\sigma(2))}\otimes\cdots\otimes\phi_{s(\sigma(n))},$$

summing over all permutations of the set $\{1, 2, \ldots, n\}$, where $\text{sign}(\sigma)$ is the parity of σ. Then $\{\Phi_{\mathbf{s}}\}_{\mathbf{s}\in\mathcal{B}_n}$ is an orthonormal basis of the subspace of $\mathcal{H}^{\otimes n}$

consisting of antisymmetric vectors. The n-particle fermionic *pure states* are the rank-one projections of the form

$$(85) \qquad P_\Psi(x) = \langle x, \Psi \rangle \, \Psi,$$

where

$$(86) \qquad \Psi = \sum_{s \in \mathcal{B}_n} c_s \Phi_s$$

is an antisymmetric unit vector in $\mathcal{H}^{\otimes n}$. Finally, an n-particle fermionic density operator D is a (countable) convex combination of projections of the form (85), i.e.,

$$(87) \qquad D = \sum \lambda_j P_{\Psi_j}$$

with $\lambda_j \geq 0$ and $\sum_j \lambda_j = 1$.

THEOREM 5.2. *If D is an n-particle fermionic density operator then* $\left\| Tr^{(n-1)} D \right\| \leq \frac{1}{n}$.

Proof. Because of the convexity of the norm and the linearity of $Tr^{(n-1)}$, and the form (87) of the n-particle fermionic density operators, it suffices to prove (5.2) for fermionic pure states.

To this end, let Ψ be as in (86). Since $Tr^{(n-1)}(P_\Psi)$ is a compact self-adjoint operator, there exists a unit vector $\phi_0 \in \mathcal{H}$ such that

$$(88) \qquad \left\| Tr^{(n-1)}(P_\Psi) \right\| = \left\langle Tr^{(n-1)}(P_\Psi)(\phi_0), \phi_0 \right\rangle.$$

Extend $\{\phi_0\}$ to an orthonormal basis $\{\phi_b\}_{b \in B}$ of $\mathcal{H}$.

By definition of the partial trace, $\left\langle Tr^{(n-1)}(P_\Psi)(\phi_0), \phi_0 \right\rangle$ equals

$$(89) \qquad \sum_{r(1)\ldots r(n-1) \in B} \left\langle P_\Psi(\phi_0 \otimes \phi_{r(1)} \otimes \cdots \otimes \phi_{r(n-1)}), \phi_0 \otimes \phi_{r(1)} \otimes \cdots \otimes \phi_{r(n-1)} \right\rangle.$$

From (88) and (89)

$$(90) \qquad \begin{aligned} \left\| Tr^{(n-1)}(P_\Psi) \right\| &= \sum_{r(1)\ldots r(n-1)} \left\| P_\Psi(\phi_0 \otimes \phi_{r(1)} \otimes \cdots \otimes \phi_{r(n-1)}) \right\|^2 \\ &= \sum_{r(1)\ldots r(n-1)} \sum_{s} |c_s|^2 \left| \langle \phi_0 \otimes \phi_{r(1)} \otimes \cdots \otimes \phi_{r(n-1)}, \Phi_s \rangle \right|^2. \end{aligned}$$

Now $\langle \phi_0 \otimes \phi_{r(1)} \otimes \cdots \otimes \phi_{r(n-1)}, \Phi_s \rangle = 0$ unless $s = \{0, r(1), \ldots, r(n-1)\}$, in which case

$$\langle \phi_0 \otimes \phi_{r(1)} \otimes \cdots \otimes \phi_{r(n-1)}, \Phi_s \rangle = \frac{1}{\sqrt{n!}}.$$

It follows from (90) that

$$\left\| Tr^{(n-1)}(P_\Psi) \right\| \leq \sum_{s} |c_s|^2 \frac{(n-1)!}{n!} = \frac{1}{n}.$$

□

Acknowledgement. This research was supported by the European Union TMR network "Asymptotic methods in kinetic equations" (ERB FMRX CT97 0157) and by the Austrian START project "Nonlinear Schrödinger and quantum Boltzmann equations" of N.J.M. Also, F.G. was supported by the Institut Universitaire de France. The first three authors thank the ESI in Vienna for its hospitality.

REFERENCES

[BGM1] C. Bardos, F. Golse, and N.J. Mauser, "Weak coupling limit of the N-particle Schrödinger equation", to appear in Mathematical Analysis and Applications (2000).

[BM]   O. Bokanowski and N.J. Mauser "Local approximation for the Hartree-Fock exchange potential: a deformation approach" *Math. Models Methods Appl. Sci.*, vol. 9, No. 6, 941–961 (1999).

[BH]   W. Braun and K. Hepp, "The Vlasov Dynamics and its fluctuation in the 1/N limit of interacting particles", Comm. Math. Phys. **56** (1977), 101–113.

[CH]   T. Cazenave and A. Haraux, *Introduction aux problèmes d'évolution semi-linéaires*, Ellipses, Paris (1990).

[C]   C. Cercignani, *Ludwig Boltzmann*, Oxford Univ. Press (1998).

[GMMP] P. Gérard, P.A. Markowich, N.J. Mauser, and F. Poupaud, *Homogenization Limits and Wigner Transforms*, Comm. Pure and Appl. Math. **50** (1997), 321–377.

[GV]   J. Ginibre and G. Velo, "On a class of nonlinear Schrödinger equations with nonlocal interaction", Math. Zeitschr. **169** (1980), 109–145.

[Ho]   L. Hörmander, *The analysis of linear partial differential operators*, vol. 1, Springer Verlag, Berlin-New York (1983).

[K]   T. Kato, *Perturbation Theory for Linear Operators*, Springer-Verlag, (1966).

[K2]   T. Kato, "Fundamental properties of Hamiltonian Operators of Schrödinger type ", Trans. Amer. Math. Soc. **70** (1951), 195–211.

[LL3]   L. Landau and E. Lifshitz, *Quantum Mechanics*, Course in Theoretical Physics, vol. 3, Pergamon Press, Oxford - New York - Toronto(1974).

[Le]   J. Leray, "Sur le mouvement d'un liquide visqueux emplissant l'espace", Acta Math. **63** (1934), 183–248.

[LY]   E. Lieb and H.-T. Yau, "The Chandrasekhar Theory of Stellar Collapse as the Limit of Quantum Mechanic", Commun. Math. Phys. **112** (1987), 147–174.

[LP]   P.L. Lions and T. Paul, "Sur les mesures de Wigner", Revista Math. Iberoamericana **9** (1993), 553–617.

[MM]   P. Markowich and N. Mauser, "The Classical Limit of a Self-consistent Quantum-Vlasov Equation in 3-D", Math. Mod. and Meth. in the Appl. Sci. **9** (1993), 109–124.

[M6]   N.J. Mauser, "The Schrödinger-Poisson-Xα model", to appear in Appl. Math. Lett. (2000)

[NS]   H. Narnhofer and G.L. Sewell, "Vlasov Hydrodynamics of a Quantum Mechanical Model", Commun. Math. Phys. **79** (1981), 9–24.

[Nir]   L. Nirenberg, "An abstract form of the nonlinear Cauchy-Kowalewski Theorem", J. of Diff. Geom. **6** (1972), 561–576.

[Nis]   T. Nishida, "A note on a theorem of Nirenberg", J. of Diff. Geom. **12** (1978), 629–633.

[RS2]   M. Reed and B. Simon, *Methods of Modern Mathematical Physics*, vol. 2, Academic Press, 4th ed. (1987).

[Sew]   G. Sewell, "Quantum Plasma Model with Hydrodynamical Phase Transition", Helv. Phys. Acta **67** (1994), 4–19.

[Sp1]   H. Spohn, "Kinetic Equations from Hamiltonian Dynamics", Rev. Mod. Phys. **52** no. 3 (1980), 600–640.

[Sp2]   H. Spohn, "Quantum Kinetic Equations", in *On the Three Levels*, M. Fannes et al. eds., Plenum Press, New York (1994), 1–10.

[Sp3]   H. Spohn, "Boltzmann Hierarchy and Boltzmann equation", in *Kinetic theories and the Boltzmann equation*, C. Cercignani ed., LNM no. 1048, Springer-Verlag Berlin (1984).

[Th]   W. Thirring, *A course in Mathematical Physics*, vol. 4, Springer-Verlag (1983).

TAKING ON THE MULTISCALE CHALLENGE

LEONARD J. BORUCKI*

Abstract. Numerous physical modeling problems in the semiconductor industry involve large scale differences. In one class of equipment modeling problems, the goal is to relate macroscopic reactor level behavior to the growth of material on features that are six orders of magnitude smaller. In another class of problems, properties calculated at the atomic level are used to construct models that apply over much larger length scales. Several examples are described from electroplating, chemical vapor deposition, deposition of metallization, reliability, and chemical-mechanical polishing.

1. Introduction. Multiscale problems abound in the semiconductor industry. A typical problem occurs during the deposition of the many layers of materials that are laid down on a silicon wafer in the process of manufacturing integrated circuits. Processes of this kind include chemical vapor deposition (CVD), physical vapor deposition (PVD), plasma-enhanced CVD, and electroplating. Each such process is performed in a special chamber or tool that is large enough to accommodate one or more wafers and therefore is on the order of 0.1 m to 1 m in size. On the other end of the scale, each wafer may contain millions of microelectronic components, such as transistors, that contain critical features on the order of 0.1 μm or less. The challenge is to be able to fabricate microscopic structures that are as uniform as possible across the entire 0.2-0.3 m diameter of a wafer using only the macroscopic controls that are provided with each tool, such as the temperature, gas composition, pressure and flow rate. Since each wafer may be worth several thousand dollars and each tool may cost several million dollars to purchase and install, there is considerable incentive to optimize the operation of a tool on the macroscopic to obtain the greatest possible yield of components and circuits on the microscopic level.

2. Multiscale problems in deposition processes. An example deposition process, electroplating, is illustrated in Fig. 1. Fig. 1 shows a cross section of a typical cup electroplater, which is used to plate copper that eventually becomes the wires or interconnects between devices in some kinds of integrated circuits. The wafer is held inverted with its front surface submerged in a plating bath that fills a shallow cup. The plating bath flows slowly up from the bottom of the cup, impinges on the wafer, and spills over the lip of the cup. A copper source at the bottom of the cup is held at a different electrical potential than the wafer in order to drive metal ions in the bath to deposit onto the wafer. The wafer may have an array of die, or chips patterned on its front surface. Each die in turn may contain millions of features. Many factors can produce nonuniformity in the feature-scale outcome. Variations in the current density across the wafer on

---

*Motorola, Inc., 2200 West Broadway Rd., Mail Stop M360, Mesa, Arizona 85202.

the equipment scale will cause the deposited metal to be thicker where the current is higher. Typically, the plate at the edge of the wafer is thicker than at the center. Differences in the packing densities of features on a die scale, which often correspond to differences in real surface area, can also lead to nonuniformities through local metal ion and additive depletion. Thus, both the size and arrangement of die on the wafer and the density of features on the die can affect the outcome of the plating process. Quantitative prediction of how individual features will fill across the wafer is not trivial. Thus, it is difficult to know *a priori* how to best plate a wafer with a given arrangement of die and features.

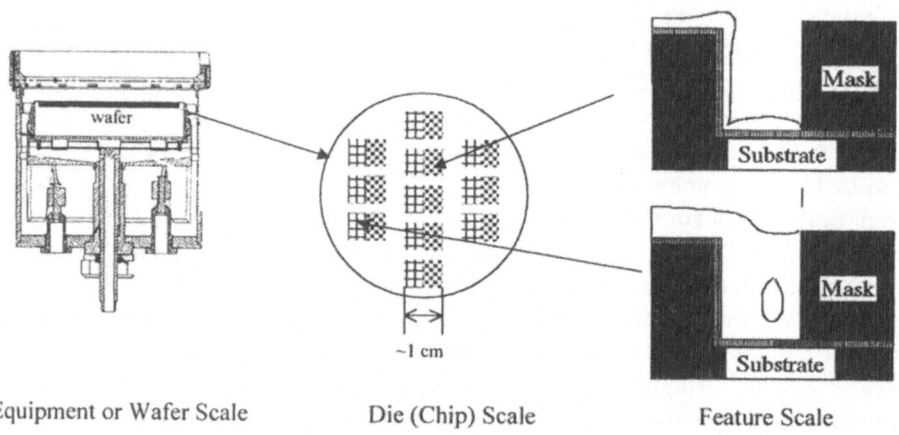

Equipment or Wafer Scale   Die (Chip) Scale   Feature Scale

FIG. 1. *A wide range of scales is typical for many semiconductor fabrication processes. In electroplating, which is used to deposit copper metallization for connecting devices in integrated circuits, a wafer whose diameter is ~0.2 m is held inverted at the top of a cup of plating fluid. The wafer may contain many identical photolithographically patterned die that may be a centimeter or less on a side. The circuits on the die are composed of transistors or other components that may have critical features on the order of 0.1 μm. Thus, spatial scales can span six orders of magnitude in electroplating. In general, the trend in the industry has been to increase the wafer size and decrease the feature size, thus exacerbating the discrepancy in scales.*

An second example (Fig. 2), involving the low pressure chemical vapor deposition of silicon dioxide from the precursor trietlyoxysilane, or TEOS, can be found in [1] and [2]. Here, TEOS in a carrier gas enters the top of an axisymmetric reactor chamber, impinges on the wafer, and then is vented through ports at the wafer perimeter. The reaction of TEOS at the wafer surface depletes the precursor in the gas phase and releases reaction byproducts such as ethanol. The gas phase precursor and byproduct concentrations depend on the local arrangement of die and on the surface area presented by the packing of features on the die (Fig. 3). The reactor and feature scale problems are coupled and must be solved simultaneously. As the features fill up, the surface area changes, forcing changes on the reactor scale. In [1] and [2], the Navier-Stokes and continuity equations were solved on the reactor scale using a commercial finite element computational fluid dynamics code. At each die location, species fluxes for the reactor scale model were computed using the feature scale ballistic transport and surface evolution program EVOLVE [3]. In some cases, a third mesoscopic, or die scale finite element model was inserted between the reactor and feature scale models. Solutions on the reactor, die, and feature scale were iterated until the boundary fluxes converged. This is a rather brute force approach to this kind of multiscale problem. An approach described in [4] uses homogenization theory to connect the reactor and feature scale models. It applies when the pressure is high enough so that the mean free path for gas molecule collisions is much smaller than the feature size .

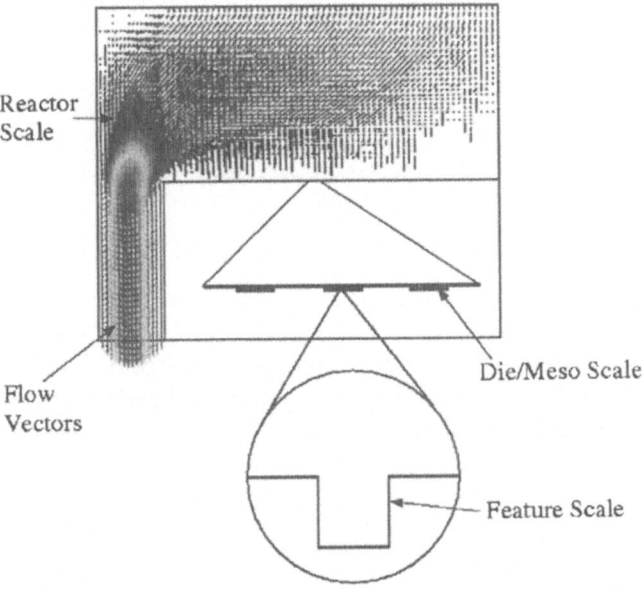

FIG. 2. *Multiscale problems occur in chemical vapor deposition [1]. Here, gas containing triethyloxysilane, a precursor for silicon dioxide deposition, enters at the top of an axisymmetric reactor, impinges on a wafer and then is vented around the wafer perimeter. The wafer again contains die with features that may have different packing densities.*

3. Multiscale modeling from the atomic level. The fundamental motivation in the above two examples is the need to connect profile evolu-

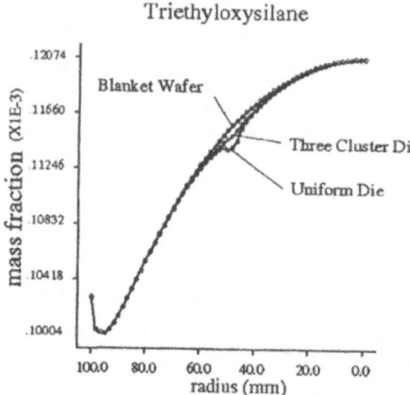 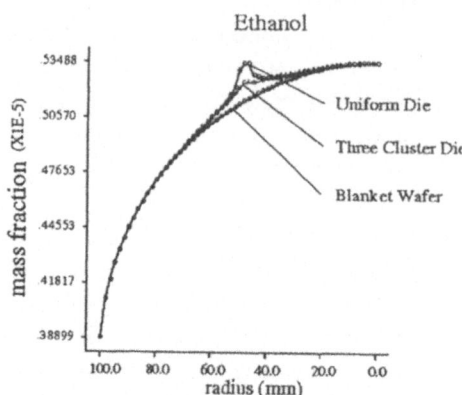

FIG. 3. *Comparison of the mass fractions of the reactant triethyloxysilane in Fig. 2 and the reaction product ethanol as a function of wafer radius on a blanket (unpatterned) wafer, a wafer with a die that contains a uniform population of features, and a die that contains three clusters of features. As the total surface area of the die increases, the reactant fraction decreases and the reaction product fraction increases. Feedback between the reactor, die and feature scales is evident.*

tion on a feature scale with reactor-level controls that act at a spatial scale six orders of magnitude larger. However, in some applications the feature scale profile is at the high end of the modeling spectrum. Improvements in numerical methods for first principles calculations have made it possible to predict, understand, and quantify basic properties of many systems and are therefore now an important tool in the development of higher level models. One advantage of using *ab initio* methods is that fitting of basic parameters, such as formation and migration energies, can sometimes be avoided. By using a combination of computed parameters and experimental data, it is often possible to develop models that are more detailed, more physically correct and more predictive than by experiment alone.

We outline an example of this kind of model development from the deposition of metallization. Wang *et. al.* [5] used the Embedded Atom Method (EAM), which relies on semi-empirical potentials derived from a database of experimental measurements, to calculate activation energies for various diffusion mechanisms on Cu (100), (110) and (111) facets, including hopping between adjacent facets of a crystal. These were in general agreement with more accurate and costly density functional theory (DFT)

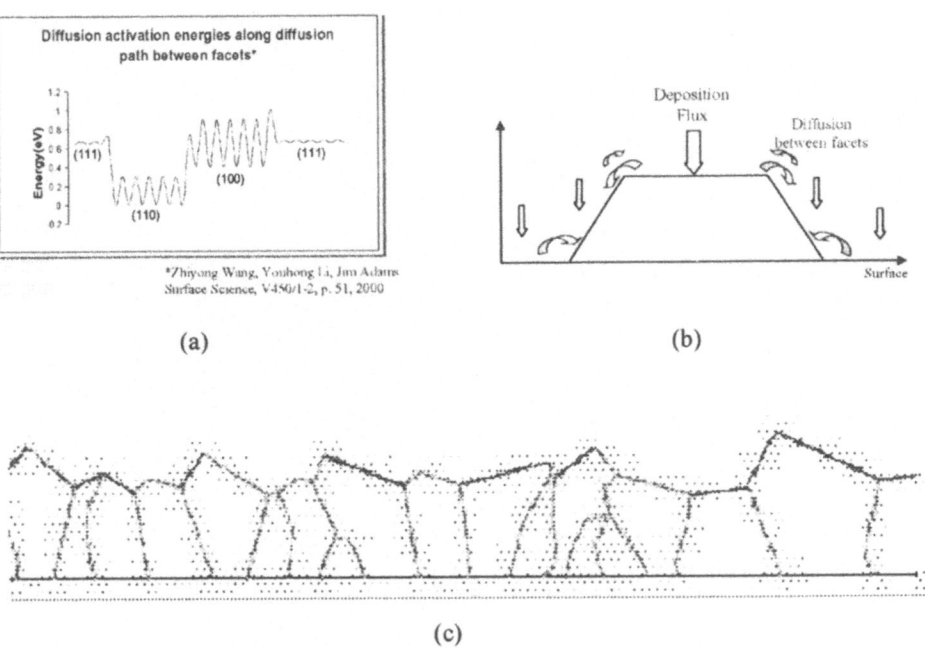

FIG. 4. *An example of the use of atomistic calculations to build a feature scale model [5]. (a) Embedded atom potentials are used to calculate activation energies for diffusion on and between crystallographic planes. This involves structures with at most a few thousand atoms. (b) The calculated energies are used to build an event table for a kinetic lattice Monte Carlo (KLMC) algorithm that simulates the growh of much larger structures over much larger times. (c) The KLMC code is used to predict how the grain structure of a deposited film depends on the isotropy of the deposition flux.*

estimates and were performed on collections of at most a few thousand atoms. The calculated activation energies were then used to build an event table for a Kinetic Lattice Monte Carlo algorithm that was used to simulate facet growth on a much larger spatial scale ($\sim$0.1 μm) and a much longer time scale than would be possible with EAM alone (Fig. 4). The KLMC code was then used to simulate the grain structure (morphology and orientation) that would occur under different assumptions about the flux distribution of deposited Cu atoms. At this point, the KLMC grain growth model could be connected with a reactor scale plasma model capable of predicting the energies and directions of incoming copper ions and neutrals. Although there are still many things missing from this combined model, such as detailed kinetics of nucleation on the substrate, this illustrates that it is currently possible to assemble potentially useful models that span atomic to macroscopic spatial and time scales.

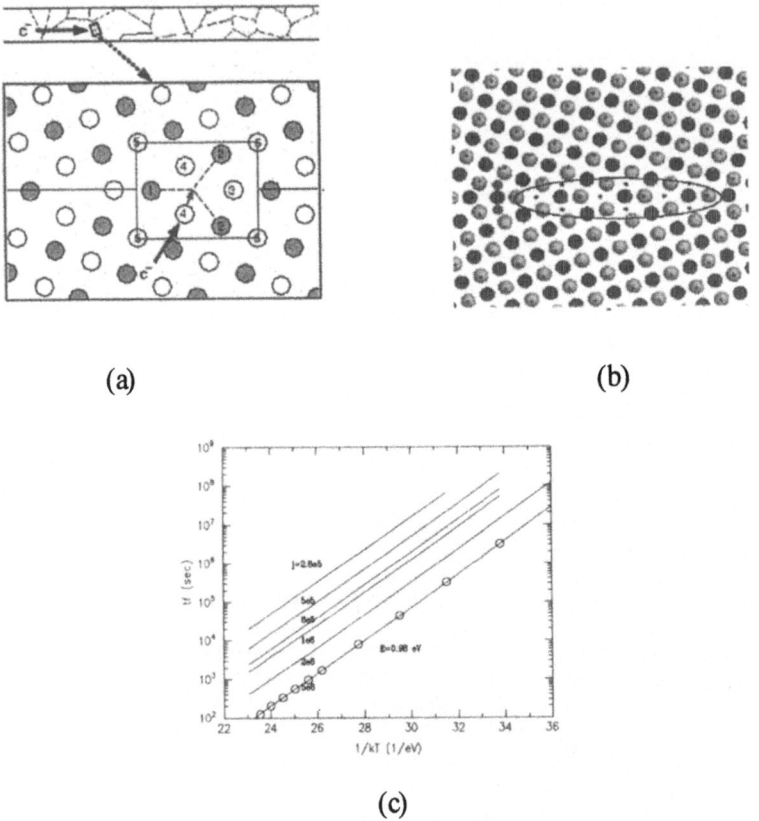

(a)

(b)

(c)

FIG. 5. *An example of multiscale modeling in metallization reliability. (a) The Embedded Atom Method was used to examine vacancy and interstitial formation energies and diffusion paths in symmetric tilt grain boundaries in alumunum. (b) The method was also used to quantify the morphologies and formation energies of small voids at grain boundaries. (c) Some of the calculated parameters were then used in a continuum model for vacancy transport and stress evolution during electromigration to predict failure times (taken to be the time to the first nucleation event) in aluminum. The calculated activation energy for failure, 0.98 eV, was about 0.2 eV higher than experimental data.*

We illustrate another application of multiscale modeling in metallization reliability. One of the major sources of failure of the wires, or interconnects, that are used in microelectronic circuits is electromigration. Electromigration occurs when electrons in the metal collide with atoms, transferring momentum and creating a slight directional bias in diffusional jumps. This leads to a slow flow of mass in the direction of the electron flow and the eventual depletion of atoms in some areas and accumulation in other areas. Depletion and accumulation are associated with a buildup of tensile stress or compressive stress that can lead to voiding or hillocking, which are major causes of failure. While failure rates can be measured in a few hours or months under high temperature, high current accelerated test conditions, there is considerable doubt about the methods that are used to project ten year failure rates under actual use conditions. Liu *et. al.* [6],[7] constructed detailed minimum energy models consisting of a few thousand atoms for various symmetric tilt grain boundaries in aluminum. The Embedded Atom Method was then used to examine vacancy and interstitial formation energies and diffusion paths for each atom in the grain boundary unit cell. This led to some interesting insights. For example, it was discovered that in some grain boundaries, formation and migration energies for interstitials from some cell atoms are small enough that interstitial atoms may sometimes make a significant contribution to atomic migration. EAM also provided qualitative and quantitative insight into why the addition of a small amount of copper to aluminum increases the resistance of the material to electromigration [7]. In Richards *et. al.* [8], EAM backed up by DFT was used to quantify the morphologies and formation energies of small voids in aluminum and copper grain boundaries. Contrary to what happens in bulk crystal, it was found the minimum energy configuration of small voids in grain boundaries is not necessarily open. Instead, vacancies tend to line up on every other lattice plane within the grain boundary, producing a flat, porous structure (Fig. 5(b)). Presumably, a grain boundary void would transition to an open structure if it became large enough. Some of the calculated parameters in these studies were subsequently used in a continuum model for vacancy transport and stress evolution during electromigration to predict failure times (taken to be the time to the first nucleation event) in aluminum [9]. The calculated activation energy for failure, 0.98 eV, was about 0.2 eV higher than experimental data (Fig. 5(c)). While this is not accurate enough for lifetime projections, it does show that multiscale, atomic-based modeling holds some promise for addressing this kind of problem.

4. A multiscale problem in chemical-mechanical polishing.

We describe a final example from chemical-mechanical polishing (CMP). During so-called back end processing of an integrated circuit, layers of interconnect metallization are deposited, patterned and etched, and then covered with an insulating dielectric material such as silicon dioxide from

TEOS. The insulating material inherits any topography variations that are present in the underlying metallization. If these variations are not removed before the next layer of metal is deposited, each successive layer becomes rougher and therefore harder to pattern using photolithography.

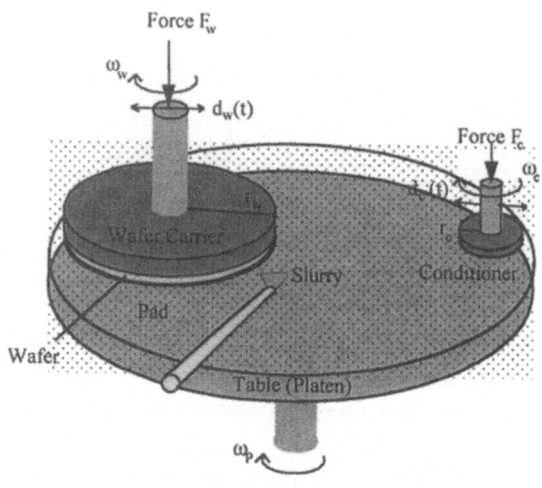

FIG. 6. *Schematic of a chemical-mechanical polishing tool.*

CMP is now the procedure of choice for planarizing successive layers of dielectric. In this process, the wafer with its still rough dielectric coating is held inverted by a wafer carrier and pressed against a soft open cell polyurethane polishing pad (Fig. 6). Both the pad and wafer are rotated in the same direction and at about the same speed, producing an approximately constant sliding speed between the wafer and pad surface. A chemically reactive slurry containing very fine abrasive particles is simultaneously sprayed in front of the wafer. The combined chemical and abrasive action of the layer of slurry entrained between the wafer and pad removes the high spots on the dielectric layer. However, the details of the planarization mechanism at work in CMP are not fully understood; for example, the source of systematic wafer scale polish rate nonuniformities [10] (Fig. 7) is not quantitatively clear.

Levert [11] and Shan [12] measured an unexpected subambient, or suction fluid pressure at the leading edge of the interface between an instrumented stainless steel fixture and a commercial polishing pad (Fig 8). Often, the fluid pressure at the trailing edge was also above ambient. It is possible that variations in the fluid pressure influence the solid contact

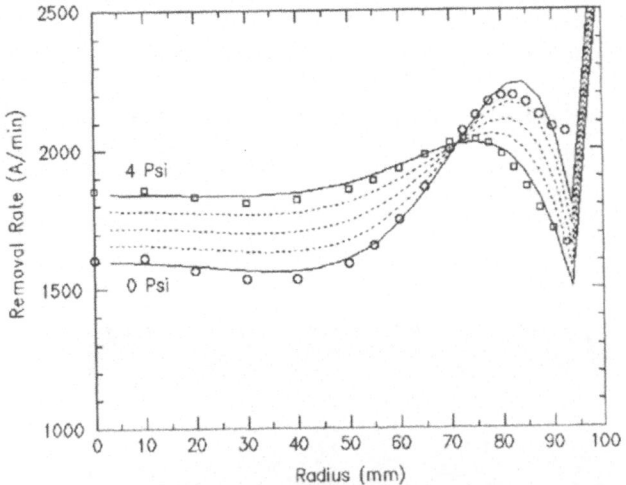

FIG. 7. *Nonuniform silicon dioxide CMP removal rates.*

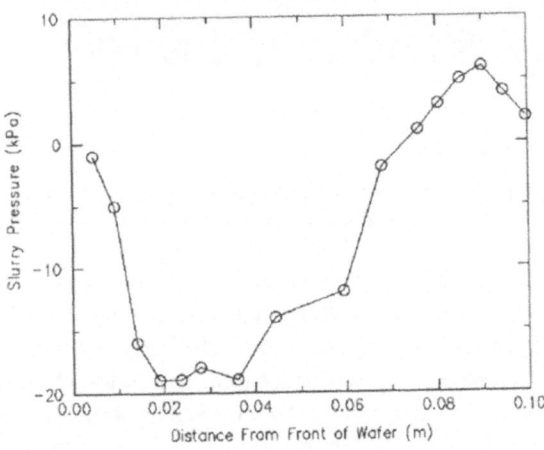

FIG. 8. *The fluid pressure measured under a stainless steel CMP polishing fixture [12].*

pressure between the pad asperities and the wafer, thereby affecting the polish rate uniformity. The fluid pressure distribution in turn seems to depend on the structure of the surface of the pad (Fig. 9), which contains pores whose average diameter is about 30 μm. The connection between the fluid pressure and the surface roughness is provided through an averaged form of the Reynolds equation-proposed by Patir and Cheng [13],

$$(1) \quad \frac{\partial}{\partial x}\left(\phi_x \frac{h^3}{12\mu}\frac{\partial p_f}{\partial x}\right) + \frac{\partial}{\partial y}\left(\phi_y \frac{h^3}{12\mu}\frac{\partial p_f}{\partial y}\right) = \frac{V}{2}\frac{\partial \bar{h}_T}{\partial x} + \frac{V}{2}\sigma\frac{\partial \phi_s}{\partial x}.$$

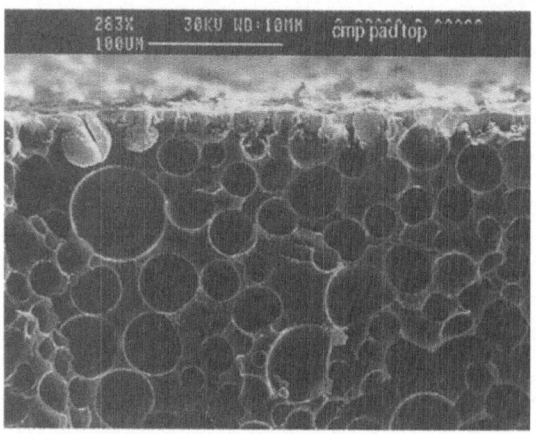

FIG. 9. *The surface of a polyurethane pad used in oxide CMP contains numerous pores that are formed when the void-filled pad is cut or conditioned.*

In (1), ϕ_x and ϕ_y are called pressure flow factors and ϕ_s is called the shear flow factor. The shear flow factor is a measure of additional flow transport due to the roughness of the sliding surfaces, in this case the pad. Patir and Cheng proposed functional forms for the pressure and flow factors that provided a good fit to numerical solutions for surfaces with randomly generated Gaussian roughness. Their funcational forms depend on the standard deviation of the roughness of each surface and their average separation. Thus, wafer scale polish rate variations in oxide CMP may be a multiscale phenomenon that depends on the distribution of surface roughness on the pad. Recently, Kirill [14] used the method of homogenization to calculate the pressure and shear flow factors in the case where the pad has either bidirectional sinusoidal roughness or has a surface consisting of a regular array of hemispherical cavities. The results were in general numerical agreement with Patir and Cheng, although for sinusoidal roughness, the functional forms were different.

REFERENCES

[1] T.P. MERCHANT, M.K. GOBBERT, T. CALE, AND L. BORUCKI, "Multiple scale integrated modeling of deposition processes," Thin Solid Films, **365**: 368–375 (2000).

[2] M.K. GOBBERT, T.P. MERCHANT, L.J. BORUCKI, AND T.S. CALE, "A Multiscale Simulator for Low Pressure Chemical Vapor Deposition," IMA Preprint Series, **#1466**, Feb. 1997.

[3] T.S. CALE AND V. MAHADEV, "Feature Scale Transport and Reaction during Low-Pressure Deposition Processes," Thin Solid Films, **22**: 176–276 (1996).

[4] A. FRIEDMAN, "Free Boundary Problems in Science and Technology," Notices of the AMS, **47**(8): 854–861 (2000).

[5] Z. WANG, Y. LI, AND J. ADAMS, "Kinetic Lattice Monte Carlo Simulation of Facet Growth," Surface Science, V. 450/1-2, pp. 51–63 (2000).

[6] C.-L. LIU, X.-Y. LIU, AND L.J. BORUCKI, "Defect Generation and Diffusion Mechanisms in Al and Al-Cu," Mat. Res. Soc. Symp. Proc., **516**: 189–192 (1998).

[7] X.-Y. LIU, C.-L. LIU, AND L.J. BORUCKI, "A New Investigation of Copper's Role in Enhancing Al-Cu Interconnect Electromigration Restance from an Atomistic View," *Acta. Mater.*, **47**(11): 3227–3231 (1999).

[8] D. RICHARDS, J. ADAMS, AND L. BORUCKI, "Modeling Nucleation and Growth of Voids During Electromigration," Proceedings of the Second International Conference on Modeling and Simulation of Microsystems, San Juan, Puerto Rico, April 1999, pp. 455–458.

[9] M.E SARYCHEV, YU. V. ZHITNIKOV, L. BORUCKI, C.-L. LIU, AND T.M. MAKHVI-LADZE, "General model for mechanical stress evolution during electromigration," J. Appl. Phys., **86**(6): 3068–3075 (15 Sept. 1999).

[10] C. SRINIVASA-MURTHY, D. WANG, S.P. BEAUDOIN, T. BIBBY, K. HOLLAND, AND T.S. CALE, "Stress distribution in chemical-mechanical polishing," Thin Solid Films, **308–309**: 533–537 (31 Oct 1997).

[11] J.A. LEVERT, "Surface Mechanics of Chemical Mechanical Polishing for Integrated Circuit Planarization," Ph.D. Dissertation, Georgia Institute of Technology, 1997.

[12] L. SHAN, "Fluid Pressure Distribution at the Interface Between a Compliant and Hard Surfaces," Masters Dissertation, Georgia Institute of Technology, September 1998.

[13] N. PATIR AND H.S. CHANG, "Application of Average Flow Model to Lubrication Between Rough Sliding Surfaces," Transactions of the ASME, **101**: 220–230, April 1979.

[14] D. KIRILL, "Homogenized Reynolds Equation for Rough Surfaces," private communication.

NONRESONANT SMOOTHING
FOR COUPLED WAVE + TRANSPORT EQUATIONS
AND THE VLASOV-MAXWELL SYSTEM

FRANÇOIS BOUCHUT*, FRANÇOIS GOLSE†, AND CHRISTOPHE PALLARD‡

Abstract. Consider a system (see (1.1) below) consisting of a linear wave equation coupled to a transport equation. Such a system is called *nonresonant* when the maximum speed for particles governed by the transport equation is less than the propagation speed in the wave equation. Velocity averages of solutions to such nonresonant coupled systems are shown to be more regular than those of either the wave or the transport equation alone. This smoothing mechanism is reminiscent of the proof of existence and uniqueness of C^1 solutions of the Vlasov-Maxwell system by Glassey-Strauss [9] for time intervals on which particle momenta remain uniformly bounded. Applications of our smoothing results to solutions of the Vlasov-Maxwell system are discussed.

Key words. Wave equation, transport equation, regularity of solutions, Vlasov-Maxwell system.

AMS(MOS) subject classifications. 35B65, 35B34, 35L05, 35Q75, 82C40, 82D10.

1. Nonresonant coupled wave + transport systems. Consider a coupled system consisting of a linear wave equation and a transport equation, of the form

$$(1.1) \qquad \begin{aligned} \Box_{t,x} u &= f, \\ (\partial_t + v(\xi) \cdot \nabla_x) f &= P(t, x, \xi, D_\xi) g, \end{aligned}$$

where $\Box_{t,x} = \partial_t^2 - \Delta_x$. The unknowns in that system are the real-valued functions $u \equiv u(t, x, \xi)$ and $f \equiv f(t, x, \xi)$, while the source term in the right-hand side of the transport equation involves a given real-valued function $g \equiv g(t, x, \xi)$. The notation $P(t, x, \xi, D_\xi)$ designates a (smooth) linear differential operator in the variable ξ only, while $v \equiv v(\xi)$ is a smooth $\mathbf{R}^D$-valued vector field on $\mathbf{R}^M$.

The system (1.1) is posed for all $(t, x, \xi) \in \mathbf{R}_+^* \times \mathbf{R}^D \times \mathbf{R}^M$. Associated to this system are the initial conditions

$$(1.2) \qquad \begin{aligned} u_{|t=0} &= u_I, \\ \partial_t u_{|t=0} &= u_I', \\ f_{|t=0} &= f_I, \end{aligned}$$

---

*C.N.R.S. and D.M.A., Ecole Normale Supérieure, 45 rue d'Ulm, 75230 Paris cedex 05 France; E-mail: fbouchut@dma.ens.fr.

†Institut Universitaire de France and D.M.A., Ecole Normale Supérieure, 45 rue d'Ulm, 75230 Paris cedex 05 France; E-mail: golse@dma.ens.fr.

‡Ecole Normale Supérieure de Cachan, 61 avenue du Président Wilson, 94235 Cachan Cedex; E-mail: pallard@dptmath.ens-cachan.fr.

where the functions u_I, u'_I, f_I, together with g, are the data of the Cauchy problem (1.1)-(1.2).

The subject matter of this work is the local regularity of averages with respect to ξ of the unknown u, namely of functions of the form

$$(1.3) \qquad \rho_\chi \equiv \rho_\chi(t,x) = \int u(t,x,\xi)\chi(\xi)d\xi ,$$

where χ is an arbitrary test function in $C_c^\infty(\mathbf{R}^M)$.

One possible approach to this problem would be

- to first establish the regularity of velocity averages of the solution f of the transport equation

$$\int f(t,x,\xi)\chi(\xi)d\xi ;$$

- and since averaging in ξ commutes with the d'Alembert $\square_{t,x}$ operator, to infer the regularity of ρ_χ from the classical energy estimate for the wave equation

$$\square_{t,x}\rho_\chi = \int f\chi d\xi ,$$

the regularity of its right hand side obtained at the previous step and that of the initial data u_I, u'_I.

Step 1 in this procedure is by now classical in kinetic theory: for smooth, generic v's,

$$(1.4) \quad \int f\chi d\xi \in H_{loc}^{\frac{1}{2(m+1)}}(\mathbf{R}_+ \times \mathbf{R}^D) \text{ if } g \text{ and } f \in L_{loc}^2(\mathbf{R}_+ \times \mathbf{R}^D \times \mathbf{R}^D)$$

where m is the order of the differential operator $P(t,x,\xi,D_\xi)$ involved in the right-hand side of the transport equation in (1.1). This gain of regularity was observed for the first time in [12], [13] for $m = 0$ and [6] for $m \in \mathbf{N}^*$ and is referred to as smoothing by Velocity Averaging. The precise condition on a smooth vector field v required for (1.4) to hold is that

$$(VA) \quad \sup_{\epsilon>0} \sup_{(\omega,k)\in\mathbf{R}\times\mathbf{R}^D} \frac{1}{\epsilon}\text{meas}\,(\{\xi \in \text{supp}\,\chi \,|\, |\omega + v(\xi)\cdot k|$$
$$\leq \epsilon\sqrt{\omega^2 + |k|^2}\}) < +\infty .$$

The classical energy estimate for the wave equation (see [14], formula (6.3.1)) finally implies that

$$(1.5) \quad \rho_\chi \in H_{loc}^{1+\frac{1}{2(m+1)}}(\mathbf{R}_+ \times \mathbf{R}^D) \text{ if } g \text{ and } f \in L_{loc}^2(\mathbf{R}_+ \times \mathbf{R}^D \times \mathbf{R}^M),$$

provided that

$$u_I \in L^2(\mathbf{R}^M; H_{loc}^{1+\frac{1}{2(m+1)}}(\mathbf{R}_+ \times \mathbf{R}^D)), \quad u'_I \in L^2(\mathbf{R}^M; H_{loc}^{\frac{1}{2(m+1)}}(\mathbf{R}_+ \times \mathbf{R}^D)) .$$

However, this method fails to predict the exact amount of regularity on ρ_χ for a large class of systems (1.1), namely those for which

$$(NR) \qquad v_M := \sup_{\xi \in \text{supp } \chi} |v(\xi)| < 1.$$

The relevance of this condition comes from physical considerations. Various kinetic models describe the coupling of particle transport with a background electromagnetic field. For massive particles with uniformly bounded momenta (i.e. with momenta in the support of χ), the maximum speed of transport is less than the speed of light (normalized here to 1), with a uniform bound as in (NR).

In order to gain some intuition on the role of this non-resonance condition (NR) in the regularity problem for ρ_χ as in (1.3), we propose the following line of reasoning in the case where $P(t, x, \xi, D_\xi)$ is the identity (or equivalently, $m = 0$). To avoid unnecessary complications, we also assume that the initial conditions u_I and u'_I are smooth.

Under assumption (NR), the characteristic manifold of the wave operator

$$\text{Char } (\square_{t,x}) = \{(t, x, \omega, k) \in T^*(\mathbf{R}_+^* \times \mathbf{R}^D) \mid \omega^2 - |k|^2 = 0\}$$

and that of the transport operator

$$\text{Char } (\partial_t + v(\xi) \cdot \nabla_x) = \{(t, x, \omega, k) \in T^*(\mathbf{R}_+^* \times \mathbf{R}^D) \mid \omega + v(\xi) \cdot k = 0\}$$

intersect at the zero section:

$$\text{Char } (\square_{t,x}) \cap \text{Char } (\partial_t + v(\xi) \cdot \nabla_x) = \{(t, x, 0, 0) \mid t > 0, \ x \in \mathbf{R}^D\}.$$

Consider a point $(t_0, x_0, \omega, k) \in T^*(\mathbf{R}_+^* \times \mathbf{R}^D)$ such that $(\omega, k) \neq (0, 0)$. Then

- either $(t_0, x_0, \omega, k) \notin \text{Char } (\square_{t,x})$, and thus u has two derivatives more than f microlocally at point (t_0, x_0, ω, k);
- or $(t_0, x_0, \omega, k) \notin \text{Char } (\partial_t + v(\xi) \cdot \nabla_x)$, and thus u has one derivative more than $w = (\partial_t + v(\xi) \cdot \nabla_x) u$ microlocally at point (t_0, x_0, ω, k); then in the scale of L^2-based Sobolev spaces, $w = (\partial_t + v(\xi) \cdot \nabla_x) u$ has one derivative more than g at (t_0, x_0, ω, k), independently of whether this point belongs to $\text{Char } (\square_{t,x})$ or not, by the usual energy estimate for the wave equation $\square_{t,x} w = g$ satisfied by w.

This little argument suggests that, for an arbitrary *fixed* ξ, if the differential operator $P(t, x, \xi, D_\xi) = Id$ (or more generally is of order 0) and $|v(\xi)| < 1$, then $u(\cdot, \cdot, \xi) \in H^2_{loc}(\mathbf{R}_+ \times \mathbf{R}^D)$ as soon as both f and $g \in L^2_{loc}(\mathbf{R}_+ \times \mathbf{R}^D)$[1].

---

[1] An analogous important observation was communicated by S. Klainerman to the second author: under integration along a time-like curve, solutions of the wave equation gain in regularity relative to the space variables. (This is a natural amplification of Proposition 2.7 in [15].)

This gain of regularity is not only better than (1.5) even in the case of $m = 0$, but also relies on a completely different mechanism, as witnessed by the fact that this smoothing effect occurs *pointwise in* ξ. At variance, the former procedure relies fundamentally on smoothing the solution of the transport equation by averaging in ξ, as implied by condition (VA). It also completely separates the roles of both the transport and wave equations in (1.1), while the new mechanism for smoothing described above is based on the *joint* properties of the transport and wave equations in (1.1).

Below we call this mechanism "nonresonant smoothing" in view of its analogy with the classical envelope theory for the Mathieu equation (see [2] §17 or D. Pesme's contribution in [5]).

The microlocal argument above fails however to indicate what happens in the important case where the differential operator in the right hand side of (1.1) has order $m > 0$; it also fails in the case where the regularity is measured in L^p-based Sobolev spaces for $1 \le p \le +\infty$, with $p \ne 2$ and in the case where the space dimension is 3. This is of course the most relevant case in view of physical applications (see the next section).

The outline of the paper is as follows:

- Section 2 states the main results on nonresonant smoothing, together with explicit counterexamples showing that our statements are sharp;
- Section 3 explains how the relativistic Vlasov-Maxwell (RVM) system can be put in the form (1.1);
- Section 4 discusses applications to the smoothness of solutions to the (RVM) system.

2. Nonresonant smoothing: main results. As suggested in the previous section, the most direct way to measure nonresonant smoothing is in L^2-based Sobolev spaces. Indeed these are natural spaces for the classical energy estimate of the wave equation.

THEOREM 2.1. *Let f and $g \in L^2_{loc}(\mathbf{R}^*_+ \times \mathbf{R}^D \times \mathbf{R}^M)$, and assume that the initial data $f_I \in L^2_{loc}(\mathbf{R}^D \times \mathbf{R}^M)$, that $u'_I \in L^2_{loc}(\mathbf{R}^M{}_\xi; H^1_{loc}(\mathbf{R}^D))$ while $u_I \in L^2_{loc}(\mathbf{R}^M{}_\xi; H^2_{loc}(\mathbf{R}^D))$. Let $P(t, x, \xi, D_\xi)$ be a linear differential operator of order $m \in \mathbf{N}$ on $\mathbf{R}^M{}_\xi$ with smooth coefficients. Pick $\chi \equiv \chi(\xi)$ be a test function in $C^m_c(\mathbf{R}^M)$ and let $v \equiv v(\xi)$ be in $C^m(\mathbf{R}^M)$ and satisfy the nonresonant condition (NR).*

Then, if (1.1)–(1.2) hold, the ξ-average

$$\rho_\chi(t, x) = \int u(t, x, \xi)\chi(\xi)d\xi$$

belongs to $H^2_{loc}(\mathbf{R}^*_+ \times \mathbf{R}^D)$.

The key argument in the proof of theorem 2.1 is that some well chosen combinations of the wave operator $\Box_{t,x}$ and of the transport operator $\partial_t + v(\xi) \cdot \nabla_x$ are elliptic in the variables t and x.

LEMMA 2.1. *For $\chi \in C_c^m(\mathbf{R}^M)$, let $v \equiv v(\xi)$ in $C^m(\mathbf{R}^M)$ satisfy the nonresonant condition (NR), and let $\lambda \in \mathbf{R}$. The two following conditions are equivalent:*

- *λ satisfies the condition*

(2.1) $$v_M^2 < \lambda < 1, \quad where \ v_M = \sup_{\xi \in supp \ \chi} |v(\xi)| \, ;$$

- *for each $\xi \in supp \ \chi$, the second order differential operator*

(2.2) $$Q_\xi^\lambda = \lambda \square_{t,x} - (\partial_t - v(\xi) \cdot \nabla_x)(\partial_t + v(\xi) \cdot \nabla_x)$$

   is elliptic.

When λ verifies any of these conditions, the symbol q_ξ^λ of the operator Q_ξ^λ satisfies the following uniform ellipticity estimates: for all $m \in \mathbf{N}$

(2.3) $$\sup_{\xi \in supp \ \chi} \sup_{\omega^2 + |k|^2 > 0} (\omega^2 + |k|^2) \left| D_\xi^m \left(\frac{1}{q_\xi^\lambda(\omega, k)} \right) \right| < +\infty.$$

The uniform ellipticity estimates (2.3) provide precisely the quantitative information missing in the little microlocal argument of the previous section and necessary to address the case where the source term of the transport equation in (1.1) effectively involves ξ derivatives. Notice that one could also use the operator $\lambda \square_{t,x} - (\partial_t + v(\xi) \cdot \nabla_x)^2$ instead of Q_ξ^λ.

Once lemma 2.1 is established, the proof of theorem 2.1 is based upon controlling $Q_\xi^\lambda u$ by the usual energy estimate for the wave equation. Finally, the uniform ellipticity estimates (2.3) are used to control the various contributions to the ξ-average ρ_χ after integrating by parts to bring all ξ-derivatives to bear on either χ or $1/q_\xi^\lambda$.

There is an analogous statement in space dimension 3, with L^2 and H^s replaced by L^p and $W^{s,p}$, for $1 < p < 2$ and $2 < p < \infty$. The proof is essentially the same as in the L^2 case, except for the arguments that involve the energy estimate for the wave equation. These are replaced by the fact that the elementary solution of the wave operator expressing u in terms of f is proportional to the uniform measure on the unit sphere S^2, to which one can apply the corollary to theorem 7 of [7]. One eventually finds that $\rho_\chi \in W_{loc}^{s,p}(\mathbf{R} \times \mathbf{R}^D)$ where $s = 1 + 2 \inf(\frac{1}{p}, \frac{1}{p'})$. Because this approach relies in the end on L^p estimates for the elliptic operator Q_ξ^λ, the cases $p = 1$ or $p = +\infty$ require a different treatment based on the commutation of the Lorentz boosts $L_j = x_j \partial_t + t \partial_{x_j}$, $j = 1, 2, 3$ with $\square_{t,x}$: this part bears some definite analogy with one of the key techniques in [9].

The result stated in theorem 2.1 is sharp, unlike that in (1.5).

First, the H^2 regularity is optimal. Indeed, let $f \equiv f(t,x)$ be in $H_{loc}^1(\mathbf{R}_+ \times \mathbf{R}^D)$ and such that the solution of

$$\square_{t,x} u = f, \quad u_{|t=0} = \partial_t u_{|t=0} = 0$$

belongs to $H^2_{loc}(\mathbf{R}_+ \times \mathbf{R}^D)$ and no better. Thus the ξ-average $\rho_\chi = u \int \chi d\xi$ clearly belongs to $H^2_{loc}(\mathbf{R}_+ \times \mathbf{R}^D)$ and no better. Finally, f solves the transport equation with right hand side $g = (\partial_t + v(\xi) \cdot \nabla_x)f \in L^2_{loc}(\mathbf{R}_+ \times \mathbf{R}^D \times \mathbf{R}^M)$. For such an f, one can take $f(t, x) = -A'(x_1 - t)$ where $A \in H^2_{loc}(\mathbf{R})$, and $u(t,x) = \frac{1}{2}tA(x_1 - t) - \frac{1}{4}\int_{x_1-t}^{x_1+t} A(z)dz$.

Second, the nonresonant condition (NR) cannot be dispensed with. Otherwise, ρ_χ is in general less regular than H^2, as shown by the following example. Suppose that $v(\xi) = (1, 0, \ldots, 0)$ for all ξ, and let f be of the form $f(t, x, \xi) = -B'(x_1 - t)$ for some $B \in H^1_{loc}(\mathbf{R})$. Clearly, one has $(\partial_t + v(\xi) \cdot \nabla_x)f = 0$, $f_{|t=0} = -B' \in L^2_{loc}(\mathbf{R}^D)$. Define $u(t,x) = \frac{1}{2}tB(x_1 - t) - \frac{1}{4}\int_{x_1-t}^{x_1+t} B(z)dz$; this function u satisfies $\Box_{t,x} u = f$, with $u_{|t=0} = \partial_t u_{|t=0} = 0$. However, one has $\rho_\chi = u \int \chi d\xi \in H^1_{loc}(\mathbf{R}_+ \times \mathbf{R}^D)$ and in general no better. Notice that here (1.5) does not hold since (VA) fails to be satisfied.

One final comment: the first counterexample above shows that the smoothing mechanism of Velocity Averaging cannot improve upon nonresonant smoothing when condition (NR) holds[2]. However, the same mechanism as in Velocity Averaging helps when condition (NR) fails, if it is known that the set of ξ-s for which (NR) is not verified is of small measure in some sense. This situation occurs in the relativistic Vlasov-Maxwell (RVM) system when only a few particles reach large momenta. We shall explain in Section 4 below how this last idea helps in studying the regularity of weak solutions of (RVM) as in [6] with smooth initial data.

We refer to [3] for further discussion of nonresonant smoothing and proofs of all the results in this section.

3. Kinetic formulation of Maxwell's equations; applications to the RVM system.
By a "kinetic formulation of Maxwell's equations", we mean a representation of the four components of the electromagnetic potential as moments of a single, scalar potential which depends of course on t and x but also on an extra variable ξ. The moments mentioned above are ξ-averages of this potential, like ρ_χ in (1.3).

Maxwell's system of equations in the vacuum reads

(3.1)
$$\begin{aligned}
\partial_t E - \nabla_x \wedge B &= -j, \\
\nabla_x \cdot E &= \rho, \\
\partial_t B + \nabla_x \wedge E &= 0, \\
\nabla_x \cdot B &= 0,
\end{aligned}$$

---

[2]This is consistent with the fact that Velocity Averaging lemmas are used neither in the Pfaffelmoser proof of global existence of classical solutions to the Vlasov-Poisson system (see [8]), nor in the corresponding Glassey-Strauss argument for the relativistic Vlasov-Maxwell system, which both deal with solutions having bounded support in the momentum variable. Recently R. Glassey confirmed to the second author that any attempt to use the Velocity Averaging method in order to simplify the arguments in [9] or extend their validity had not been successful yet.

where the unknowns $E \equiv E(t, x)$ and $B \equiv B(t, x)$ are respectively the electric and magnetic field, while the current $j \equiv j(t, x)$ and charge density $\rho \equiv \rho(t, x)$ are given. The system (3.1) is well-posed on $\mathbf{R}_+ \times \mathbf{R}^3$ in some appropriate class of functions once initial conditions are prescribed, as follows:

$$E_{|t=0} = E_I, \qquad B_{|t=0} = B_I, \tag{3.2}$$

where E_I and B_I are compatible with the second and fourth equations in (3.1) and provided that ρ and j satisfy the continuity equation

$$\partial_t \rho + \nabla_x \cdot j = 0. \tag{3.3}$$

Suppose now that, instead of the macroscopic quantities ρ and j, one is given a microscopic, phase-space density of charges $f(t, x, \xi)$, as in the kinetic theory of gases. In other words, $f(t, x, \xi)$ is the density of (like) charged particles (electrons or ions) which, at time t, occupy position x and have momentum ξ. The macroscopic density of charge and the current are given in terms of the microscopic density f by the formulas

$$\rho(t, x) = \int f(t, x, \xi) d\xi, \quad j(t, x) = \int f(t, x, \xi) v(\xi) d\xi, \tag{3.4}$$

where the velocity of particles with momentum ξ is expressed as

$$v(\xi) = \frac{\xi}{\sqrt{1 + |\xi|^2}} \tag{3.5}$$

in dimensionless variables. In kinetic theory, the continuity equation (3.3) is usually implied by a transport equation on f, of the form

$$\partial_t f + v(\xi) \cdot \nabla_x f = S, \quad \text{with} \quad \int S d\xi = 0. \tag{3.6}$$

In order to satisfy (3.1)-(3.2), we first choose a vector field $A_I \equiv A_I(x)$ such that

$$\nabla_x \wedge A_I = B_I, \quad \nabla_x \cdot A_I = 0, \tag{3.7}$$

and define $A^{(I)} \equiv A^{(I)}(t, x)$ by

$$\Box_{t,x} A^{(I)} = 0,$$
$$A^{(I)}_{|t=0} = A_I, \tag{3.8}$$
$$\partial_t A^{(I)}_{|t=0} = -E_I.$$

Solve then for $u \equiv u(t, x, \xi)$ the Cauchy problem for the wave equation

$$\Box_{t,x} u = f,$$
$$u_{|t=0} = 0, \tag{3.9}$$
$$\partial_t u_{|t=0} = 0.$$

Elementary computations based on the continuity equation (3.3) and the uniqueness of solutions to the Cauchy problem (3.1)-(3.2) show that

$$(3.10) \qquad \phi = \int u d\xi \,, \quad A = A^{(I)} + \int u v(\xi) d\xi$$

are respectively the scalar and vector potentials satisfying the wave equations

$$\Box_{t,x}\phi = \rho \,, \quad \Box_{t,x}A = j \,,$$

the Lorentz gauge condition

$$(3.11) \qquad \partial_t \phi + \nabla_x \cdot A = 0 \,,$$

and giving the electromagnetic field by the formulas

$$(3.12) \qquad \begin{aligned} E &= -\partial_t A - \nabla_x \phi = -\partial_t A^{(I)} - \partial_t \int u v(\xi) d\xi - \nabla_x \int u d\xi \,, \\ B &= \nabla_x \wedge A = \nabla_x \wedge A^{(I)} + \nabla_x \wedge \int u v(\xi) d\xi \,. \end{aligned}$$

Thus Maxwell's system of equations (3.1)-(3.2) can be replaced by the single scalar wave equation (3.9) with the continuity equation implied by (3.6).

This kinetic formulation of Maxwell's system of equations is of course very natural when the electromagnetic field is the self-consistent field of a plasma. This is precisely the situation described by the relativistic Vlasov-Maxwell system. In this case, the source term S in (3.6) is the term modeling the acceleration by the Lorentz force.

$$(3.13) \qquad \begin{aligned} \partial_t f + v(\xi) \cdot \nabla_x f &= -(E + v(\xi) \wedge B) \cdot \nabla_\xi f \,, \\ \partial_t E - \nabla_x \wedge B &= -j_f \,, \\ \nabla_x \cdot E &= \rho_f \,, \\ \partial_t B + \nabla_x \wedge E &= 0 \,, \\ \nabla_x \cdot B &= 0 \,, \end{aligned}$$

with $v(\xi)$ as in (3.5) and the notations

$$(3.14) \qquad \rho_f = \int f(t, x, \xi) d\xi \,, \quad j_f = \int f(t, x, \xi) v(\xi) d\xi \,.$$

This system for the unknown $(f, E, B) \equiv (f(t, x, \xi), E(t, x), B(t, x))$ is posed in $\mathbf{R}_+ \times \mathbf{R}_x^3 \times \mathbf{R}_\xi^3$ and is completed by the initial conditions

$$(3.15) \qquad f_{|t=0} = f_I \,, \quad E_{|t=0} = E_I \,, \quad B_{|t=0} = B_I \,.$$

The main results known to this date on (RVM) are

- the global existence of weak (and even renormalized) solutions, proved by R. DiPerna and P.-L. Lions [6];
- existence and uniqueness of classical solutions under the assumption that supp $f(t, x, \cdot)$ is bounded for each $t > 0$, proved by R. Glassey and W. Strauss [9].

Subsequently, the global existence and uniqueness of classical solutions to (RVM) was established in [10] under the weaker assumption that the macroscopic energy density satisfy

$$(3.16) \qquad \int \sqrt{1 + |\xi|^2} f d\xi \in L^\infty_{loc}(\mathbf{R}_+; L^\infty(\mathbf{R}^3)).$$

Finally, R. Glassey and W. Strauss established the global existence and uniqueness of classical solutions to (RVM) for small (in some sense) initial data in [11], by proving that (3.16) holds for such initial data.

The main open problem on (RVM) is to prove (or disprove) the same result as in [9] without assuming (3.16) or the support condition for all $t > 0$:

"Let f_I, E_I and B_I be compactly supported and C^∞. Does there exist a unique global C^∞ solution to the Cauchy problem (3.13)-(3.15)?"

The system (RVM) can be somewhat simplified by using the kinetic formulation of the Maxwell equation. It becomes

$$(3.17) \qquad \begin{aligned} \partial_t f + v(\xi) \cdot \nabla_x f &= \nabla_\xi \cdot [-(E + v(\xi) \wedge B)f], \\ \Box_{t,x} u &= f, \end{aligned}$$

where (E, B) are given in terms of u by (3.12). The initial conditions are

$$(3.18) \qquad f_{|t=0} = f_I, \quad u_{|t=0} = \partial_t u_{|t=0} = 0.$$

The formulation (3.17)-(3.18)-(3.12) of (RVM) is the main reason for considering coupled wave + transport systems as (1.1). It greatly simplifies the formulas in [9] representing the electromagnetic field in terms of the acceleration part in the transport equation of (RVM). Indeed, these formulas occupy 13 of the 32 pages in [9] and their complexity somewhat hinders a complete understanding of the key arguments in this otherwise carefully written paper.

Finally, let us mention that the functions u and $v(\xi)u$ with u as in (3.17) are natural physical quantities. They can be viewed as the Liénard-Wiechert potentials (see [17] §63) distributed under the initial microscopic density f_I.

4. Regularity of solutions of the (RVM) system. This section expands on the idea introduced in the last paragraph of Section 2, namely merging the techniques of Velocity Averaging with nonresonant smoothing. We concentrate on the important example of the (RVM) system, for which we have been able to establish the following *a priori* regularity result on the electromagnetic field.

THEOREM 4.1. *Consider initial data* (f_I, E_I, B_I) *such that* $f_I \in L^\infty(\mathbf{R}^3 \times \mathbf{R}^3)$, $f_I \geq 0$ *a.e.*, E_I *and* $B_I \in H^1_{loc}(\mathbf{R}^3)$ *satisfy*

$$(4.1) \qquad \nabla_x \cdot B_I = 0, \quad \nabla_x \cdot E_I = \int f_I d\xi,$$

and the finite energy condition

$$(4.2) \qquad \iint \sqrt{1 + |\xi|^2} f_I dx d\xi + \int (|E_I|^2 + |B_I|^2) dx < +\infty$$

holds. Let (f, E, B) *be a weak solution of the (RVM) system (the existence of which is predicted by [6]). If the macroscopic energy density satisfies*

$$(4.3) \qquad \int \sqrt{1 + |\xi|^2} f d\xi \in L^p_{loc}(\mathbf{R}_+ \times \mathbf{R}^3), \quad \text{with } p \in]\tfrac{3}{2}, 2]$$

then the electromagnetic field has regularity given by

$$(4.4) \qquad E \text{ and } B \in H^s_{loc}(\mathbf{R}^*_+ \times \mathbf{R}^3), \quad \text{with } s = \frac{2p - 3}{2p + 4}.$$

Before describing the main ideas in the proof of this result, let us stress a few points.

Observe first that the condition (4.3) is indeed weaker than the condition (3.16) under which R. Glassey and W. Strauss have proved in [10] the global existence and uniqueness of a classical solution, with E and B belonging to $L^\infty_{loc}(\mathbf{R}_+; W^{1,\infty}(\mathbf{R}^3))$. Accordingly, the regularity on E and B predicted by theorem 4.1 is weaker. Besides, taking $p = \infty$ and copying the proof of theorem 4.1 would not give the $W^{1,\infty}$ control of [10], which is based on iterating twice a rather intricate procedure which is close in spirit to the mechanism of nonresonant smoothing described above. It would instead give a weaker piece of information, namely that E and $B \in H^1_{loc}(\mathbf{R}_+ \times \mathbf{R}^3)$. It could be that the regularity predicted in theorem 4.1 is not optimal and would be improved by using part of the information in [9] or [10]. However, the assumption (4.3) is not more natural than (3.16). The only natural condition on the macroscopic energy density is that

$$\int \sqrt{1 + |\xi|^2} f(t, x, \xi) d\xi \in L^\infty_t(L^1_x)$$

which is guaranteed by the conservation of energy for (RVM)[3] but we have not yet been able to use it to control the density of particles with large

---

[3] Actually, the theory of weak solutions to the (RVM) only predicts that the total energy at time t is less than or equal to that at time 0, for any positive t.

momenta in a way compatible with nonresonant smoothing as suggested in the last paragraph of Section 2.

Our second main observation on theorem 4.1 is that the Sobolev regularity index it predicts exceeds that predicted by Velocity Averaging. For example, in the case where (4.3) holds with $p = 2$, a direct application of the Velocity Averaging lemma of [6] (or Theorem 1.5.6 of [4]) would imply that

$$(4.5) \qquad \rho = \int f d\xi \text{ and } j = \int v(\xi) f d\xi \in H^{1/16}_{loc}(\mathbf{R}^*_+ \times \mathbf{R}^3)$$

which, by the classical energy estimate for Maxwell's system, entails that

$$E \text{ and } B \in H^{1/16}_{loc}(\mathbf{R}^*_+ \times \mathbf{R}^3).$$

This regularity is indeed weaker than the one predicted by theorem 4.1, in this case that E and $B \in H^{1/8}_{loc}(\mathbf{R}^*_+ \times \mathbf{R}^3)$. At variance with the Velocity Averaging method however, theorem 4.1 says nothing of the regularity of the density of charge ρ and current j.

We now proceed to describe the main steps in the proof of theorem 4.1.

First, a simple interpolation argument leads to L^2 estimates on the charge and current densities.

LEMMA 4.1. *Let* $f \equiv f(t, x, \xi)$ *be a measurable function on* $\mathbf{R}_+ \times \mathbf{R}^3 \times \mathbf{R}^3$. *Then, for each* $\alpha \in [0, 1]$, *one has*

$$(4.6) \qquad \left\| \int |f| d\xi \right\|_{L^2_{t,x}} \leq 9 \|f\|_{L^\infty_{t,x,\xi}}^{\frac{\alpha}{\alpha+3}} \left\| \int |\xi|^\alpha |f| d\xi \right\|_{L^{\frac{6}{\alpha+3}}_{t,x}}^{\frac{3}{\alpha+3}}.$$

Further, for each $R > 0$

$$(4.7) \qquad \left\| \int_{|\xi|>R} |f| d\xi \right\|_{L^2_{t,x}} \leq \frac{9}{R^{\frac{3(1-\alpha)}{\alpha+3}}} \|f\|_{L^\infty_{t,x,\xi}}^{\frac{\alpha}{\alpha+3}} \left\| \int \sqrt{1+|\xi|^2} |f| d\xi \right\|_{L^{\frac{6}{\alpha+3}}_{t,x}}^{\frac{3}{\alpha+3}}.$$

The second step in the proof of theorem 4.1 is a more accurate version of lemma 2.1

LEMMA 4.2. *Let* $v \equiv v(\xi) \in W^{1,\infty}(\mathbf{R}^3)$ *satisfy* $|v(\xi)| < 1$ *for all* $\xi \in \mathbf{R}^3$. *For each* $\lambda \in]|v(\xi)|^2, 1[$, *set* $q^\lambda_\xi(\omega, k) = \omega^2 - |v(\xi) \cdot k|^2 - \lambda(\omega^2 - |k|^2)$. *Then, for each* $\xi \in \mathbf{R}^3$,

$$(4.8) \qquad \inf_{|v(\xi)|^2 < \lambda < 1} \sup_{\omega^2 + |k|^2 > 0} \frac{\omega^2 + |k|^2}{q^\lambda_\xi(\omega, k)} = \frac{2}{(1 - |v(\xi)|^2)},$$

with the inf attained at $\lambda(\xi) = \frac{1}{2}(1 + |v(\xi)|^2)$. *Likewise,*

$$(4.9) \qquad \inf_{|v(\xi)|^2 < \lambda < 1} \sup_{\omega^2 + |k|^2 > 0} (\omega^2 + |k|^2) \left| D_\xi \frac{1}{q^\lambda_\xi(\omega, k)} \right| \leq \frac{8 \|\nabla v\|_{L^\infty}}{(1 - |v(\xi)|^2)^2}.$$

With these estimates at our disposal, the proof of theorem 4.1 follows the line of the Velocity Averaging method. The main idea, as can be seen in [12], [13] and [6], consists in splitting the momentum space in two regions:

- in the first region, defined by the inequality $|\xi| \leq R$, the speed of particles is bounded by $|v(\xi)| \leq \frac{R}{\sqrt{1+R^2}}$; hence the condition (NR) is satisfied so that nonresonant smoothing holds for the electromagnetic field created by these particles; further, the ellipticity estimates (4.8) and (4.9) control the growth of the H^1 norm of this part of the electromagnetic field as $R \to +\infty$;
- in the second region, defined by the inequality $|\xi| > R$, the control on the macroscopic energy density (4.3) together with the estimates (4.6) and (4.7) imply that the densities of charge and current created by the corresponding particles are small in L^2 as $R \to +\infty$; the L^2 norms of the corresponding fields are then controlled by the classical energy estimate for Maxwell's system of equations.

Hence the electromagnetic field can be split as the sum of a field whose H^1 norm tends to infinity with R and of a field whose L^2 norm vanishes with $1/R$. One concludes by a straightforward interpolation argument.

5. Conclusion. In this note, we have discussed a mechanism of nonresonant smoothing for a wave equation coupled to a transport equation. The main application of this idea seems to be the (RVM) system, with Maxwell's system of equations reduced to a scalar wave equation by a kinetic formulation reminiscent of the notion of Liénard-Wiechert potentials. While theorem 4.1 exemplifies the power of using the method of Velocity Averaging together with nonresonant smoothing, it is very likely that the regularity statement (4.4) is not optimal. However, we believe that the idea of using (some form of) nonresonant smoothing for the vast majority of particles with momenta below a certain threshold together with the method of Velocity Averaging to control the few particles with higher momenta might prove helpful in the question of global existence of classical solutions to the (RVM) system, still open to this date except for special cases (small data, almost neutral initial data, $2 + 1/2D$ solutions...)

Finally, this mechanism of nonresonant smoothing is by no means confined to the examples considered here, but can easily be generalized to a wide generality of systems of coupled hyperbolic equations. For example, one could think of models arising in the study of the laser-plasma interaction or Langmuir turbulence. Such models involve a wave equation for the electromagetic field and another wave equation for the acoustic disturbances in the plasma coupled by terms involving in particular the ponderomotive force: see [1], or the contribution by D. Pesme in [5]. The nonresonance condition adapted to a system of two coupled wave equations reduces to saying that the speeds of propagation in both wave equations are different, a condition obviously satisfied in the case of laser-plasma interaction, where the speed of light is to be compared to the speed of sound in the plasma. In this case, nonresonant smoothing entails a gain of 3 derivatives on one of the fields, which might be useful in the mathematical treatment of models such as considered in [1].

Acknowledgement. This work is partially supported by the European T.M.R. network "Asymptotic methods in kinetic theory", contract no. ERB- FMRXCT970157.

Note added in proof. S. Klainerman recently informed the second author of a new approach to the Glassey-Strauss theorem by himself and G. Staffilani [16]. One of the steps in their proof is similar to the integration by parts in ξ using estimate (2.3).

REFERENCES

[1] BERGÉ, L.; BIDEGARAY, B.; AND COLIN, T. *A perturbative analysis of the time envelope approximation in sttrong Langmuir turbulence.* Phys. D 95 (1997), 351–379.

[2] BOGOLIUBOV, N.N. AND MITROPOLSKY, Y.A. *Asymptotic methods in the theory of non-linear oscillations.* Translated from the second revised Russian edition. International Monographs on Advanced Mathematics and Physics Hindustan Publishing Corp., Delhi, Gordon and Breach Science Publishers, New York 1961.

[3] BOUCHUT, F.; GOLSE, F.; AND PALLARD, C. *Work in preparation.*

[4] BOUCHUT, F.; GOLSE, F.; AND PULVIRENTI, M. *Kinetic Equations and Asymptotic Theory.* L. DESVILLETTES & B. PERTHAME eds.. Series in Applied Mathematics, 4. Gauthier-Villars, Editions Scientifiques et Médicales Elsevier, Paris; North-Holland, Amsterdam, 2000.

[5] DAUTRAY, R. AND WATTEAU, J. *La fusion thermonucléaire inertielle par laser.* Eyrolles, Paris 1993.

[6] DiPERNA, R. AND LIONS, P.-L. *Global weak solutions of Vlasov-Maxwell systems.* Comm. Pure Appl. Math. 42 (1989), no. 6, 729–757.

[7] FEFFERMAN, C. AND STEIN, E.M. H^p *spaces of several variables.* Acta Math. 129 (1972), no. 3–4, 137–193.

[8] GLASSEY, ROBERT T. *The Cauchy problem in kinetic theory.* Society for Industrial and Applied Mathematics (SIAM), Philadelphia, PA, 1996

[9] GLASSEY, ROBERT T. AND STRAUSS, WALTER A. *Singularity formation in a collisionless plasma could occur only at high velocities.* Arch. Rational Mech. Anal. 92 (1986), no. 1, 59–90.

[10] GLASSEY, ROBERT T. AND STRAUSS, WALTER A. *High Velocity Particles in a Collisionless Plasma.* Math. Meth. Appl. Sci. 9 (1987), 46–52.

[11] GLASSEY, ROBERT T. AND STRAUSS, WALTER A. *Absence of Schocks in an Initially Dilute Collisionless Plasma.* Comm. Math. Phys. 113 (1987), 191–208.

[12] GOLSE, F.; PERTHAME, B.; AND SENTIS, R. *Un résultat de compacité pour les équations de transport et application au calcul de la valeur propre principale d'un opérateur de transport.* C. R. Acad. Sci. Paris Sér. I Math. 301 (1985), 341–344.

[13] GOLSE, F.; LIONS, P.-L.; PERTHAME, B.; AND SENTIS, R. *Regularity of the moments of the solution of a transport equation.* J. of Funct. Anal. 76, (1988), 434–460.

[14] HÖRMANDER, L. *Lectures on nonlinear hyperbolic differential equations.* Mathématiques & Applications [Mathematics & Applications], 26. Springer-Verlag, Berlin, 1997.

[15] KLAINERMAN, S. AND MACHEDON, M. *Finite energy solutions of the Yang-Mills equations in* $\mathbf{R}^{3+1}$. Ann. of Math. (2) 145, (1995), 39–119.

[16] KLAINERMAN, S. AND STAFFILANI, G. *A new approach to study the Vlasov-Maxwell system,* preprint.

[17] LANDAU, L. AND LIFSHITZ, E. *Cours de physique théorique. Vol. 2: Théorie des champs.* Editions Mir, Moscou, 1970.

INTEGRATED MULTISCALE PROCESS SIMULATION IN MICROELECTRONICS

TIMOTHY S. CALE*, MAX O. BLOOMFIELD*, DAVID F. RICHARDS*,
SOFIANE SOUKANE*, KENNETH E. JANSEN†, JOHN A. TICHY‡, AND
MATTHIAS K. GOBBERT§

Abstract. We discuss selected applications of integrated multiscale process simulation (IMPS) that are particularly relevant to integrated circuit fabrication. We first summarize approaches to IMPS for two processes for which the governing equations are well accepted. In these cases, models for equipment scale (meter), pattern scale (mm) and feature scale (micron) are solved simultaneously. The first approach uses regular grids, and is applied to low-pressure chemical vapor deposition (LPCVD) of silicon dioxide from tetraethoxysilane (TEOS). The second approach uses unstructured meshes, and is applied to electrochemical deposition (ECD) of copper. The goal is to develop approaches to estimate "loading" in these processes; *i.e.*, the effects of pattern density and topography on local deposition rates. This is accomplished by resolving pattern (mesoscopic, mm) scales, which are between equipment (0.1-1 m) and feature scales (0.1-1 μm). In this work, we focus on steady state simulation results. We close the discussion of deposition processes with a few thoughts on extending IMPS to the grain scale, and the conversion of discrete atomistic representations to continuum representations of islands during deposition. We end by discussing progress made towards IMPS for chemical mechanical planarization (CMP). In this example, well-accepted models or relevant simulators do not exist for any scale of the process.

1. Introduction. Wafers undergo hundreds of processes during the fabrication of integrated circuits (ICs). Each process is intended to accomplish a specific change in wafer state. Relatively simple models can be used to represent many of these processes. On the other hand, changes in wafer state desired for many processes require more complex models for reliable process development. In general, process modeling and simulation have been used to gain understanding, and have not been relied upon for quantitative predictions of changes in wafer state. Equipment scale simulation has gained acceptance as a tool to quantitatively address issues of reactor design, optimization and prediction of blanket wafer scale properties such as growth rates [1]. Feature scale simulations, on the other hand, are used to predict the evolution of surface topography and composition during processing [1-3]. The reactor scale drives the process, but the change in wafer state is the bottom line; *i.e.*, the ICs are packaged and sold. Simulators for one scale do not adequately address the other scale, resulting in limited

---

*Focus Center–New York: Rensselaer, Rensselaer Polytechnic Institute, Troy, NY 12180.

†Scientific Computation Research Center, Rensselaer Polytechnic Institute, Troy, NY 12180.

‡Mechanical Engineering, Aeronautical Engineering and Mechanics, Rensselaer Polytechnic Institute, Troy, NY 12180.

§Department of Mathematics and Statistics, University of Maryland, Baltimore County, Baltimore, MD 21250.

predictive capability for processes over patterned wafers. Integrating simulators for these scales is attractive [1], as it allows self-consistent estimates of wafer scale uniformity and feature scale evolution as functions of position. This can reduce process development and optimization time and cost. The basic difficulty in merging equipment and feature scale simulators is the disparity in length scales, which span about six orders of magnitude, from 0.1–1 m to 0.1–1 μm.

In this paper, we attempt to summarize the state of the art in integrated multiscale process simulation (IMPS) related to IC fabrication. We begin by discussing two approaches to IMPS for processes whose governing equations are well-accepted, chemical vapor deposition (CVD) and electrochemical deposition (ECD). The purpose is to demonstrate approaches to IMPS. As in many modeling and simulation efforts, the accuracy of the calculated results is limited by the accuracy of the transport and reaction kinetic models available for a given process.

We then briefly discuss our efforts to extend IMPS for deposition to the growth of islands, and the formation of polycrystalline films. Grain growth modeling is a rapidly evolving field [4], and we restrict our attention to briefly discussing our work related to IMPS for deposition processes, and the conversion from discrete atomistic to continuum representations.

We end with a discussion of IMPS for chemical mechanical planarization (CMP). After presenting our approach to wafer scale CMP modeling, we focus on the unique challenges facing IMPS for CMP.

2. IMPS for deposition. Deposition processes play a central role in IC manufacturing. After being deposited, films are often patterned using lithography and etch techniques. Sequences of deposition and etch steps can be used to fabricate device and circuit structures; *e.g.*, to form the network of conducting lines that are inherent to ICs. Deposition processes in use range from physical vapor deposition (PVD) processes such as evaporation and sputtering, to thermal CVD (at low, medium and high pressures), to plasma enhanced CVD, to ECD processes. The equations that govern transport and reaction in thermal CVD and ECD are well accepted, and these are natural processes on which to focus our IMPS studies. IMPS is realized by coupling models for different length scales; the results from larger scales are fed to the smaller scales, and the results from the smaller scales are fed back up to form a tightly coupled solution. In the following sections, two approaches are described, each with an application.

2.1. Regular grids: LPCVD of silicon dioxide. Gobbert *et al.* [5, 6] presented the first approach to true IMPS for deposition processes. The method takes advantage of homogenization, through which the wafer surface can be considered flat in models for larger spatial scales, even in the presence of patterns. They discussed a two-scale and a three-scale approach. In the two-scale approach, the reactor scale is coupled to the feature scale using representative features at the surface nodes of the mesh, if features exist at that node. Thus, the effect of having a larger deposition surface per flat wafer surface is included if the grid is fine enough to have nodes where patterns exist. In the three-scale approach, a pattern scale (a few mm wide) regular finite element grid extends from the wafer surface back into the reactor (on the order of a mm), and is used to resolve the patterned region of the wafer. As in the two-scale approach, solving a feature scale problem at appropriate surface nodes of the pattern scale grid accounts for the increased surface area per flat area due to patterned regions. This multi-grid approach not only avoids the large amount of grid resolution that is necessary for a single model, but also allows for the capture of the underlying physics by varying the model description according to the scale. Thus the model physics can be changed from continuum to the molecular flow (ballistic transport) regime at length scales where the mean free path becomes comparable to geometry dimensions; *i.e.*, the Knudsen number becomes large.

Figure 1 depicts the transition between models for three different scales, and shows a representative axisymmetric, finite element grid of the reactor and mesoscale, and indicates that feature geometry and pitch are considered. When using a two-scale model, which includes the reactor and feature scales, the region between the grid nodes of the reactor scale are implicitly assumed to have a uniform pattern corresponding to the feature density as simulated on the feature scale. This implies that feature size and pitch are uniform on a length scale associated with the inter-nodal distance, and results in a simplified representation of the prescribed pattern density. In the three-scale model, the reactor scale nodes reflect the local pattern density associated with the mesoscale. The grid nodes on the mesoscale get net fluxes that correspond to the feature density. Thus, additional information about the changes in feature density is obtained in the three-scale model that is absent in the two-scale model [6].

The initial guess of species concentrations over an element on the reactor scale is interpolated onto the mesoscale grid using the finite element basis functions. The mesoscale model is then solved in a coupled fashion with multiple feature scale simulations at each node representing a patterned area. The guesses for concentrations at the individual nodes are fed into the feature scale simulator, which returns a net flux of each species at that node. The mesoscale and feature scale models are then iterated until convergence. The net flux of each species within the converged mesoscale solution is passed to the reactor scale, along with the net fluxes associated with the flat regions of the wafer, which then provides a new guess. Gobbert *et al.* [6] provide more details on homogenization as used here, for steady state simulations. A fully coupled transient simulation is more complicated [7], because the feature scale simulator moves the surface corresponding to the time step taken in the reactor scale simulation, and return the resulting fluxes at the end of the time step. In addition, the current profile at every node has to be stored so that only an update is made to the current

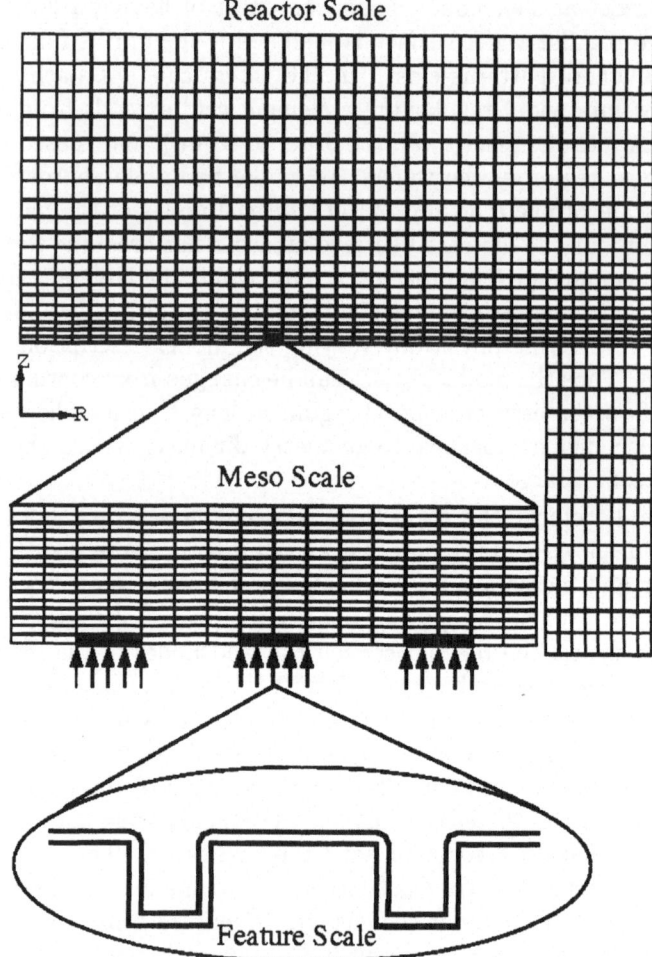

FIG. 1. *Reactor scale, mesoscale and feature scale simulation grids and transition scheme.*

profile at a subsequent time. While this is not difficult to conceptualize, formulation of the problem for arbitrary features and nodes is a non-trivial bookkeeping task. The addition of other intermediate simulation scales is straightforward, as long as the continuum equations are valid. On the feature scale, the particulate nature of the species transport is taken into account, and is "driven" by transferring the flux distribution of each species at the feature surface.

The LPCVD of SiO_2 from TEOS is used to demonstrate IMPS. The kinetic model for the gas phase involves six species involved in four reactions [8–10]. The major gas phase intermediates are triethoxysilanol, water, ethylene, and ethanol. In addition, there are six surface species involved in eight surface reactions, whose primary byproducts are water, ethylene and ethanol. The reaction mechanism and kinetics for this process have not been established, as is discussed in more detail in Ref. [10]. Our goal here is to demonstrate an approach to IMPS that works for complex kinetics, and not a particular model.

Process gases enter the water-cooled reactor at room temperature about 5 cm from the susceptor, which is held at 1000 K, and leave through an annular outlet at the bottom of the susceptor. For the simulations discussed here, argon is the inert carrier gas flowing at 2 slm, and the pressure is 0.01 atm. All the transport properties, such as mass diffusivities, thermal diffusivities, viscosity, etc. are determined using the CHEMKIN database [9], which is coupled into FIDAP [11]. The forward and reverse homogeneous reaction rates at the nodal points in the reactor volume are also computed using the CHEMKIN formulation for this process. Some of the grid nodes corresponding to elements of the wafer surface are flagged as having patterns, where the reactor scale model is coupled to either a mesoscale model (three-scale approach) or a feature scale model (two-scale approach). For nodes on unpatterned regions of the wafer, fluxes corresponding to a flat surface are returned to the reactor scale model. The local heterogeneous reaction rates (and fluxes) are computed by EVOLVE [12] using CHEMKIN. The transient simulations are started under the pseudo-steady state conditions described above. This corresponds to a situation where the reactor flow, concentration and temperature fields are stabilized at the specified operating conditions, with the patterned wafer exposed to the incoming reactive gases, but little growth has taken place to alter the original feature topography. Since the growth rates are much smaller than residence times for the flow, this is a reasonable approximation.

For most of the simulation results shown here, a single pattern about 3.3 mm wide is placed at about the halfway point of the wafer radius. The mesoscale model spans this distance, and in some cases contains three 0.4 mm wide clusters of features. The height of the mesoscale model is taken to be 1 mm, which is more than three times the mean free path for all species under the specified operating conditions. The grid for the mesoscale simulation and the feature scale geometry are also shown in Figure 1. For simulations with this single pattern, the individual features are infinite trenches of 1 μm height, 1 μm width, and a pitch of 3 μm.

Gobbert et al. [6] demonstrated the validity of the IMPS approach described, for pseudo-steady state. They showed that introducing the mesoscale model to a two-scale model does not change the computed concentrations of reactive intermediates for two extreme cases; a blanket wafer, and a uniform die. Figure 2 is a representative sample of the pseudo-steady state results from two-scale and three-scale models. It shows the mass fraction of the reaction byproduct water for three different cases, as functions of radial position on the wafer. The three cases shown in the graph correspond to the case of a blanket wafer that has no topography; a uniform

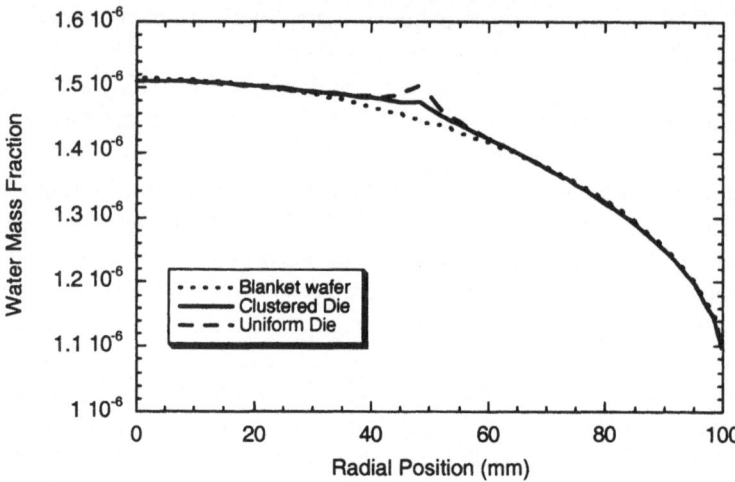

FIG. 2. *Mass fraction of water at the wafer surface as functions of radial position on the wafer at the reactor scale, for three cases; a blanket (unpatterned) wafer, a uniformly patterned die, and a die with three clusters of features, as described in the text.*

die case where the feature density extends uniformly throughout the 3.3 mm patterned area, and the case of three clusters of features across the die. The mass fraction of water increases as the feature density increases, consistent with intuition. The water mass fraction for the clustered die case falls in between the other two scenarios, because its average feature density is between those of the blanket wafer and uniform die. One point to note is that the effect is fairly local, affecting only a small region near the die. Also, no information about the effect of individual clusters can be seen at the reactor scale. Figure 3 shows that information about the individual clusters can be seen on the mesoscale; *i.e.*, variations in the mass fraction of water associated with the clustered die are observed. The spatial variations decrease in amplitude quickly as a function of distance away from the surface, and only an effective value is observed at the reactor scale. Thus the mesoscale model can capture effects that are not captured at the other scales. Since the amplitudes of the variations over each cluster at the mesoscale are about the same, the effect of variations within a die do not seem to be important for SiO_2 deposition under these conditions, although it could be important for other systems.

Figure 4 shows the mass fraction of water radially across the wafer for different patterns on the wafer, for 1 μm deep by 0.5 μm wide features and a 2.5 μm pitch. In the case of five patterns, the patterns were placed uniformly every 10 mm from the center of the wafer and are 10 mm wide. The effect of just one pattern is small and local; however, the effect of having multiple large patterns is significant. The mass fraction of water increases by 22% over the blanket wafer case. The figure also shows differences in the

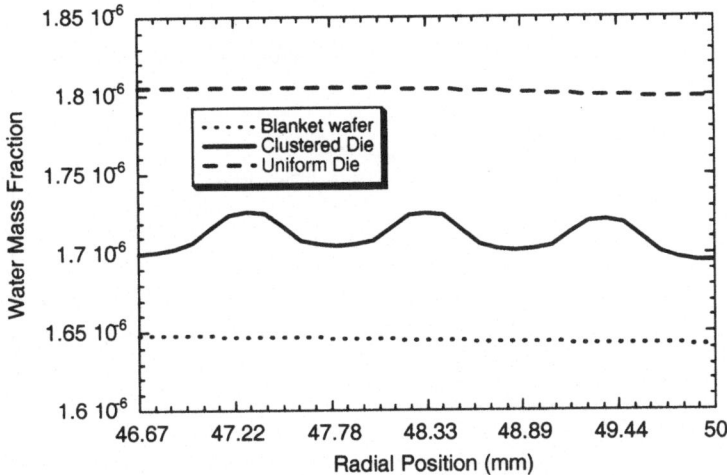

FIG. 3. *Mass fraction of water at the wafer surface as functions of radial position on the mesoscale, for three cases; a blanket (unpatterned) wafer, a uniformly patterned die, and a die with three clusters of features, as described in the text.*

amplitudes of the variations across the wafer, indicating that these loading effects depend on the position and density of the patterns on the wafer. Similar behavior is observed for other intermediates and byproducts as well, although the amplitudes are quantitatively different for each species. While a simple exposed area approximation may give a similar overall behavior for the mass fractions, it would not be able to correctly predict the differences in the amplitude of variations across the wafer. These results imply that under certain conditions and chemistries the effect of loading can be significant to the overall performance of the process.

Figure 5 shows an example of the evolution of the SiO_2 film profile in a representative feature at the center of the die on the wafer, during a transient simulation [7]. The profile contours are 100 seconds apart and the feature closes somewhere between 600 and 700 seconds. The result shows that for this particular chemistry and conditions, the TEOS deposition is conformal, and a smooth uniform film is deposited without any void formation.

2.2. Unstructured meshes: ECD of copper. We use a single, locally refined, finite-element mesh for Cu ECD IMPS. The ability to decrease the local mesh size of unstructured and semi-structured meshes in regions of high gradients or of particular interest allows a very natural transition from one scale to another. Resolution on the appropriate scale can be obtained without undue computational expense. Application of a continuum of discretization sizes is particularly advantageous if there is not a substantial change in the physics between the represented scales. For example, consider low-pressure processes. Reactor and die scale models could be

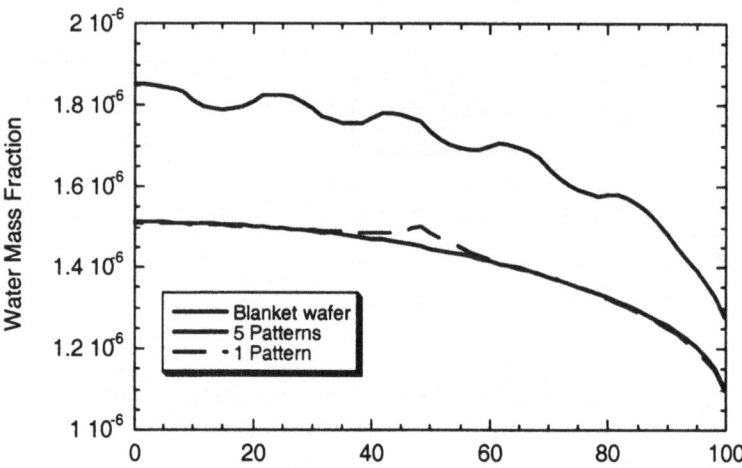

FIG. 4. *Mass fraction of water at the wafer surface vs. radial position, for three cases; an unpatterned wafer, a wafer with 5 patterns, and a wafer with one pattern (see text).*

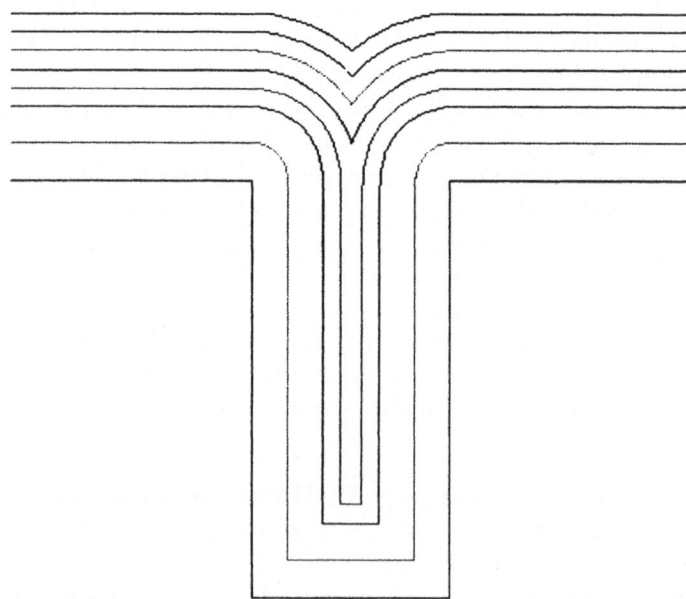

FIG. 5. *Evolution due to SiO$_2$ deposition in an aspect ratio two feature at the center node of the die on the wafer surface. Profile contours are every 100 s and extend to 900 s.*

modeled on a smoothly varying mesh, but transitions between continuum and molecular flow regimes would require a change of model [see Regular Grids: LPCVD of Silicon Dioxide]. Thus, the major advantages of having a single solution domain are lost.

To demonstrate this approach, we have investigated simple Cu ECD from Cu^{2+}(aq) on a non-uniform wafer in a deliberately generic ECD reactor; see Figure 6. We perform completely 3D modeling of an axisymmetric reactor. The deposition surface is a flat 200 mm diameter wafer with an annular die having interior and exterior radii of 50.0 mm and 53.0 mm, respectively. This die follows an ABABA sequence, with three patterned regions, each 0.6 mm wide, separated by 0.6 mm wide flat areas, for a total of five regions within the 3.0 mm wide die. Each patterned region is populated by 1 μm wide, 2 μm deep infinite trenches with a 3 μm pitch. A 10° slice of the reactor is discretized with an unstructured mesh of an appropriate length scale ($\sim$10 mm for a $\sim$0.5 m reactor). In regions near the deposition surface (within $\sim$2 cm), the length scale of the unstructured mesh is gradually decreased to $\sim$1 mm, where it meets a 'near-surface' region that consists of a semi-structured mesh with elements aligned in 12 strata over a distance of 1.5 mm; see Figure 7. The elements in this near-surface layer smoothly increase in aspect ratio to match the region of high concentration gradients immediately adjacent to the surface; the elements next to the surface have a thickness of 0.01 mm.

The mesh is further refined in patterned regions of the wafer; see Figure 8. The characteristic radial length of an element changes from 1.5 mm in the reactor bulk to 0.10 mm in the vicinity of the patterned die that has an internal length scale of 0.6 mm.

No-slip, no-penetration boundary conditions are assigned to the reactor walls, periodicity to the radial planes, natural pressure to the outlet, and a parabolic velocity profile to the inlet. The inlet conditions result in a Reynolds number of 100. To parallelize the computation, the entire reactor scale mesh is partitioned into four approximately computationally equal domains with the goal of minimizing inter-processor communication. The fully 3D, steady state fluid flow solution was then computed using PHASTA, a stabilized transient finite element fluid flow code [13], over 20 steps of 10^6 seconds each on an 8 processor SGI R10K (see Figure 6). The concentration of copper ion in the solution is tracked during this computation, but is uniform everywhere due to the concentration boundary conditions.

After the steady state flow has been computed, the flow solver is turned off, in order to compute the concentration field of the copper ion. At each node on the deposition surface, a value of the ion flux is required. This flux condition is supplied by using a previously constructed table to look up the flux that corresponds to the ion concentration and potential at that node. Once the flux is known at the surface, PHASTA solves a convective-diffusion equation, using the previously computed 3D flow field to establish

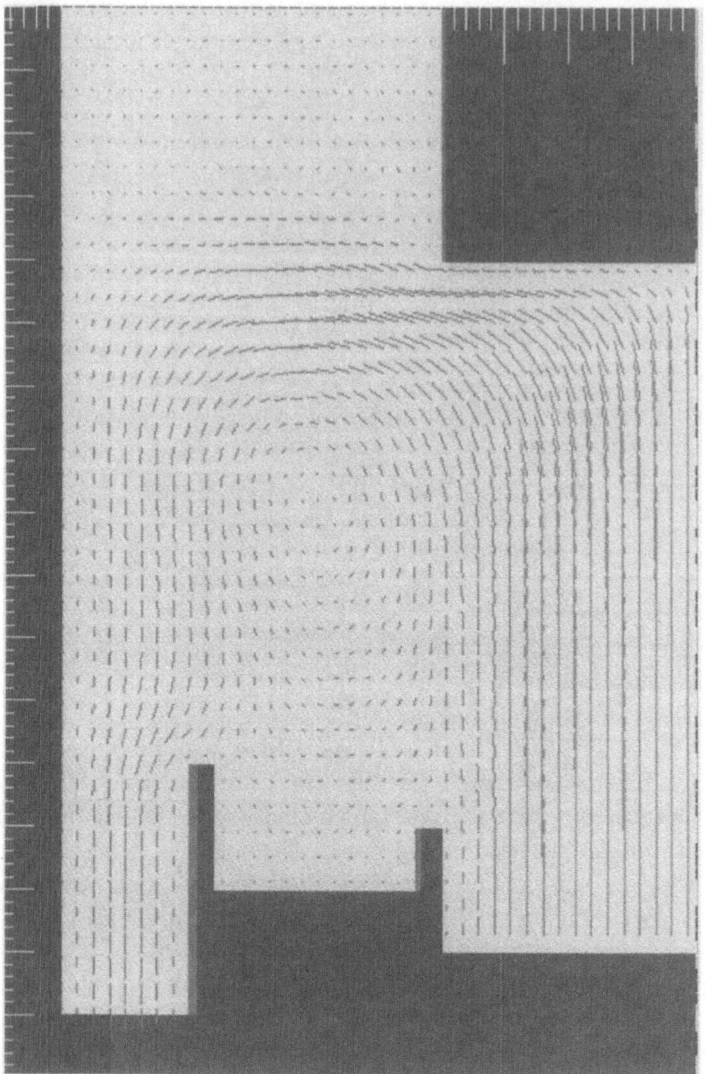

FIG. 6. *Steady state fluid flow pattern from a reactor scale simulation. The horizontal wafer surface is located in the upper right of the picture.*

a new reactor scale concentration field, which in turn is used to get new values of the fluxes. The system in then iterated to consistency. Note that we consider only the initial deposition rate profile across the wafer and features; *i.e.*, the initial wafer and pseudo-steady state conditions. Figure 9 shows an initial deposit on the features making up the pattern, performed under conditions found in the middle of one of these patterned regions.

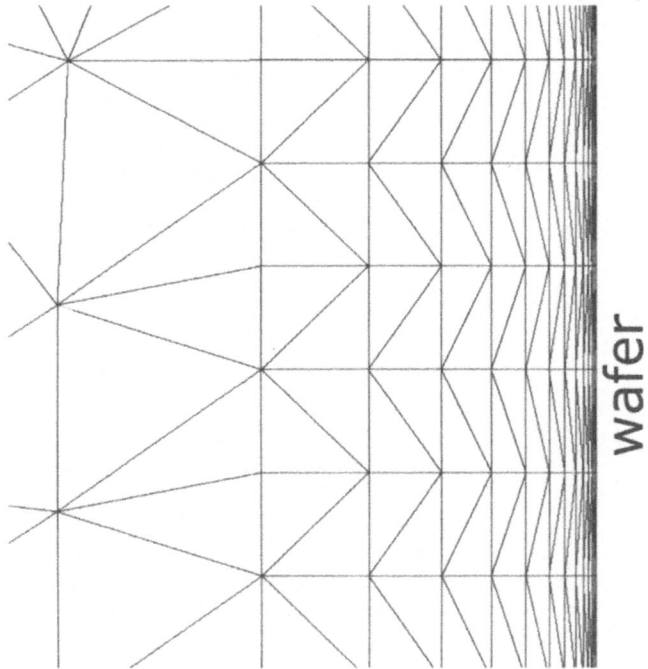

FIG. 7. *Semi-structured boundary layer mesh region, showing how elements change in aspect ratio over 12 strata.*

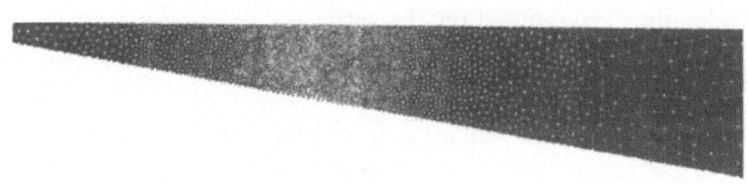

FIG. 8. *A mesh along the deposition surface, showing that it is selectively refined in areas near the patterned die.*

The lookup tables are built using results from a feature scale deposition model implemented with EVOLVE's ECD module [14]. As appropriate for our focus, the model consists solely of a charge transfer reaction governed by Butler-Volmer kinetics at the anode [14, 15]. For the conditions in the studied range, the transport of Cu^{2+} is far from the limiting current, indicating the current density is related to the potential through Ohm's law.

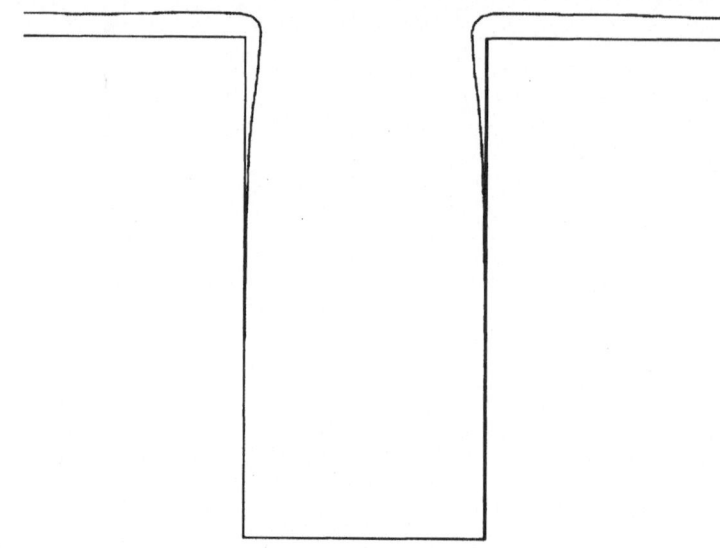

Fig. 9. *A simulated copper film profile at the start of deposition over an infinite trench is shown, assuming typical conditions observed in the patterned region.*

The fluid is taken to be uniformly conductive and in a stagnant film on the feature scale. EVOLVE uses this model to create a steady state feature scale concentration field corresponding to a given bulk ion concentration and the specified electrical potential, here 0.30 V, through an iterative finite element method on a 2D triangular mesh. From the resulting concentration field, the ion flux is then calculated. There is one table for each type of feature (or flat) represented on the surface. For this work, there are two tables, one for the flat regions of the wafer and one for the specific features under consideration.

Figure 10 shows the depletion in the reactor near the patterned die. As one might expect, the region directly next to the patterned areas are depleted more than the regions next to unpatterned areas. The concentrations directly on the pattern show a maximum depletion of about 5% relative to the reactor inlet condition of 80 mmoles/liter. This is in contrast to the minimum depletion of 3%, observed at the center of the wafer.

Depletion is uniform along the large unpatterned areas, but a very slight relative depletion can be observed between the areas on either side of the patterned region. The flat area exterior to the patterned annulus is about 0.2% lower in copper ion than the flat interior to the annulus. This incomplete recovery indicates that, just as for the case of LPCVD, that ECD can exhibit significant loading for large regions of high pattern density.

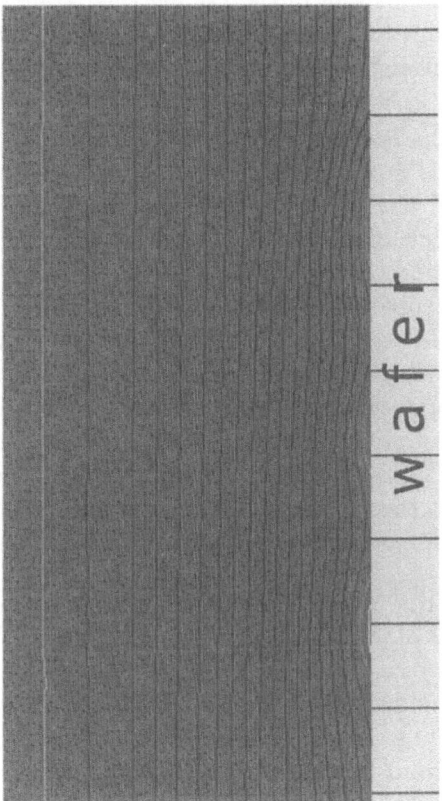

FIG. 10. *Copper ion concentration near the patterned region on the wafer.*

2.3. Discussion. The techniques and results presented here represent a starting point in investigating the interaction of multiscale phenomena for semiconductor manufacturing processes with well-accepted governing equations. While the technique is conceptually straightforward, implementation requires considerable bookkeeping to ensure seamless passage of information from one scale to another. For example, feature scale information needs to be stored at the current time and several previous times for time integration, for each node. Storage of the feature profile at earlier times is necessary to allow for non-convergence of the reactor scale solution. The number of feature scale simulations add up quickly; for each node on the wafer, for each reactor scale iteration, and for each time step. Because the reactor scale can take many iterations to converge at a particular time step, the overall time for a simulation increases rapidly. Efficiency of the process is improved by performing mesoscale and feature scale simulations only once for several reactor scale iterations. This is possible because the change in the current guesses of the reactor scale mass fractions has a relatively small impact on the resulting fluxes. While this reduces the overall solution time considerably, the optimal ratio of reactor scale iterates to lower scale simulations will depend on the particular chemistry and operating conditions.

There is also the issue of robustness of the codes in the integrated environment. This is especially true for transient simulations, where the time scale associated with one level is vastly different from the time scale of the other level. For instance, the time scale in the reactor scale simulator is associated with the flow characteristics and is on the order of seconds. The time scale on the feature scale is however the time to fill the feature, which can be a few hundred seconds. Both simulators should be robust enough to handle these widely varying time scales in situations where the codes are coupled. A robust reactor scale simulator should not give erroneous answers at long times, and the feature scale simulator should not diverge for very short time steps. In the current IMPS implementation, there is minimal coupling of the time scales in the reactor scale simulators and EVOLVE. Further discussion can be found in Refs. [6] and [7].

3. Developments in microstructure formation. As the implementation difficulties of IMPS at the reactor and feature scale are overcome for deposition processes such as those above, the next challenge will be to extend IMPS to include grain scale and atomic scale models capable of predicting wafer state even smaller length scales. Quantitative prediction of IC performance from process conditions will become a reality only as IMPS is extended to include all relevant length scales. For example, nucleation density of Cu on TaN during CVD has been shown to increase nearly 100-fold in the presence of small amounts of water vapor [16]. Greater nucleation density is reflected in the final thick film properties in the form of lower resistivity and reduced surface roughness. Even though Cu CVD on TaN may prove to be of limited practical interest, this example shows that models of feature scale film evolution should be integrated with models of atomic scale processes. Another clear advantage to including atomic scale models in IMPS is that the impact of feature or reactor scale process non-uniformity on atomic scale processes can be investigated.

To date, most atomic scale simulations of thin film nucleation and growth have been performed using Kinetic Monte Carlo (KMC) techniques. For example, Huang *et al.* [17] demonstrated KMC simulations of film growth that include multiple crystal orientations as well as overhangs and voids. However, simulated deposition of more than a few monolayers is currently impractical due the large computational demands of their method.

An IMPS framework that includes atomic scale models must implement a method to exchange the data for tens or perhaps hundreds of thousands of atoms with larger scale models. To meet this need, we are implementing a method to convert from discrete atomistic data to continuum models using finite element meshes. These meshes are created so that element faces conform as closely as possible to any identifiable material interfaces (including interfaces between islands or grains), and attributes such as material composition and crystal orientation can be mapped onto corresponding volume elements. As shown in Figure 11, the mesh can be created using different element sizes, however; some detail must be sacrificed for larger elements. The encapsulation process converts discrete atom data to a mesh representation that is suitable for continuum modeling.

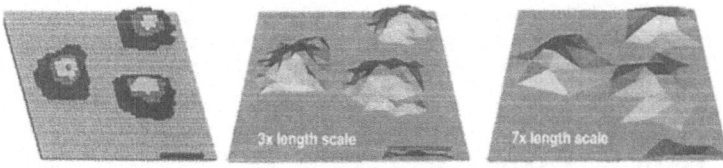

FIG. 11. *Discrete atomistic islands generated using a kinetic Monte Carlo simulation, and continuum representations of islands with different mesh sizes.*

One advantage of the conversion technique just described is that it provides a natural way to combine grain scale models and feature scale continuum thin film evolution models. As the film evolves, the mesh can be modified so that the correspondence between volume elements and material attributes is retained.

One promising method of tracking the evolution and coalescence of the islands and resulting film during deposition is through the use of level sets [20]. We have created, and continue to develop, a finite-element-based level-set tool for tracking the evolution. We use the initial encapsulation to embed the interface in a scalar field in three dimensions. Physical models of deposition can then provide the necessary information to evolve the field. The new interface can then be extracted from the updated field. Barth and Sethian [19] formulated the relevant Petrov-Galerkin finite element algorithm. Gyure *et al.* [20] have reported similar work on island growth. Figure 12 shows two stages of a sample evolution of a set of continuum islands that have been converted as described above. Here, the field is evolving under an isotropic growth. A primary advantage of this level set representation of evolution is in the tracking of the topological changes in the surface as discrete islands coalesce into a blanket film.

When this strategy is fully implemented it will be possible to predict not only the surface topography of a deposited film, but also the microstructure. The ability to predict and represent microstructure information will in turn open the door to predictions of properties such as adhesion, resistivity, electromigration resistance, stress during thermal cycling and so on [21, 22]. Models and modeling frameworks that can predict such properties will make it possible, for the first time, to supply modeling tools that meet the industrial need to predict the performance of devices based on material properties and processing parameters.

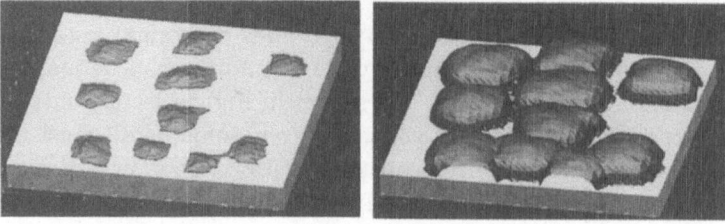

FIG. 12. *Sample level-set-based evolution of islands generated using kinetic Monte Carlo, after converting them to continuum representations as described in the text. The level-set field has been evolved under the assumption of isotropic growth.*

4. IMPS for CMP: unique challenges. Chemical-mechanical planarization (or polishing), CMP, has recently received considerable attention, because it is a cost-effective method to achieve planarized (flat) wafers during the manufacture of ICs. Desired improvements in CMP processes include; increasing wafer throughput, increasing the control of removal rate, and decreasing both polishing nonuniformity and polishing generated defects. There are a number of chemical and physical mechanisms at work in CMP, and achieving the desired improvements will require studying CMP from a number of perspectives.

In a typical CMP process, a wafer whose surface is covered with die is pressed face down into a rotating, compliant polishing pad flooded with a slurry of abrasive particles, as in Figure 13. Polishing occurs by rotating both the wafer and the larger table that holds the polishing pad and slurry.

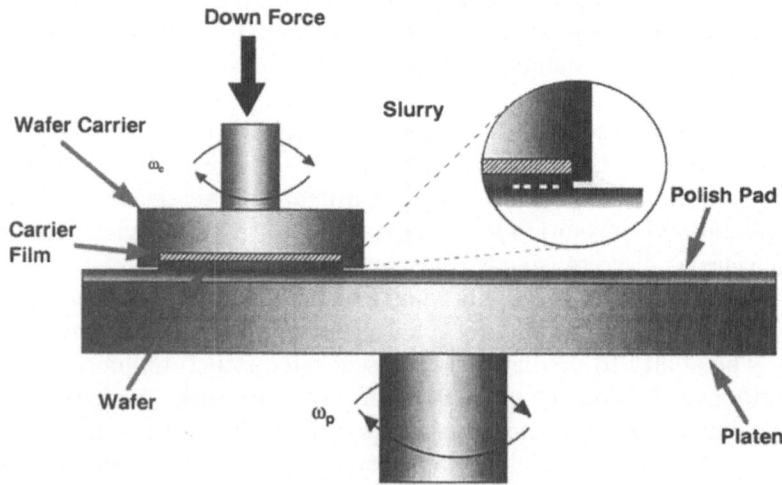

FIG. 13. *Schematic of typical rotational CMP process equipment.*

This motion drags slurry into the interface between the pad and wafer. The normal load (down-force on the wafer holder), wafer velocity, pad velocity and properties, and slurry properties are known to affect the polishing (material removal) rate as well as local and global polishing uniformity [23]. CMP mechanisms are not well enough understood to provide models and simulation tools that can predict process responses from tool design, consumables specifications and operating conditions. However, several recent modeling efforts have improved our understanding, and have helped in process development [24–28].

One primary hurdle for modeling CMP is to explain pad-wafer interactions on all of the relevant length scales. There are two distinct conceptual views on the pad-wafer contact component of material removal. The first view holds that the wafer surface is separated from the polishing pad by a hydrodynamic film of slurry, and polishing is done by collision of the slurry particles with the surface (erosion). The other view is that the hydrodynamic effect is not strong enough to separate the workpiece from the pad and the asperities of the pad rub against the wafer with trapped slurry particles in between (abrasion) [28]. In each case, one would intuitively expect a positive fluid pressure at the interface. Surprisingly, Danyluk and co-workers [29, 30] reported sub-ambient pressures in the fluid between the wafer and pad. They measured downward wafer vertical displacement (rather than hydrodynamic lift) and negative (sub-ambient or suction) pressures. Subsequent to this, Tichy et al. [28] presented a conceptual model that predicts sub-ambient and positive fluid pressures at the interface of a hard wafer and a soft porous material such as a polyurethane pad.

4.1. Differences in multiscale modeling of CMP. In the title of this section, we refer to unique challenges, and they do, indeed, exist. In the cases of LPCVD and ECD discussed above, the governing equations at the reactor scale and the mesoscale are, for lack of a better word, classical. They are the accepted balance equations of mass, momentum and species. Depending on the scale considered, certain terms of the governing PDEs may be negligible or dominant. There is general agreement that if these equations are solved with the proper boundary conditions, the process can be modeled, at least to the accuracy to which the transport and reaction sub-models for a specific process are known. Studies may be two-dimensional (2-D) or three-dimensional (3-D), with the coordinate direction normal to the surface always being an independent variable. Linkages between the scales are set up in a mathematically straightforward fashion. Fluxes obtained from the smaller scale serve as (Neumann) boundary conditions to the larger scale. Interpolated nodal point values obtained from the larger scale serve as Dirichlet boundary conditions to the smaller scale. The feature scale representation may be either molecular (as in LPCVD) or continuum based (as in ECD).

We describe the corresponding aspects of CMP modeling. We can specify a wafer scale, a mesoscale and a feature scale; with approximately the same length magnitudes as in the LPCVD and ECD cases. Much of the similarity ends there. The descriptions of the wafer scale and feature scale are heuristic and controversial, not formal and generally accepted. *Ad hoc* models for wafer scale physics abound, although the conceptual view of Danyluk, Tichy and Cale, as described in Refs. [28, 31, 32] is attaining consensus. In our view, the mesoscale can be described by a statistical model (not a PDE), which uses an "asperity height" distribution function. The feature scale model, which must describe the material removal process, is also conceptual and highly heuristic at this point, the details of which are surely open to controversy.

4.2. CMP wafer-scale model. The simplified conceptual model proposed below appears to capture the basic physics of the wafer-pad contact mechanics and slurry fluid flow of the CMP process. In particular, the model can predict the counter-intuitive experimental determination of suction fluid pressure below the pad introduced above. If the model continues to successfully predict observed trends, better predictions are expected as model features and parameter values are refined.

The essential description of the mechanical problem is that the wafer is subject to an externally applied normal load and undergoes a displacement into a compliant pad surface. The pad surface is very rough by usual tribology standards — for the purposed of this paper we assume that both the roughness/pore height and the spacing of the asperity peaks are about 50 μm [28]. The pad material is porous but largely impermeable, except near the surface. The process of 'conditioning' (roughening) the pad surface exposes the rough pore structure. Both the contact forces of the pad asperities and the slurry fluid forces balance the externally applied normal download. The solid contact pressure, while treated in the model as continuously varying, is visualized as an ensemble of discrete contact points, transmitted through the asperities. Thus, we assume the fluid and solid forces are both distributed over the same area without influencing each other (as in the formal construct of mixture theory [33]). Due to the small size of the slurry particles, and the relatively small solids fraction of the slurry, the fluid (water-based) is assumed to be Newtonian. The fluid is assumed to flow through a film of varying height, which is essentially the effective height of the compressed asperities.

We propose the following model to explain the interfacial pressure profile. Relative motion of the interface produces a shear stress due to rubbing, which produces an applied moment on the wafer. The applied normal load and moment produces an asymmetric bowl-shaped contact stress [34, 35]. The stress shape leads to a thinner fluid film around the fixture (wafer) edge and a thicker film at the center due to the non-uniform deformation of pad asperities. This spatial distribution of fluid film thickness, together with the pad-wafer relative motion, generates a fluid pressure in accord with Reynolds equation of lubrication theory.

Figure 14 is a schematic of the model, and the solid line that runs beneath the wafer shows a profile of the mean position of the asperities, $h(x, y)$. This curve diverges and converges as a result of the relative motion of the fixture and the rebound of the asperities. The mesoscale model described below relates the contact stress of asperities to the local mean separation of the surfaces.

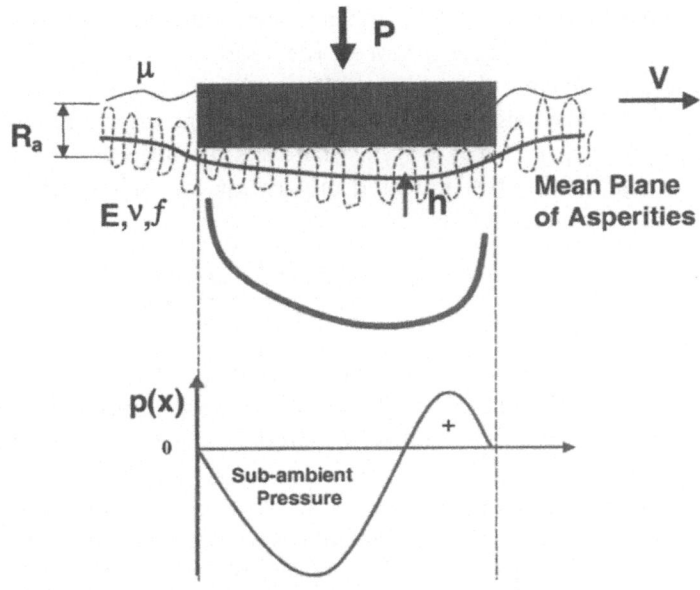

FIG. 14. *Schematic of wafer-scale model and fluid pressure at the wafer/pad interface.*

Assuming we know the film thickness $h(x, y)$, the fluid flow problem is addressed by assuming a continuous fluid flow at the interface. The Reynolds equation of hydrodynamic lubrication can be used to calculate the interfacial fluid pressure [36], viz

$$(4.1) \qquad \frac{\partial}{\partial x}\left(h^3 \frac{\partial p}{\partial x}\right) + \frac{\partial}{\partial y}\left(h^3 \frac{\partial p}{\partial y}\right) = 6\mu \left(V_x \frac{\partial h}{\partial x} + V_y \frac{\partial h}{\partial y}\right),$$

where μ is the liquid viscosity and (V_x, V_y) are the relative sliding velocities. Coordinates x and y are in the plane of the wafer, while z is normal. The simplified model presented in Ref. [30] uses one-dimensional analysis, $V_y = \partial V/\partial y = 0$. For the case of the wafer rotating at the same speed as the platen (normal industrial practice), it is easy to show that the motion is linear, $V_y = 0$. The formulas of contact mechanics (convolution integrals) and lubrication hydrodynamics reduce the 3-D equations of elasticity and hydrodynamics to 2-D equations (or reduce a 2-D problem to 1-D).

4.3. CMP mesoscale model. Assuming Hertzian contact of individual asperities, the equivalent liquid film thickness can be calculated using the Greenwood and Williamson [37] contact model for curved surfaces:

$$(4.2) \qquad \sigma(x, y) = \frac{4E}{3(1 - \nu^2)} \eta \sqrt{R_{asp}} \int_{h(x,y)}^{\infty} [z - h(x, y)]^2 \, \varphi(z) \, dz,$$

where, E is the elastic modulus of the pad, ν is Poisson's ratio, η and R_{asp} are density (number/area) and average radius of the asperities, respectively, h is the distance between the rigid flat and the mean plane of asperities (equivalent fluid film thickness), and $\varphi(z)$ is the distribution function of asperity heights. For demonstration, an exponential distribution is assumed, $\varphi(z) = \exp(z/s)/s$, where s is the RMS of the pad surface roughness. Performing the indicated integration and solving the resulting equation for h, we obtain

$$(4.3) \qquad\qquad h(x, y) = s \ln \left(\frac{\eta E \sqrt{\pi R_{asp} s^3}}{(1 - \nu^2)\sigma(x, y)} \right).$$

4.4. CMP feature-scale heuristic model. The model is conceptualized along the following lines. The mesoscale model implies an ensemble of asperities of constant tip radius, and uniform density distribution, but distributed over various heights from the mean surface according to an asperity height distribution function $\varphi(z)$. At a given global normal load F_z and asperity height distribution function, some pad asperity tips do not touch the opposing wafer surface, others barely touch and are slightly deformed, others are highly deformed.

We model and parameterize the contact of such an individual asperity with the rigid wafer surface, Figure 15. This modeling is highly idealized with many possible variations, both conceptually (*e.g.*, shape of the asperity) and in implementation (material model, conditions at the pad-wafer interface, *etc.*). This modeling effort is in progress, and some details are sketchy.

To set up this configuration as a boundary value problem, we must specify the domain of the problem, the constitutive equations, and the boundary conditions. We assume the asperity shape is a hemispherical cap. The constitutive equation for the polymer pad material should probably be some form of nonlinear elasticity, such as hyper-elasticity [38], the strains for a highly deformed asperity being of order one. Though the strain is not small, as required for linear elastic theory, it is also doubtful that the pad material flows plastically. The boundary conditions would be a specified normal displacement for the region of wafer-pad contact (based on asperity height) and a tangential displacement to produce a specified friction coefficient. The boundary would be stress-free at the non-contacting

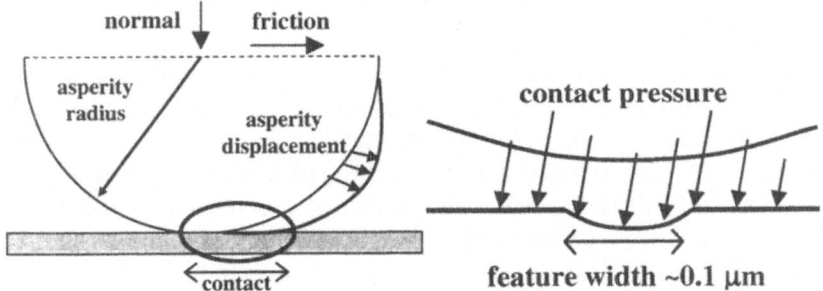

FIG. 15. *Schematic of feature scale asperity contact model.*

portion, and specified stress where the idealized asperity connects to the pad in bulk (assumed uniform).

The computations would be highly nonlinear in that the contact region is not known *a priori*. Calculations would yield local asperity contact stress σ_i (for the ith asperity). The ensemble average of the σ_i is the continuum wafer-scale contact stress $\tilde{\sigma}$. The fractional real area of contact is small ($< 10\%$) for all typical tribological conditions.

Finally, we need a model for material removal. For the local asperity contact region, the linear material removal (volume rate per area) is abrasion, and we use a *locally* applied Preston's equation,

$$(4.4) \qquad \left(\frac{\dot{V}}{A}\right)_i = k_{ab}\,\frac{\sigma_i}{E_Y}\,U,$$

where k_{ab} is the empirical abrasive wear coefficient, E_Y is Young's modulus and U is the relative speed. Presumably, (4.4) is more accurate than a corresponding global version. For the non-contacting surface of the wafer we use an erosive wear model in which we envision slurry particles moving relative to the wafer surface

$$(4.5) \qquad \frac{\dot{V}}{A} = k_{er}\dot{\gamma}D_{particle},$$

where k_{er} is the empirical erosion wear coefficient, $\dot{\gamma}$ is the shear rate, and $D_{particle}$ is the particle diameter. At a given instant of time, a small fraction of the surface area of the wafer is undergoing abrasion, but the abrasion removal rate is much larger than erosion.

4.5. CMP wafer-scale modeling results. The results discussed below make use of the mesoscale model to obtain the effective film thickness from the contact stress. The solid contact pressure and fluid pressures are calculated using the commercial finite element analysis program ANSYS [39]. There are many aspects to ANSYS contact analyses. As a starting point, we have chosen "perfect-sticking" to simulate the solid pressure —

the normal and tangential displacements of the pad and wafer are identical at the interface.

The analysis consists of a rigid, cylindrical indenter impinging on the center of the top surface of a flat, cylindrical, piece of the complete elastic pad. Thus, only the region of the pad that is near the wafer is considered in the model. The bottom surface of the pad is completely fixed. Sliding occurs only in the x-axis. The rigid wafer is displaced into the pad (negative z-axis), and tangentially (x-axis), and rotated about the y-axis by an angle α such that the leading edge is down. It is interesting that a nearly imperceptible tilt of the wafer has a profound effect in causing the large suction pressures. A top view of the 3-D finite element mesh used in ANSYS is shown in Figure 16. The mesh is denser near the outer radius of the wafer, because singular behavior exists near the region of a rigid indenter (wafer) with a sharp edge.

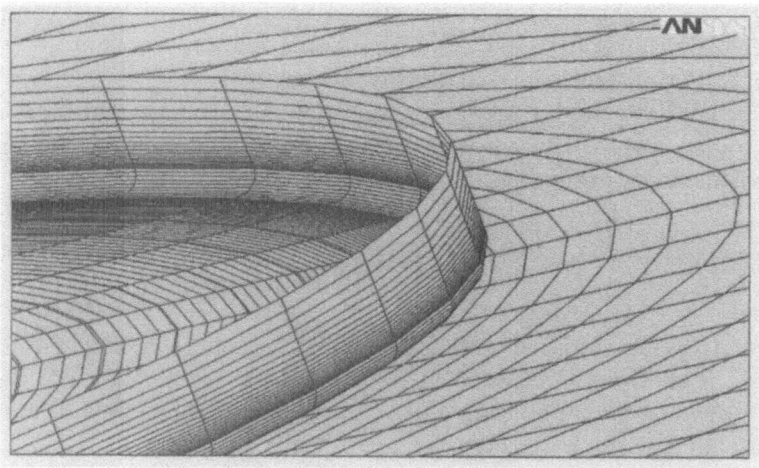

FIG. 16. *View of 3-D finite element mesh for wafer-pad contact analysis.*

The resulting contact stress is calculated over the pad-wafer contact area. We use experimentally obtained parameter values as for roughness. Reynolds equation is solved on the contact domain using ANYSYS with $p = 0$ applied to the boundary. ANYSYS does not have a Reynolds equation solver. Equation (4.1) is a generalized Poisson's equation — identical in form to the heat conduction equation. ANYSYS does allow solution for the temperature field, so through suitable selection of the variables and parameters ANYSYS can solve Reynolds equation. One example of a calculated suction pressure distribution below the pad is shown in Figure 17. There is a very small region of positive pressure at the trailing edge.

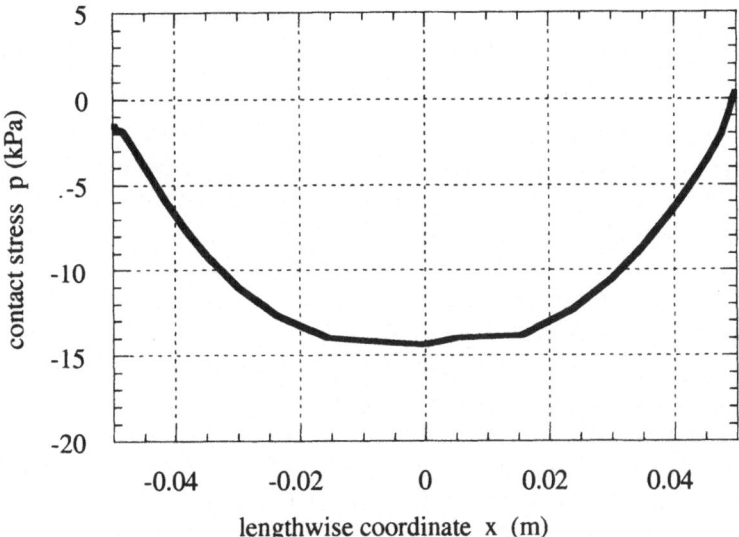

FIG. 17. *Profile of fluid pressure below pad, along diameter parallel to sliding direction.*

4.6. Comments and future research. We expect that the outcome of the feature-scale (individual asperity) model will be the local contact stress σ_i and contact area A_i, for a specified normal load and friction coefficient on the asperity, as well as normal displacement and pad properties. The local volumetric removal of wafer material is then calculated from (4.4).

Linking the feature-scale asperity contact analysis (in progress) to the wafer scale model occurs through the statistical mesoscale, described above. The evolution of the wafer surface geometry, due to material removal means that time varying (but quasi-steady) conditions must also be considered. Both the wafer and feature scale analyses are computationally intensive. Clearly, the computational challenges are formidable.

Toward the goal of a user process model, a major complication is that many sets of conditions will need to be run for sensitivity analysis, and comparison to known behavior. For the multi-scale model described there are numerous parameter values, *e.g.*, speed, viscosity, elasticity parameters, asperity parameters (density, tip radius), friction coefficient, material removal parameters, *etc.* Many different model implementations can be foreseen, *e.g.*, different sliding and wafer support configurations, rheological properties of the slurry, *etc.* At the feature scale and mesoscale the asperity models could be varied, *e.g.*, varying tip radius, differing asperity frequency distribution functions.

Such difficulties and complications imply that considerable use must be made of design of experiment and response surface modeling methodologies.

We feel that with suitable physical insight to useful approximations, and some development work on multi-scale methodologies, we are headed down a path that will lead to useful modeling of CMP.

5. Conclusions. Techniques that can predict feature scale behavior from equipment scale operating conditions are very relevant to the semiconductor industry, as they can reduce development time as well as expedite process optimization. Within the IMPS framework, the equipment scale information is fed to mesoscale and feature scale simulators to obtain the effects at the evolving surface. The simulators are iterated to achieve self-consistent solutions at all length scales. While the framework of this methodology is established for CVD and ECD, there is scope for improvement in both the efficiency and robustness of this approach. One way to dramatically improve efficiency is to parallelize the simulations, which involves additional coding for resource management. The approach demonstrated is clearly extendable to other processes.

Though we can model CVD and ECD processes, from the reactor scale down to the continuum film evolution level, the information required for accurate predictions of deposition rates and film profile evolution for any given process is generally not known. Thus, there is still a large amount of work being done in this area, notably in the area of reaction chemistry. Unfortunately, the examples used points out a major weakness in such quests; they may yield complicated kinetic models that are hard to deal with (calibrate). Simpler, physically based 'engineering' transport and reaction sub-models will continue to play a dominant role in making process decisions.

We also discussed how discrete atomistic representations of islands, perhaps from a KMC deposition simulation, can be converted to continuum representations more suitable for level set based evolution.

Acknowledgments. T.S. Cale, D.F. Richards, and S. Soukane acknowledge support from MARCO, DARPA, NYSSTF, SRC and NSF. K.E. Jansen and M.K. Gobbert acknowledge support from NSF. The authors acknowledge F. Shakib of Acusim Software.

REFERENCES

[1] International Technology Roadmap for Semiconductors, 1999 edition; (http://public.itrs.net/Files/1999_SIA_Roadmap/Home.htm).
[2] T.S. CALE, T.P. MERCHANT, L.J. BORUCKI, AND A.H. LABUN, *Topography simulation for the virtual wafer fab*, Thin Solid Films, **365**(2), 2000, pp. 152–175.
[3] T.S. CALE, B.R. ROGERS, T.P. MERCHANT, AND L.J. BORUCKI, *Deposition and Etch Processes: Continuum Film Evolution in Microelectronics*, J. Comp. Mat. Sci., **12**, 1998, pp. 333–353.
[4] Proceedings of the Fifth IUMRS International Conference on Advanced Materials, J. Comp. Aided Mats. Des., **6**(2–3), 1999; Proceedings of the Sixth International IUMRS Conference on Advanced Materials, J. Comp. Mat. Sci., in press.

[5] M.K. GOBBERT, C.A. RINGHOFER, AND T.S. CALE, *Mesoscopic scale modeling of microloading during low pressure chemical vapor deposition*. J. Electrochem. Soc., **143**(8), 1996, pp. 2624–2631.

[6] M.K. GOBBERT, T.P. MERCHANT, L.J. BORUCKI, AND T.S. CALE, *A multiscale simulator for low pressure chemical vapor deposition*, J. Electrochem. Soc., **144**(11), 1997, pp. 3945–3951.

[7] T.P. MERCHANT, M.K. GOBBERT, T.S. CALE, AND L.J. BORUCKI, *Multiple scale integrated modeling of deposition processes*, Thin Solid Films, **365**(2), 2000, pp. 368–375.

[8] M.E. COLTRIN, P. HO, H.K. MOFFAT, AND R.J. BUSS, *Chemical kinetics in chemical vapor deposition: growth of silicon dioxide from tetraethoxysilane (TEOS)*, Thin Solid Films, **365**(2), 2000, pp. 251–263.

[9] R.J. KEE, F.M. RUPLEY, E. MEEKS, AND J.A. MILLER, *CHEMKIN-III: A Fortran chemical kinetics package for the analysis of gas-phase chemical and plasma kinetics*, Sandia National Laboratories, Livermore, CA, 1996.

[10] A.H. LABUN, H. MOFFAT, AND T.S. CALE, *Mechanistic feature scale profile simulation of SiO_2 LPCVD by TEOS pyrolysis*, J. Vac. Sci. Technol., B**18**(1), 2000, pp. 267–272.

[11] FIDAP 7.6, Fluent Inc., 500 Davis St. Suite 600, Evanston, IL 60201, 1996.

[12] EVOLVE is a deposition, etch, and reflow process simulator developed under the direction of T.S. Cale. Copyright 1990–2000 by Timothy S. Cale.

[13] C.H. WHITING AND K.E. JANSEN, *A stabilized finite element method for the incompressible Navier-Stokes equations using a hierarchical basis*, International Journal of Numerical Methods in Fluids, **35**, 2001, pp. 93–116.

[14] S. SOUKANE AND T.S. CALE, *Proceedings of the Seventeenth International VLSI Multilevel Interconnection Conference*, T. Wade, ed., IMIC, 2000, pp. 260.

[15] J.O. DUKOVIC, *Feature-scale simulation of resist-patterned electrodeposition*, IBM J. Res. Develop., **37**(2), 1993, pp. 125–141.

[16] D. YANG, J. HONG, AND T.S. CALE, *Effects of process variables on Cu(TMVS)(hfac) sourced copper CVD films*, in: Advanced Metallization Conference 1999, M.E. Gross, T. Gessner, N. Kobayashi, and Y. Yasuda, eds., MRS, 2000, pp. 207–211.

[17] G.H. GILMER, H. HUANG, T.D. DE LA RUBIA, J. DALLA TORRE, AND F. BAUMANN, *Lattice Monte Carlo models of thin film deposition*, Thin Solid Films, **365**(2), 2000, pp. 189–200.

[18] J.A. SETHIAN, *Fast level set methods and fast marching methods. Evolving interfaces in computational geometry, fluid mechanics, computer vision and materials science*, second ed., Cambridge University Press, Cambridge, 1999.

[19] T.J. BARTH AND J.A. SETHIAN, *Numerical schemes for the Hamilton-Jacobi and level set equations on triangulated domains*, J. Comp. Phys., **145**(1), 1998, pp. 1–40.

[20] M.F. GYURE, C. RATSCH, B. MERRIMAN, R.E. CAFLISCH, S. OSHER, J.J. ZINCK, AND D.D. VVEDENSKY, *Level-set methods for the simulation of epitaxial phenomena*, Phys. Rev. E, **58**(6), 1998, pp. R6927–R6930.

[21] D. MAROUDAS, M.N. ENMARK, C.M. LEIBIG, AND S.T. PANTELIDES, *Theory and computer simulation of microstructure evolution in polycrystalline metallic thin films*, Proceedings of Fourth International Symposium on Process Physics and Modeling in Semiconductor Devices, 1996, pp. 249–260.

[22] C.S. HAU-RIEGE AND C.V. THOMPSON, *The effects of microstructural transitions at width transitions on interconnect reliability*, J. Appl. Phys., **87**(12), 2000, pp. 8467–8472.

[23] S.P. MURARKA, J.M. STEIGERWALD, AND R.J. GUTMANN, *Inlaid copper multilevel interconnections using planarization by chemical-mechanical polishing*, MRS Bulletin, **18**(6), 1993, pp. 46–51.

[24] R.S. SUBRAMANIAN, L. ZHANG, AND S.V. BABU, *Transport phenomena in chemical mechanical polishing*, J. Electrochem. Soc., **146**(11), 1999, pp. 4263–4272.

[25] W.-T. TSENG, Y.-H. WANG, AND J.-H. CHIN, *Effects of film stress on the chemical mechanical polishing process*, J. Electrochem. Soc., **146**(11), 1999, pp. 4273–4280.

[26] S.R. RUNNELS, R. MICELI, AND I. KIM, *Validation of a large area three-dimensional erosion simulator for chemical mechanical polishing*, J. Electrochem. Soc., **146**(12), 1999, pp. 4619–4625.

[27] C.-H. YAO, D.L. FEKE, K.M. ROBINSON, AND S. MEIKLE, *Modeling of chemical mechanical polishing processes using a discretized geometry approach*, J. Electrochem. Soc., **147**(4), 2000, pp. 1502–1512.

[28] J. TICHY, J. LEVERT, L. SHAN, AND S. DANYLUK, *Contact mechanics and lubrication hydrodynamics of chemical mechanical polishing*, J. Electrochem. Soc., **146**(4), 1999, pp. 1523–1528.

[29] J.A. LEVERT, A.R. BAKER, F.M. MESS, R.F. SALANT, S. DANYLUK, AND L. COOK, STLE Tribology Trans., **41**, 1998, p. 593.

[30] J.A. LEVERT, *Interface Mechanics of Chemical Mechanical Polishing for Integrated Circuit Planarization*, Ph.D. Thesis, Georgia Institute of Technology, 1997.

[31] L. SHAN, J.A. LEVERT, J. TICHY, AND S. DANYLUK, *Interfacial fluid mechanics and pressure prediction in chemical mechanical polishing*, J. of Trib., **122**(3), 2000, pp. 539–543.

[32] J. TICHY, C. CLUTZ, AND T. CALE, *CMP pad displacement and slurry flow characteristics' finite element analysis*, in *Proceedings of the Fifth International Chemical Mechanical Polish for ULSI Multilevel Interconnection Conference*, IMIC, 2000, pp. 222–228.

[33] A.E. GREEN AND P.M. NAGHDI, *The flow of fluid through an elastic solid*, Acta Mechanica, **9**(3–4), 1970, pp. 329–340.

[34] K.L. JOHNSON, *Contact Mechanics*, Cambridge University Press, Cambridge, 1985.

[35] C. SRINIVASA-MURTHY, D. WANG, S.P. BEAUDOIN, T. BIBBY, K. HOLLAND, AND T.S. CALE, *Stress distribution in chemical mechanical polishing*, Thin Solid Films, **308–309**, 1997, pp. 533–537.

[36] B.J. HAMROCK, *Fundamentals of Fluid Film Lubrication*, McGraw-Hill, 1994.

[37] J.A. GREENWOOD AND J.B.P. WILLIAMSON, Proc. Roy. Soc. London, A**295**, 1966, p. 300.

[38] W.M. LAI, D. RUBIN, AND E. KREMPL, *Continuum Mechanics*, third edition, Pergamon, 1993.

[39] ANSYS, Faculty/Research Release 5.6; ANSYS, Inc, Canonsburg PA, 1999.

CONSTITUTIVE RELATIONS FOR VISCOLEASTIC FLUID MODELS DERIVED FROM KINETIC THEORY

P. DEGOND*, M. LEMOU*, AND M. PICASSO†

Abstract. Constitutive relations for the stress tensor in viscoselastic fluids are derived from the kinetic theory of polymeric liquids. The Fokker-Planck equation corresponding to the so-called dumbbells theory for diluted solutions of polymers is considered. The fluid models are derived for time-dependent, nonhomogeneous, and nonpotential flows, thus extending the results of [2]. A comparison with Oldroyd-B, FENE and FENE-P fluids is presented in the frame of the plane Couette flow. This paper is an abriged version of [8].

1. Introduction. In this paper the kinetic theory corresponding to a diluted solution of polymeric liquids is considered. According to [2], the simplest model to account for non interacting polymer chains is the so-called dumbbells model. A dumbbell is made out of two beads connected with an elastic spring, and is characterized at time t by the position of its center of mass $X(t)$ and its elongation $Q(t)$. When a dumbbell is placed into a given incompressible velocity field $v(x,t)$, three forces are acting on each of the beads. The first force is nothing but the drag force and is proportional to the difference between the bead velocity and the velocity of the surrounding fluid particles; the second force is the elastic force due to the spring stiffness; the third force corresponds to thermal agitation and is modeled using Brownian motion. Writing Newton's equations on the beads yields a set of stochastic differential equations (see [2, 15, 1] for more details) for $X(t)$ and $Q(t)$. The deterministic restatement of the problem is the following. Let $f(x,q,t)$ be the probability density corresponding to the stochastic processes $X(t)$ and $Q(t)$. The quantity $f(x,q,t)dxdq$ thus represents the probability at time t to find a dumbbell located between x and $x + dx$ having elongation between q and $q + dq$. Assuming that the elastic force $F : \mathcal{D} \subset \mathbb{R}^d \to \mathbb{R}^d$ ($d = 2$ or 3) of the spring derives from a potential U: $F(q) = U'(q^2/2)q$, then $f(t,x,q)$ has to satisfy the following Fokker-Planck equation

$$(1.1) \qquad \frac{\partial f}{\partial t} + v \cdot \nabla_x f + \nabla_q \cdot \left((\nabla v)qf \right) = \frac{1}{2\lambda} \nabla_q \cdot \left(\nabla_q f + U'qf \right),$$

Here $\nabla v_{ij} = \partial v_i / \partial x_j$ is the velocity gradient, F is the elastic force due to the spring, and λ characterizes the elastic property of the fluid. According to [2], the extra-stress τ due to the polymer chains can be modelled as

$$(1.2) \qquad \tau = kT(C - nI), \quad \text{with} \quad n(x,t) = \int_{\mathcal{D}} f(x,q,t)dq,$$

*Laboratoire MIP, Université Paul Sabatier, 118 route de Narbonne, 31062 Toulouse Cedex, France.

†Département de Mathématiques, Ecole Polytechnique Fédérale de Lausanne, 1015 Lausanne, Switzerland.

I the unit tensor and C is defined by

$$(1.3) \quad C = \int_{\mathcal{D}} fU'q \otimes q \, dq, \quad \text{that is} \quad C_{ij}(x,t) = \int_{\mathcal{D}} f(x,q,t)U'(q^2/2)q_iq_j dq.$$

k is Boltzmann's constant, T the absolute temperature and n the density of the polymer chains. In the case when the springs are Hookean, $\mathcal{D} = \mathbb{R}^d$, $F(q) = q$, $U(s) = s$, then the above kinetic model corresponds exactly to an Oldroyd-B fluid. Indeed, multiplying (1.1) by q_iq_j and integrating over q yields

$$(1.4) \qquad \qquad \lambda\frac{\delta\tau}{\delta t} + \tau = nkT(\nabla v + \nabla v^T),$$

where $\delta/\delta t$ is the upper convected time derivative defined by

$$(1.5) \qquad \qquad \frac{\delta\tau}{\delta t} = \frac{\partial\tau}{\partial t} + v \cdot \nabla\tau - (\nabla v)\tau - \tau(\nabla v)^T,$$

Relation (1.4) is nothing but Oldroyd-B constitutive equation. In practical situations, an Oldroyd-B fluid is too simple to describe experimental phenomena such as shear thinning, i.e. the fact that viscosity decreases with increasing shear rate. A more interesting model [2, 15] is the FENE (Finitely Extensible Nonlinear Elastic) model for which the force law of the spring F is defined on $\mathcal{D} = \mathcal{B}(0, \sqrt{b})$ by

$$(1.6) \quad F(q) = \left(1 - \frac{q^2}{b}\right)^{-1} q, \quad \text{thus} \quad U(s) = \ln\left(1 - \frac{2s}{b}\right)^{-b/2},$$

which prevents the elongation q from reaching values greater than $\sqrt{b}$. The FENE model has no corresponding constitutive equation and is implemented by means of expensive Brownian simulations, see for instance [15, 14, 3, 12, 9] for details. However, approximate constitutive equations are available for FENE dumbbells and a nonlinear constitutive equations can be obtained [2],. This model predicts shear thinning and is justified in the limit when b is large (i.e. when the dumbbells are almost Hookean). Another possible approach is to consider the case when the elastic property of the fluid is small and to expand the probability density f in powers of λ. The corresponding calculations are available in [2] for steady-state, homogeneous, potential flows.

The goal of this paper is to extend these calculations to general time-dependent, nonhomogeneous and nonpotential flows, using Chapman-Enskog expansion techniques. In the sequel, eq. (1.1) is written in operator form as follows:

$$(1.7) \qquad Tf + Bf = \frac{1}{\lambda} Af,$$

where T is the transport operator and B the operator corresponding to the drag force exerted on the beads

$$(1.8) \qquad Tf = \frac{\partial f}{\partial t} + v \cdot \nabla_x f, \quad Bf = \nabla_q \cdot \left((\nabla v) q f \right),$$

A is the operator due to Brownian motion and springs stiffnesses

$$(1.9) \qquad Af = \frac{1}{2} \nabla_q \cdot \left(\nabla_q f + U' q f \right).$$

The structure of the paper is the following. In the next section, some properties of A are discussed. In section 3, Chapman-Enskog expansion techniques are applied to eq. (1.7). In section 4,, our model is compared to the kinetic FENE model, to the FENE-P approximation, and to the Oldroyd-B model in the frame of the plane Couette flow. Finally, in section 5, we give a conclusion and some perspectives. The present paper is an abriged version of [8] in which the details of the proofs can be found.

2. Properties of the diffusion operator A. In this section we give some basic mathematical properties of the operator A defined by (1.9). More precisely, we prove that this operator is self-adjoint and coercitive on some functional space. As we will see, such properties are useful to justify the analysis of the asymptotics $\lambda \to 0$ in the kinetic equation (1.7). First, we remark that A may also be expressed in the following form

$$(2.1) \qquad Af = \frac{1}{2} \nabla_q \cdot \left(M \nabla_q \left(\frac{f}{M} \right) \right),$$

with M being the normalized Maxwellian

$$(2.2) \qquad M(q) = \frac{e^{-U(q^2/2)}}{\int_{\mathcal{D}} e^{-U} dq}.$$

The formulation (2.1) of operator A suggests to introduce the following functional space

$$(2.3) \qquad \mathcal{E} = \{f, \quad s.t. \quad \frac{f}{\sqrt{M}} \in L^2(\mathcal{D})\},$$

endowed with the scalar product:

$$(2.4) \qquad <f, g> = \int_{\mathcal{D}} fg \frac{1}{M} dq$$

for all f and g in $\mathcal{E}$. The associated norm will be denoted by $\|.\|$. With these notations, we can state the following result

PROPOSITION 2.1.
1. $-A$ is a positive and self-adjoint operator on $\mathcal{E}$.
2. The Null-space of A in $\mathcal{E}$ is:

$$(2.5) \qquad N(A) = \mathbb{R}M = \{cM, c \in \mathbb{R}\}$$

with M defined by (2.2).
1. If, in addition, U is strictly uniformly convex, that is:

$$(2.6) \qquad \exists \alpha > 0,\ s.t.\ \forall q \in \mathcal{D},\ \forall x \in \mathbb{R}^d,\ \left(\nabla^2\, U(q^2/2)\right) x \cdot x \geq \alpha x^2,$$

then $-A$ is coercitive on the orthogonal (in $\mathcal{E}$) of its Null-space:

$$(2.7) \qquad < -Af, f > + \alpha < f, M >^2 \geq \alpha \|f\|^2, \qquad \forall f \in \mathcal{E}\ s.t.\ Af \in \mathcal{E}.$$

Proof. The first two assertions are immediate consequences of the following weak formulation of the operator A:

$$(2.8) \qquad < Af, g > = -\frac{1}{2} \int_{\mathcal{D}} \nabla_q \left(\frac{f}{M}\right) \cdot \nabla_q \left(\frac{g}{M}\right) M(q) dq.$$

The third property comes from a Poincare-like inequality proved by Brascamp and Lieb in [4]. □

The results of Proposition 2.1 will be usefull to perform the Chapman-Enskog expansion in the next section, since the inverse of the operator A is needed. For a given function $g \in \mathcal{E}$, the equation $Af = g$ has a solution $f \in \mathcal{E}$ if and only if g is in $N(A)^{\perp}$, indeed, A is a closed operator and then the range of A is equal to $N(A)^{\perp}$. Furthermore, the solution is unique in $N(A)^{\perp}$. Finally, in all this paper, we will denote by Π the orthogonal projection onto $N(A)$. Precisely, Π is the following linear operator:

$$(2.9) \qquad \Pi f(q) = \left(\int_{\mathcal{D}} f(q) dq\right) M(q),$$

where M is given by (2.2).

3. Approximate solutions by Chapman-Enskog expansions. As mentioned in the introduction of this paper, the system of equations for the constraints C_{ij}, defined by (1.3), which is derived from the kinetic equations (1.1) is not closed, unless Hookean dumbbells are considered ($U(s) = s$). From a physical point of view, the Hookean potential is too simple and does not lead to a realistic description of the fluid. In this section we propose a closure strategy for a large class of "physically-admissible" potentials. This strategy allows us to extend the results of [2] to unsteady, nonhomogeneous and non-potential flows. It is based on the Chapman-Enskog expansions that is explained below. We refer to [5–7, 10, 13] for more details about this procedure and for general presentations about hydrodynamic limits of kinetic theories.

We shall say that $\tilde{f}^\lambda$ is an order n approximate solution to equation (1.1) or (1.7) if and only if $\tilde{f}^\lambda$ is a solution to:

$$(3.1) \qquad T\tilde{f}^\lambda + B\tilde{f}^\lambda = \frac{1}{\lambda}A\tilde{f}^\lambda + O(\lambda^n).$$

We now make the "Ansatz" that such approximate solutions can be expanded in powers of λ:

$$(3.2) \qquad \tilde{f}^\lambda = f_0 + \lambda f_1 + \lambda^2 f_2 + \dots + \lambda^n f_n.$$

The successive corrections f_0, f_1, ... may depend on λ but remain of order $O(1)$ in λ. This is because we want f_0 to have the same macroscopic quantities (here the density) as the original distribution function f up to order n terms in λ. This is satisfied by following the Chapman-Enskog procedure. We first insert expansion (3.2) in (3.1) and ontain for $n = 2$:

$$(3.3) \qquad \begin{aligned} (T + B)f_0 + &\lambda(T + B)f_1 + \lambda^2(T + B)f_2 \\ &= \frac{1}{\lambda}\left\{A(f_0) + \lambda A(f_1) + \lambda^2 A(f_2)\right\} + O(\lambda^2), \end{aligned}$$

this implies, since $\Pi A = 0$, that

$$(3.4) \qquad \Pi(T + B)f_0 = O(\lambda).$$

Thus the quantity $\Pi(T + B)f_0$ can be subtracted from the zero-th order term and added to the first order one. We then obtain by identification:

$$(3.5) \qquad \begin{aligned} A(f_0) &= 0, \\ A(f_1) &= (I - \Pi)(T + B)f_0, \\ A(f_2) &= (T + B)f_1 + \frac{1}{\lambda}\Pi(T + B)f_0. \end{aligned}$$

From the definition of Π, $(I - \Pi)(T + B)f_0 \in N(A)^\perp$, thus the second equation of (3.5) has a unique solution $f_1 = A^{-1}(I - \Pi)(T + B)f_0$ where A^{-1} being the inverse of the restriction of A to $N(A)^\perp$. However, for the third equation to have a solution, one has to write a solvability condition:

$$(3.6) \qquad \begin{aligned} \Pi(T+B)f_0+\lambda\Pi(T+B)f_1 &= \Pi(T+B)f_0+\lambda\Pi A^{-1}(I-\Pi)(T+B)f_0 \\ &= 0. \end{aligned}$$

The Chapman-Enskog method can be pursued to higher orders, following the same process as for the order 2. The order $n \geq 3$ expansion leads to the following recursive expressions for f_p:

$$(3.7) \qquad \begin{cases} f_0 \in N(A) \\ f_{p+1} = A^{-1}(I - \Pi)(T + B)f_p, \quad \text{for } p \in \mathbb{N}, \ 0 \leq p \leq n - 1, \end{cases}$$

where A^{-1} is the inverse of the restriction of A to $N(A)^{\perp}$. We recall that Π is the orthogonal projection onto $N(A)$ with respect to the scalar product (2.4) and is expressed by (2.9). The solvability condition is now:

(3.8)
$$\Pi(T + B)f_0 + \lambda\Pi(T + B)A^{-1}(I - \Pi)(T + B)f_0 +$$
$$\sum_{i=2}^{n-1} \lambda^i\Pi(T + B)\left[A^{-1}(I - \Pi)(T + B)\right]^i f_0 = 0.$$

3.1. Explicit computations and constitutive relations for the constraint. We need a few notations. For any function $\phi(q)$, we define $< \phi >_M$ by

$$< \phi >_M = \int_{\mathcal{D}} \phi(q)M(q)dq,$$

where M is the normalized Maxwellian (2.2). We define the three tensors γ, ω and γ_1 by following the notations of [2]:

(3.9) $\gamma = \dfrac{1}{2}(\nabla v + \nabla v^T), \quad \omega = \dfrac{1}{2}(\nabla v - \nabla v^T), \quad \text{and} \quad \gamma_1 = \gamma\omega - \omega\gamma,$

and set $(\gamma : \gamma)$ the contracted product between γ and itself, that is the trace of the tensor product $\gamma\gamma$. We can now state the main result of this section.

PROPOSITION 3.1. *The solutions f_0, f_1, and f_2 of (3.7) are explicitly given by:*

(3.10)
$$\begin{cases} f_0 = n_0 M, \\[2mm] f_1 = n_0(\gamma q \cdot q)M, \\[2mm] f_2 = n_0 M\left[\alpha(q^2/2)\left(T(\gamma) + \gamma_1\right)q \cdot q + \dfrac{1}{2}(\gamma q \cdot q)^2 \right. \\[3mm] \qquad\qquad \left. - \dfrac{1}{d(d + 2)}(\gamma : \gamma) < q^4 >_M\right]. \end{cases}$$

Here $\alpha(s)$ satisfies the following ordinary differential equation:

(3.11) $2s\alpha''(s) + (d + 4 - 2sU'(s))\alpha'(s) - 2U'(s)\alpha(s) = 2,$

and $n_0(x, t)$ satisfies the continuity equation:

(3.12)
$$\frac{\partial n_0}{\partial t} + v \cdot \nabla_x n_0 = 0.$$

Finally, $T(\gamma)$ is the transport operator applied to γ.

Proof. Since $Af_0 = 0$, then, from Proposition 2.1, there exists $n_0(x, t)$ such that $f_0 = n_0 M$. Now, from (3.7), f_1 is a solution to the following equation:

(3.13) $Af_1 = n_0\nabla_q \cdot \left(\gamma q M(q)\right).$

A solution to this last equation is given by

(3.14)
$$f_1 = n_0(\gamma q \cdot q)M(q).$$

Now, we have to determine the solution $f_2 \in N(A)^{\perp}$ to the following equation:

(3.15)
$$Af_2 = (I - \Pi)(T + B)f_1.$$

We know that $f_1 \in N(A)^{\perp}$, then $\Pi T f_1 = T \Pi f_1 = 0$. Furthermore, $\Pi B = 0$, then after replacing f_1 by its expression (3.14):

(3.16)
$$Af_2 = T\left(n_0(\gamma q \cdot q)\right)M + n_0\nabla_q \cdot \left((\gamma q \cdot q)\gamma qM(q)\right)$$
$$+ n_0\nabla_q \cdot \left((\gamma q \cdot q)\omega qM(q)\right).$$

or equivalently:

(3.17) $\quad Af_2 = T\left(n_0(\gamma q \cdot q)\right)M + n_0(\gamma_1 q \cdot q)M + n_0\nabla_q \cdot \left((\gamma q \cdot q)\gamma qM(q)\right).$

Then a solution f_2 of (3.15) has the form:

$$f_2 = \left[\alpha\left(\frac{q^2}{2}\right)T\left(n_0(\gamma q \cdot q)\right) + n_0\alpha\left(\frac{q^2}{2}\right)(\gamma_1 q \cdot q) + \frac{1}{2}n_0(\gamma q \cdot q)^2 + a\right]M.$$

with α satisfying (3.11). To obtain the value of a, it suffices to write that $f_2 \in N(A)^{\perp}$ or equivalently $\Pi f_2 = 0$, we get:

$$a = -\frac{n_0}{2}\int_{\mathcal{D}}(\gamma q \cdot q)^2 M = -\frac{n_0}{d(d+2)}(\gamma : \gamma) < q^4 >_M .$$

It remains to prove the continuity equation (3.12). We recall that $f_0 \in N(A)$ and that $f_p \in N(A)^{\perp}$ for $p \geq 1$. Thus, in the solvability condition (3.8), the first term is the only non-vanishing term. We get $\Pi T(n_0 M) = 0$, which is equivalent to the continuity equation (3.12). $\qquad \square$

3.2. Constitutive relations. Now we give constitutive relations for the constraint. In the previous proposition, we have computed the first three terms in expansion (1.1). This will lead (by direct computations of the moments) to constitutive relations for C or τ up to terms of order λ^3. In this section we will show how to derive these relations in a simple way. Furthermore, the method presented below yields constitutive relations up to the order λ^4, while only f_0, f_1, and f_2 are known. First, we introduce the notations:

(3.18)
$$C_n(x,t) = \int_{\mathcal{D}}(q \otimes q)U'(q^2/2)f_n(x,q,t)dq,$$

in which the functions f_n are the terms of the Chapman-Enskog expansion and are defined by (3.2) and (3.7). We also denote by τ_n the following quantities:

$$(3.19) \qquad \tau_0 = kT(C_0 - n_0 I), \quad \text{and} \quad \tau_n = kTC_n \text{ for } n \geq 1.$$

Since $f_0 = n_0 M$, we find that $C_0 = n_0 I$, so that $\tau_0 = 0$. Thus, the Chapman-Enskog expansion (3.2) to the order 0 in λ yields

$$(3.20) \qquad C = n_0 I + O(\lambda), \quad \text{or} \quad \tau = O(\lambda).$$

Then, constitutive relations of higher orders in λ require the computation of the contributions C_n or τ_n from the known functions f_n. Precisely, we have the following result:

PROPOSITION 3.2. *The constitutive relation up to the order n is given by:*

$$(3.21) \qquad \lambda \frac{\delta \tau}{\delta t} + \tau = \lambda T_1 + \lambda^2 T_2 + \dots + T_n \lambda^n + O(\lambda^{n+1}),$$

where $(T_n)_{n \geq 1}$ are given by

$$(3.22) \qquad T_n = \frac{\delta}{\delta t} \tau_{n-1} + \tau_n,$$

and are linked to the functions $(f_n)_{n \geq 1}$ by the relations:

$$(3.23) \qquad \begin{aligned} T_1 &= -kT \frac{\delta}{\delta t} \left(\int_{\mathcal{D}} (q \otimes q) f_0 \, dq \right), \\ T_n &= kT \frac{\delta}{\delta t} \left(\int_{\mathcal{D}} (U' - 1)(q \otimes q) f_{n-1} \, dq \right), \quad \text{for all } n > 1. \end{aligned}$$

In particular,

$$(3.24) \qquad \left\{ \begin{aligned} T_1 &= \frac{2 n_0 kT}{d} < q^2 >_M \gamma, \\ T_2 &= \frac{2 n_0 kT}{d(d+2)} < q^4 (U' - 1) >_M \frac{\delta \gamma}{\delta t} \\ &= \frac{2 n_0 kT}{d(d+2)} < q^4 (U' - 1) >_M \left(T(\gamma) + \gamma \omega - \omega \gamma - 2\gamma^2 \right), \\ T_3 &= \frac{\delta}{\delta t} \left[\frac{n_0 kT}{d(d+2)} \left(2 \frac{\delta \gamma}{\delta t} + 4\gamma^2 \right) < \alpha q^4 (U' - 1) >_M + \right. \\ &\quad \frac{n_0 kT}{d(d+2)(d+4)} < q^6 (U' - 1) >_M \left(4\gamma^2 + (\gamma : \gamma) I \right) - \\ &\quad \left. \frac{n_0 kT}{d^2(d+2)} < q^4 >_M < q^2 (U' - 1) >_M (\gamma : \gamma) I \right]. \end{aligned} \right.$$

Remember that T is the transport operator, γ and ω are defined in 3.9, and α is a solution to (3.11).

Proof. First, we recall that $f_n \in N(A)^\perp$ for $n \geq 1$, thus $\int_\mathcal{D} f_n dq = 0$ and we have from (3.18):

$$C_n = \int_\mathcal{D} \left[U'(q \otimes q) - \frac{<q^2 U'>_M}{d} I \right] f_n dq.$$

According to (3.7), we replace f_n by:

$$f_n = A^{-1}(T + B)f_{n-1},$$

and use the self-adjointness of A (see Proposition (2.1)) in the Hilbert space $\mathcal{E}$ defined in (2.3) endowed with the scalar product (2.4). We then obtain:

$$C_n = \int_\mathcal{D} A^{-1} \left[\left(U'(q \otimes q) - \frac{<q^2 U'>_M}{d} I \right) M \right] (T + B)f_{n-1} M^{-1} dq.$$

To compute $A^{-1} \left[\left(U'(q \otimes q) - \frac{<q^2 U'>_M}{d} I \right) M \right]$, we use the following lemma:

LEMMA 3.1. *Let $\beta(q^2/2)$ be a regular function. Formal solutions F of*

$$(3.25) \qquad AF = \left(\beta(q^2/2)(q \otimes q) - \frac{<q^2 \beta>_M}{d} I \right) M(q),$$

are of the form

$$(3.26) \qquad F = \left[\alpha_1(q^2/2)(q \otimes q) + \alpha_2(q^2/2)I \right] M,$$

with α_1 and α_2 being functions of $q^2/2$ that satisfy the following ODE equations:

$$(3.27) \quad \begin{cases} 2s\alpha_1''(s) + (d + 4 - 2sU'(s))\alpha_1'(s) - 2U'(s)\alpha_1(s) = 2\beta(s) \\ 2s\alpha_2''(s) + (d - 2sU'(s))\alpha_2'(s) + 2\alpha_1(s) = -2\dfrac{<q^2 \beta>_M}{d}. \end{cases}$$

We then obtain the desired result after some computations.   □

REMARK. In the case of the Hookean potential, we have $U' = 1$, and then we check that the obtained constitutive relations for the constraint reduce to the Oldroyd-B equation at any order of truncation. This is consistant with the fact that the Hookean potential leads to an exact closed system of equations for the constraint which is nothing else than the Oldroyd-B equation itself.

4. Numerical results. We now focus on the steady-state plane Couette flow, for which a great number of numerical results are available, see for instance [11, 18]. A diluted solution of polymer is considered between two infinite planes. According to Fig. 1, the upper plane remains fixed,

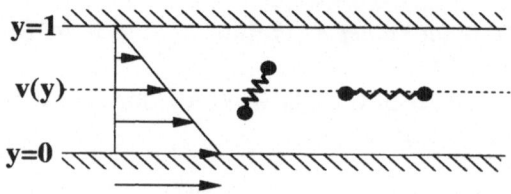

FIG. 1. *The plane Couette flow and Hookean dumbbells.*

while the lower plane moves at constant horizontal speed. Let $v(y)$ be the horizontal velocity field, $\tau_{xy}(y)$ the xy component of the extra-stress due to the polymer chains. The horizontal component of momentum conservation yields

$$-\mu\frac{d^2v}{dy^2}(y) = \frac{d\tau_{xy}}{dy}(y),$$

where μ is the fluid viscosity. Oldroyd-B constitutive equation for $\tau_{xy}(y)$ reduces to

$$\tau_{xy}(y) = n_0kT\lambda\frac{dv}{dy}(y).$$

Thus the velocity field $v(y)$ is linear, the extra-stress $\tau_{xy}(y)$ is constant and the ratio between τ_{xy} and $n_0kT\lambda\, dv/dy$ - the so-called dimensionless elastic viscosity - is one. From the experimental point of view (see for instance the results reported in [2]), this is not the case and shear thinning occurs. Indeed, the dimensionless elastic viscosity decreases with increasing values of $\lambda\, dv/dy$, this being predicted when using FENE or FENE-P dumbbells.

The aim of this section is therefore to compare the dimensionless elastic viscosity between FENE, FENE-P dumbbells and the model (3.21) up to the order 4.

4.1. The model (3.21) at order 2. Selecting the FENE dumbbells for $d = 2$ and from (3.24), we obtain

$$(4.1) \qquad \mathcal{T}_1 = n_0kT\frac{b}{b+4}(\nabla v + \nabla v^T).$$

In the frame of the plane Couette flow, γ is given by:

$$\gamma = \frac{1}{2}\frac{dv}{dy}\begin{pmatrix} 0 & 1 \\ 1 & 0 \end{pmatrix},$$

and the xy component of (4.1) is written:

$$(4.2) \qquad \tau_{xy} = \frac{b}{b+4}n_0kT\lambda\frac{dv}{dy} + O(\lambda^2).$$

Thus, the dimensionless elastic viscosity is $b/(b+4)$ at order 2.

4.2. The model (3.21) at order 3. In the case of the plane Couette flow, a simple calculation from (3.24) shows that T_2 does not contribute to the xy component of the extra-stress τ, nor on the dimensionless elastic viscosity.

4.3. The model (3.21) at order 4. With FENE dumbbells, the function α satisfying (3.11) is given by

$$\alpha(s) = 2\frac{s-b}{2b+d+4},$$

so that, with $d = 2$ and after some computations, we obtain the third order correction in the constitutive relation for the constraint:

$$
\begin{aligned}
T_3 &= \frac{\delta}{\delta t}\left[\frac{n_0 kT\left(\frac{dv}{dy}\right)^2 b^2}{(b+3)(b+4)^2(b+6)(b+8)}\begin{pmatrix}16b^2+110b+192 & 0\\ 0 & 2b(2b+7)\end{pmatrix}\right]\\
&= -\frac{2b^3(2b+7)}{(b+3)(b+4)^2(b+6)(b+8)}n_0 kT\left(\frac{dv}{dy}\right)^3\begin{pmatrix}0 & 1\\ 1 & 0\end{pmatrix}.
\end{aligned}
$$

Finally, the xy component of the extra-stress is, at order 4:

$$\tau_{xy} = n_0 kT\lambda\frac{dv}{dy}\left(\frac{b}{b+4} - \frac{2b^3(2b+7)}{(b+3)(b+4)^2(b+6)(b+8)}\lambda^2\left(\frac{dv}{dy}\right)^2\right),$$

and the dimensionless elastic viscosity is

$$\frac{b}{b+4} - \frac{2b^3(2b+7)}{(b+3)(b+4)^2(b+6)(b+8)}\lambda^2\left(\frac{dv}{dy}\right)^2.$$

Note that the dimensionless elastic viscosity goes to one when b goes to infinity, that is to say when FENE dumbbells approach Hookean dumbbells.

In Fig. 2 we have compared the dimensionless elastic viscosity at order 4 to the one predicted by FENE Brownian simulations and FENE-P deterministic computations, as described in [3]. The velocity gradient dv/dy was kept equal to one, while λ was varying from 0.1 to 5. Clearly, when λ is small (less than one), the order 4 model accurately predicts the dimensionless elastic viscosity of FENE dumbbells. However, when λ becomes larger than one, the discrepancy becomes important and FENE-P dumbbells should be preferred.

5. Conclusion and perspectives. In this paper, we have considered a diluted solution of polymer, and extended the expansions of [2] to time-dependent, nonhomogeneous, nonpotential flows. Chapman-Enskog techniques were used and numerical results in the frame of the steady-state plane Couette flow validate the approach. However, the following points still need to be addressed. Firstly, the time-dependent behaviour of

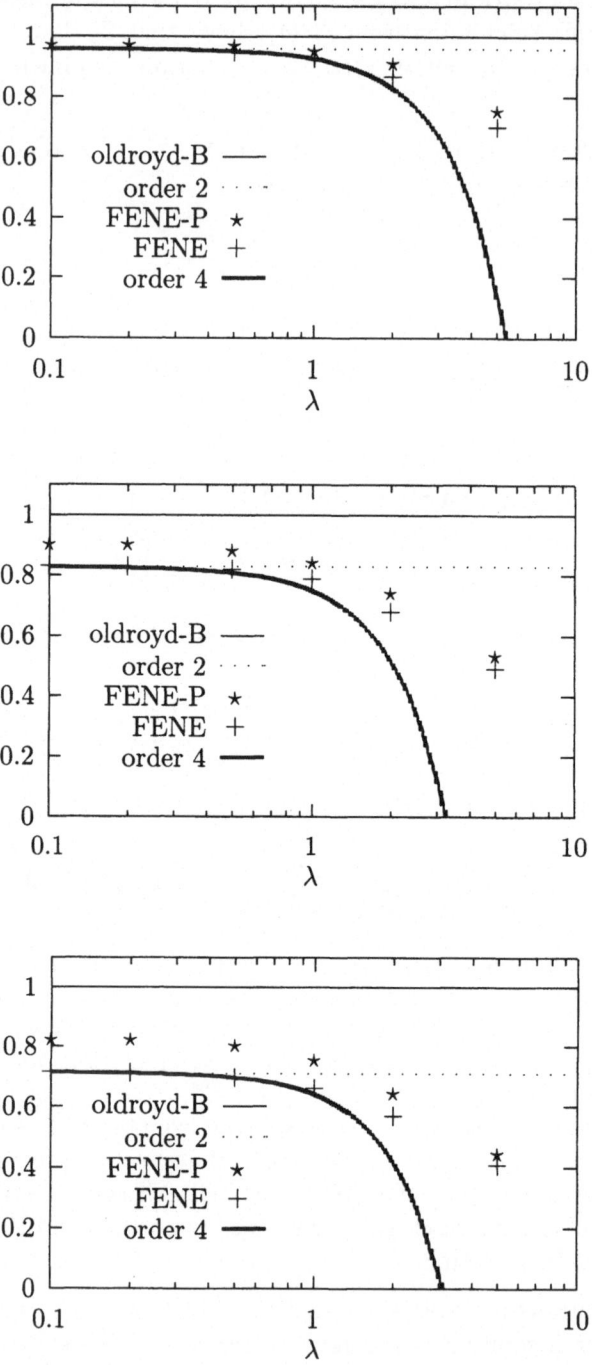

FIG. 2. *Steady-state plane Couette flow. Dimensionless elastic viscosity as a function of λ for stochastic FENE dumbbells, deterministic FENE-P dumbbells and the expansion at order 4. Top figure: $b = 100$, middle figure: $b = 20$, bottom figure: $b = 10$.*

the model has to be studied. Secondly, these expansions have been carried on for the dumbbell model and should be extended to the Rouse chain model in which several beads and springs are connected. A last important question is to investigate the stability of the model (at least the linear stability). Indeed, it is well known that the Chapman-Enskog expansion at high orders could generate instable models. Stabilization procedures similar to those introduced by Slemrod [17, 16] have to be applied. This is the subject of a future work.

REFERENCES

[1] A. BHAVE, R. ARMSTRONG, AND R. BROWN, *Kinetic theory and rheology of dilute, nonhomogeneous polymer solutions*, J. Chem. Phys., **95**: 2988–3000 (1991).

[2] R. BIRD, C. CURTISS, R. ARMSTRONG, AND O. HASSAGER, *Dynamics of polymeric liquids, Vol.* **2**, *kinetic theory*, John Wiley & Sons, New-York, 1987.

[3] J. BONVIN AND M. PICASSO, *Variance reduction methods for CONNFFESSIT-like simulations*, J. Non-Newtonian Fluid Mech., **84**: 191–215 (1999).

[4] H.J. BRASCAMP AND H. LIEB, *On extensions of the Brunn-Minkowski and Précopa-Leindler theorems, including inequalities for log-concave functions, and with an application to the diffusion equation*, Journal of Functional Analysis, **22**: 366–389 (1976).

[5] R. CAFLISH, *The fluid-dynamic limit of the nonlinear Boltzmann equation*, Comm. Pure Appl. Math., **33**: 651–666 (1980).

[6] C. CERCIGNANI, R. ILLNER, AND M. PULVIRENTI, *The mathematical theory of dilute gases*, Springer-Verlag, Berlin, 1994.

[7] S. CHAPMAN, *The kinetic theory of simple and composite gases: viscosity, thermal conduction and diffusion*, Proc. Roy. Soc. (London), **93**: 1–20 (1916).

[8] P. DEGOND, M. LEMOU, AND M. PICASSO, *Viscoleastic fluid models derived from kinetic equations for polymers*, to appear in SIAM. Appl. Maths.

[9] X. GALLEZ, P. HALIN, G. LIELENS, R. KEUNINGS, AND V. LEGAT, *The adaptive lagrangian particle method for macroscopic and micro?macro computations of time-dependent viscoelastic flows*, Comp. Meth. Applied Mech. Engrg., **180**: 345–364.

[10] H. GRAD, *Principles of the kinetic theory of gases*, Vol. **12** of Handbuch der Physik, Springer-Verlag, Berlin, 1958, pp. 205–294.

[11] M. HERRCHEN AND H.-C. ÖTTINGER, *A detailed comparison of various FENE dumbbell models*, J. Non-Newtonian Fluid Mech., **68**: 17–42 (1997).

[12] M. HULSEN, A. VAN HEEL, AND B. VAN DEN BRULE, *Simulation of viscoelastic flows using brownian configuration fields*, J. Non-Newtonian Fluid Mech., **70**: 79–101 (1997).

[13] L. LANDAU AND E. LIFSHITZ, *Physical kinetics*, in Theoretical Physics, Vol. **10**, Akademie-Verlag, Berlin, 1983.

[14] M. LASO AND H. ÖTTINGER, *Calculation of viscoelastic flow using molecular models*, J. Non-Newtonian Fluid Mech., **47**: 1–20 (1993).

[15] H.-C. ÖTTINGER, *Stochastic processes in polymeric fluids*, Springer-Verlag, Berlin, 1996.

[16] M. SLEMROD, *Constitutive relations for monoatomic gases based on generalized rational approximation to the sum the chapman-enskog expansion*, Arch. Rational Mech. Anal., **150**: 1–22 (1999).

[17] ——, *Constitutive relations for Rivlin-Ericksen fluids based on generalized rational approximation*, Arch. Rational Mech. Anal., **146**: 73–93 (1999).

[18] A. VAN HEEL, M. HULSEN, AND B. VAN DEN BRULE, *On the selection of parameters in the FENE-P model*, J. Non-Newtonian Fluid Mech., **75**: 253–271 (1998).

DISPERSIVE/HYPERBOLIC HYDRODYNAMIC MODELS FOR QUANTUM TRANSPORT (IN SEMICONDUCTOR DEVICES)*

CARL L. GARDNER† AND CHRISTIAN RINGHOFER†

Abstract. The smooth quantum hydrodynamic (QHD) model is derived from a moment expansion of the Wigner-Boltzmann equation, using a quantum Maxwellian to close the moments. The smooth QHD model reproduces the original $O(\hbar^2)$ QHD model for small $\hbar^2$. Both QHD models have hyperbolic, dispersive, and parabolic modes. Numerical simulations of a resonant tunneling diode are presented, using a steady-state conservative upwind method.

1. Introduction. Modern VLSI chips incorporate millions of semiconductor devices (transistors, diodes, optical devices, etc.). To predict the performance of the VLSI circuits, the current-voltage (I-V) characteristics of the semiconductor devices are required. Semiconductor device simulation codes provide a way of predicting I-V curves as device parameters—semiconductor material, size, doping, geometry, etc.—are varied, without the necessity of fabricating the device first. Thus many different designs for devices and circuits can be explored efficiently by computer simulations, and promising designs may be singled out for actual fabrication and testing.

Quantum semiconductor devices like resonant tunneling diodes and transistors, HEMTs, MODFETs, and superlattice devices are becoming of increasing importance in state-of-the-art VLSI chips. These devices rely on quantum tunneling of charge carriers through potential barriers for their operation. Advanced microelectronic applications include multiple-state logic and memory devices and high frequency oscillators and sensors.

A fundamental approach to modeling these tunneling devices would be through the Wigner-Boltzmann transport equation (the quantum generalization of the classical Boltzmann transport equation), but simulations of the Wigner-Boltzmann equation are computationally expensive since the Wigner distribution function $f_W(\mathbf{x}, \mathbf{p}, t)$—the quantum analog of the Boltzmann distribution function—is a function of four variables plus time even for a two-dimensional approximation of a three-dimensional device.

Thus a hydrodynamic approximation to the Wigner-Boltzmann transport equation, where the density, velocity, and temperature of a charge carrier are functions of $\mathbf{x}$ and t, offers enormous computational speedups in simulating devices. This approach has worked well in modeling semiclassical semiconductor devices by means of the classical hydrodynamic model.

---
*Research supported in part by the National Science Foundation under grant DMS-9706792.

†Department of Mathematics, Arizona State University, Tempe, AZ 85287.

91

We will discuss two versions of the quantum hydrodynamic (QHD) model: the original $O(\hbar^2)$ QHD model [1, 2] and the "smooth" QHD model [3]. The QHD models include quantum transport effects like particle tunneling through potential barriers and particle buildup in quantum wells. Our approach will be to present and derive the smooth QHD model, relating it to the original QHD model which can be derived from it in the limit that $\hbar^2/(mT_0l^2) \ll 1$, where m is the effective mass of the charge carrier in the semiconductor, T_0 is the ambient temperature, and l is a typical length scale in the problem.

The smooth quantum hydrodynamic model was developed to rigorously handle discontinuities in the potential energy V which occur at heterojunction barriers in quantum semiconductor devices. To derive the model, we constructed a "quantum Maxwellian" as an $O(\beta V)$ $(\beta = 1/T_0)$ solution to the Bloch equation, which governs thermal equilibrium in quantum statistical mechanics. Then the smooth QHD model is derived from a moment expansion of the Wigner-Boltzmann equation, using the quantum Maxwellian to close the moment expansion.

We present the derivation of the smooth QHD model in Section 2, and derive the $O(\hbar^2)$ model from the smooth model in Section 3. In Section 4 we give the mathematical classification of the smooth QHD model. Section 5 discusses various conservative upwind numerical methods which have been used to simulate the QHD models. In Section 6 we present simulations of a resonant tunneling diode using the QHD models, and make some comparisons with another type of simulator NEMO.

2. Derivation. The smooth QHD model is based on a moment expansion of the Wigner-Boltzmann equation with a Fokker-Planck collision term. The moment expansion is closed using a quantum Maxwellian which involves a smoothing of the classical potential energy.

Feynman [6] introduced into quantum statistical mechanics a smoothing of the classical potential as a way of partly accounting for the long range effects of quantum mechanics. However his smoothed potential

$$(1) \qquad V_{a^2}(x) = \int \frac{dy}{\sqrt{2\pi a^2}} \exp\left\{-\frac{(x-y)^2}{2a^2}\right\} V(y), \qquad a^2 \propto \beta \hbar^2/m$$

"fails in its present form when the [classical] potential has a very large derivative as in the case of hard-sphere interatomic potential" or heterojunction potential barriers in semiconductor devices.

The smooth QHD model involves a smoothing of the classical potential over both space and temperature—the extra degree of smoothing is just enough to insure that the gradient of the smooth effective potential in the momentum transport equation (3) below is continuous.

The smooth QHD model reproduces the original $O(\hbar^2)$ QHD model for $\hbar^2\beta/(ml^2) \ll 1$ and recovers classical electrogasdynamics as $\hbar^2\beta/(ml^2) \to 0$. We will use the shorthand $O(\hbar^2)$ for $O(\hbar^2\beta/(ml^2))$ throughout.

The smooth QHD equations have the same form as classical electrogasdynamics:

$$(2) \qquad \frac{\partial n}{\partial t} + \frac{\partial}{\partial x_i}(nu_i) = 0$$

$$(3) \qquad \frac{\partial}{\partial t}(mnu_j) + \frac{\partial}{\partial x_i}(mnu_iu_j - P_{ij}) = -n\frac{\partial V}{\partial x_j} - \frac{mnu_j}{\tau_p}$$

$$(4) \qquad \frac{\partial W}{\partial t} + \frac{\partial}{\partial x_i}(u_iW - u_jP_{ij} + q_i) = -nu_i\frac{\partial V}{\partial x_i} - \frac{(W - W_0)}{\tau_w}$$

$$(5) \qquad \nabla \cdot (\epsilon \nabla V_P) = e^2(N - n).$$

Scattering is modeled by the standard relaxation time approximation, with momentum and energy relaxation times τ_p and τ_w. Explicit factors of $\hbar^2$ appear in the fourth and higher moment equations. In the first three moments, $\hbar^2$ appears in the expressions for the stress tensor, energy density, and heat flux.

The stress tensor and energy density are

$$(6) \qquad P_{ij} = -nT\delta_{ij} - \frac{\hbar^2 n}{4mT_0}\frac{\partial^2 \overline{V}}{\partial x_i \partial x_j}$$

$$(7) \qquad W = \frac{3}{2}nT + \frac{1}{2}mnu^2 + \frac{\hbar^2 n}{8mT_0}\nabla^2\overline{V}$$

where the "quantum potential"

$$(8) \qquad \begin{aligned} \overline{V}(\beta, \mathbf{x}) = {} & \int_0^\beta \frac{d\beta'}{\beta}\left(\frac{\beta'}{\beta}\right)^2 \int d^3x' \left(\frac{2m\beta}{\pi(\beta - \beta')(\beta + \beta')\hbar^2}\right)^{3/2} \\ & \times \exp\left\{-\frac{2m\beta}{(\beta - \beta')(\beta + \beta')\hbar^2}(\mathbf{x}' - \mathbf{x})^2\right\} V(\mathbf{x}'). \end{aligned}$$

This expression is derived in Eqs. (28) and (29) using the quantum Maxwellian density matrix defined in Eqs. (18) and (19). Note that the potential energy $V = V_B + V_P$ consists of two parts: V_B, the potential energy due to the heterojunction barriers, and V_P, the potential energy from Poisson's equation.

The heat conduction term consists of a classical Fourier law plus a new quantum contribution

$$(9) \qquad \mathbf{q} = -\kappa n\nabla T - \frac{\hbar^2 n}{8m}\nabla^2\mathbf{u}, \quad \kappa = \kappa_0\mu_0 T_0$$

where μ_0 is the low-field mobility of the charge carriers. The quantum contribution may be derived through a Chapman-Enskog expansion for the moments of the Wigner-Boltzmann equation [7], or through an analysis [8] of the transport equations (2)–(4) in terms of the density matrix expressed as a sum over mixed state wavefunctions.

To derive the transport equations, we will work in physical space and use the density matrix $\rho(\mathbf{R}, \mathbf{s}, t)$, where

$$(10) \qquad \mathbf{R} = \frac{1}{2}(\mathbf{x} + \mathbf{y}), \quad \mathbf{s} = \mathbf{x} - \mathbf{y},$$

instead of its Fourier transform, the Wigner distribution function $f_W(\mathbf{x}, \mathbf{p}, t)$. The correspondence between the $(\mathbf{x}, \mathbf{p})$ phase space and the $(\mathbf{R}, \mathbf{s})$ space is given by Table 1:

TABLE 1

Phase space	Fourier transform
$f_W(\mathbf{x}, \mathbf{p}, t)$	$\rho(\mathbf{R}, \mathbf{s}, t)$
$n(\mathbf{x}, t) = \displaystyle\int d^3p \; f_W(\mathbf{x}, \mathbf{p}, t)$	$n(\mathbf{x}, t) = \rho(\mathbf{R}, 0, t)$
$W(\mathbf{x}, t) = \displaystyle\int d^3p \; \frac{p^2}{2m} f_W(\mathbf{x}, \mathbf{p}, t)$	$W(\mathbf{x}, t) = \displaystyle\lim_{s \to 0} -\frac{\hbar^2}{2m} \nabla_s^2 \rho(\mathbf{R}, \mathbf{s}, t)$

The classical Maxwell-Boltzmann (M) distribution function is

$$(11) \qquad f_M(\mathbf{x}, \mathbf{p}) = \frac{1}{(2\pi\hbar)^3} \exp\left\{ -\beta\left(\frac{p^2}{2m} + V(\mathbf{x}) \right) \right\}.$$

The corresponding density matrix is the Fourier transform

$$(12) \qquad \rho_M(\mathbf{R}, \mathbf{s}) = \left(\frac{m}{2\pi\beta\hbar^2} \right)^{3/2} \exp\left\{ -\frac{m}{2\beta\hbar^2} s^2 - \beta V(\mathbf{R}) \right\}.$$

Our quantum Maxwellian (QM) density matrix (derived below in Eqs. (18) and (19)) takes the form

$$(13) \qquad \rho_{QM}(\mathbf{R}, \mathbf{s}) = \left(\frac{m}{2\pi\beta\hbar^2} \right)^{3/2} \exp\left\{ -\frac{m}{2\beta\hbar^2} s^2 - \beta\tilde{V}(\mathbf{R}, \mathbf{s}) \right\}.$$

In Figures 1 and 2 we show the classical Maxwellian vs. our quantum Maxwellian (in $(\mathbf{R}, \mathbf{s})$ space) for a 0.1 eV potential step for electrons in GaAs at 300 K. Note that the moments through the energy density will require second derivatives of the density matrix, so the classical Maxwellian gives divergent results in the presence of a potential jump.

In thermal equilibrium, ρ satisfies the Bloch equation

$$(14) \qquad \frac{\partial \rho}{\partial \beta} = -\frac{1}{2}(H_x + H_y)\rho = \frac{\hbar^2}{4m}\left(\nabla_x^2 + \nabla_y^2 \right)\rho - \frac{1}{2}\left[V(\mathbf{x}) + V(\mathbf{y}) \right]\rho$$

where $H_x = -\hbar^2\nabla_x^2/2m + V(\mathbf{x})$. The Green's function for the Bloch equation (6D heat equation) is

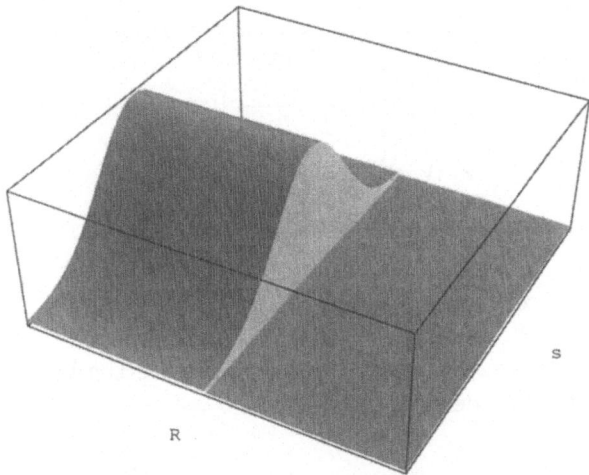

FIG. 1. *Classical Maxwellian for a 0.1 eV potential step for electrons in GaAs at 300 K.*

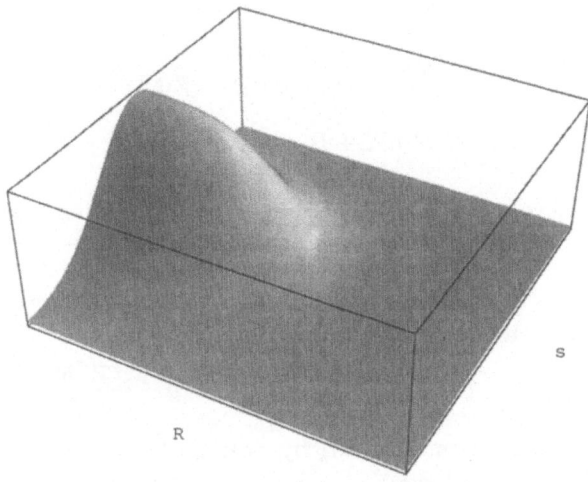

FIG. 2. *Smooth quantum Maxwellian for a 0.1 eV potential step for electrons in GaAs at 300 K.*

$$G(\beta, \mathbf{x}, \mathbf{y}; \beta', \mathbf{x}', \mathbf{y}') =$$

(15) $$\left(\frac{m}{\pi(\beta-\beta')\hbar^2}\right)^3 \exp\left\{-\frac{m}{(\beta-\beta')\hbar^2}\left[(\mathbf{x}-\mathbf{x}')^2+(\mathbf{y}-\mathbf{y}')^2\right]\right\}\theta(\beta-\beta')$$

with initial condition $\rho(0, \mathbf{x}, \mathbf{y}) = \delta^{(3)}(\mathbf{x} - \mathbf{y})$. We obtain

$$
\begin{aligned}
\rho(\beta, \mathbf{x}, \mathbf{y}) \;=\; &\rho_0(\beta, \mathbf{x}, \mathbf{y}) - \frac{1}{2}\int_0^\beta d\beta' \int d^3x' d^3y' \left(\frac{m}{\pi(\beta-\beta')\hbar^2}\right)^3 \\
&\times \exp\left\{-\frac{m}{(\beta-\beta')\hbar^2}\left[(\mathbf{x}-\mathbf{x}')^2 + (\mathbf{y}-\mathbf{y}')^2\right]\right\} \\
&\times \left[V(\mathbf{x}') + V(\mathbf{y}')\right]\rho(\beta', \mathbf{x}', \mathbf{y}')
\end{aligned}
$$
(16)

where the free-particle ($V = 0$) density matrix is

$$
\rho_0(\beta, \mathbf{x}, \mathbf{y}) = \left(\frac{m}{2\pi\beta\hbar^2}\right)^{3/2} \exp\left\{-\frac{m}{2\beta\hbar^2}(\mathbf{x}-\mathbf{y})^2\right\}.
$$
(17)

Next solve for ρ iteratively: setting $\rho = \rho_0$ inside the integral yields the quantum Maxwellian density matrix

$$
\rho_{QM}(\beta, \mathbf{R}, \mathbf{s}) \approx \left(\frac{m}{2\pi\beta\hbar^2}\right)^{3/2} \exp\left\{-\frac{m}{2\beta\hbar^2}s^2 - \beta\tilde{V}(\beta, \mathbf{R}, \mathbf{s})\right\}
$$
(18)

where

$$
\begin{aligned}
\tilde{V}(\beta, \mathbf{R}, \mathbf{s}) \;=\; &\frac{1}{2\beta}\int_0^\beta d\beta' \int d^3X \left(\frac{2m\beta}{\pi(\beta-\beta')(\beta+\beta')\hbar^2}\right)^{3/2} \\
&\times \exp\left\{-\frac{2m\beta}{(\beta-\beta')(\beta+\beta')\hbar^2}X^2\right\} \\
&\times \left[V\left(\mathbf{X}+\mathbf{R}+\frac{\beta'}{2\beta}\mathbf{s}\right) + V\left(\mathbf{X}+\mathbf{R}-\frac{\beta'}{2\beta}\mathbf{s}\right)\right].
\end{aligned}
$$
(19)

Dynamic evolution of ρ (including departures from thermal equilibrium) is governed by the Wigner-Boltzmann equation

$$
\begin{aligned}
i\hbar\frac{\partial\rho}{\partial t} + \frac{\hbar^2}{m}\nabla_R\cdot\nabla_s\rho \;=\; &\left[V\left(\mathbf{R}+\frac{\mathbf{s}}{2}\right) - V\left(\mathbf{R}-\frac{\mathbf{s}}{2}\right)\right]\rho \\
&- i\hbar\frac{1}{\tau}\mathbf{s}\cdot\nabla_s\rho - i\frac{m(2w_0/3)}{\hbar\tau}s^2\rho.
\end{aligned}
$$
(20)

Scattering is modeled in Eq. (20) by Fokker-Planck terms (the last two terms on the right-hand side) in order to produce relaxation time scattering terms in the QHD conservation laws (3) and (4) for momentum and energy [9]. The constants multiplying the Fokker-Planck terms yield the relaxation times $\tau_p = \tau$ and $\tau_w = \tau/2$ in the QHD equations. In general, though, τ_w will not have any simple relationship to τ_p.

In $(\mathbf{x}, \mathbf{p})$ space, averages of functions of momentum $\mathbf{p}$ are given by

$$
\langle\chi\rangle = \int d^3p\, \chi(\mathbf{p}) f_W(\mathbf{x}, \mathbf{p}).
$$
(21)

In $(\mathbf{R}, \mathbf{s})$ space,

$$
\langle\chi\rangle = \lim_{s\to 0} \chi\left(\frac{\hbar}{i}\nabla_s\right)\rho(\mathbf{R}, \mathbf{s})
$$
(22)

which can be verified using

$$(23) \qquad \rho(\mathbf{R}, \mathbf{s}) = \int d^3 p \; f_W(\mathbf{R}, \mathbf{p}) e^{i\mathbf{p} \cdot \mathbf{s}/\hbar}.$$

The moment expansion is obtained by multiplying the Wigner-Boltzmann equation with powers of $\hbar \nabla_s / i$ (1, $\hbar \nabla_s / i$, and $-\hbar^2 \nabla_s^2 / 2m$) and taking the limit $\mathbf{s} \to 0$:

$$(24) \qquad \frac{\partial n}{\partial t} + \frac{1}{m} \frac{\partial \langle p_i \rangle}{\partial x_i} = 0$$

$$(25) \qquad \frac{\partial \langle p_j \rangle}{\partial t} + \frac{\partial}{\partial x_i} \left\langle \frac{p_i p_j}{m} \right\rangle = -n \frac{\partial V}{\partial x_j} - \frac{\langle p_j \rangle}{\tau}$$

$$(26) \qquad \frac{\partial}{\partial t} \left\langle \frac{p^2}{2m} \right\rangle + \frac{\partial}{\partial x_i} \left\langle \frac{p_i p^2}{2m^2} \right\rangle = -\frac{\langle p_i \rangle}{m} \frac{\partial V}{\partial x_i} - \frac{\langle p^2/2m \rangle - W_0}{\tau/2}.$$

We now assume that the momentum-shifted

$$(27) \qquad \mathbf{p} = m\mathbf{u} + \mathbf{p}'$$

ρ_{QM} approximates the actual ρ well enough for average values in the conservation laws to approximate actual values.

Calculating the stress tensor and the energy density,

$$(28) \quad P_{ij} = \lim_{\mathbf{s} \to 0} \frac{\hbar^2}{m} \frac{\partial^2 \rho_{QM}}{\partial s_i \partial s_j} \approx -nT\delta_{ij} - \frac{\hbar^2 n}{4mT_0} \frac{\partial^2 \overline{V}}{\partial x_i \partial x_j}$$

$$(29) \quad W = \frac{1}{2} mnu^2 - \lim_{\mathbf{s} \to 0} \frac{\hbar^2}{2m} \nabla_s^2 \rho_{QM} \approx \frac{3}{2} nT + \frac{1}{2} mnu^2 + \frac{\hbar^2 n}{8mT_0} \nabla^2 \overline{V}$$

we obtain the smooth QHD equations (2)–(4) with the quantum potential (8).

The heat conduction term (9) vanishes for our quantum Maxwellian, but appears in the Chapman-Enskog expansion. Since heat conduction is important in charge transport in semiconductor devices, we have included this term in the classical (with $\hbar \to 0$) and quantum hydrodynamic models and simulations.

The smooth QHD equations contain at worst a step function discontinuity. To see this, we define the 1D smooth effective potential in the momentum conservation equation (3) as the most singular part of $V - P_{11}$:

$$(30) \qquad U = V + \frac{\hbar^2}{4mT_0} \frac{d^2 \overline{V}}{dx^2}.$$

As can be proved using Fourier transforms [3], the smooth effective potential is smoother by two degrees than the classical potential V; i.e., if V has

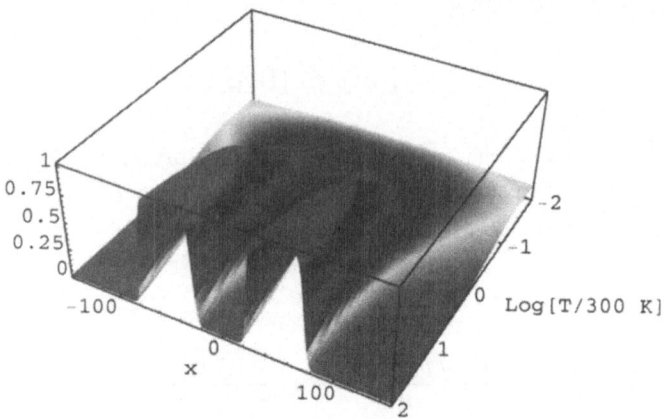

FIG. 3. *Smooth effective potential for electrons in GaAs for 50 Å wide unit potential double barriers and 50 Å wide well as a function of x and $\log_{10}(T/300 \text{ K})$.*

a discontinuity, then U is once differentiable. (There is a step function discontinuity in the momentum conservation equation from $(dn/dx)(d^2\overline{V}/dx^2)$ which is cancelled by other terms in Eq. (3).)

The double integration (over both space and temperature) provides sufficient smoothing so that the P_{11} term in the smooth effective potential cancels the leading singularity in the classical potential at a barrier (see Fig. 3), leaving a residual smooth effective potential with a lower potential height in the barrier region. This cancellation and smoothing makes the barriers partially transparent to the particle flow and provides the mechanism for particle tunneling in the QHD model. Note that the effective barrier height approaches zero as $T \to 0$. This effect explains in fluid dynamical terms why particle tunneling is enhanced at low temperatures. As $T \to \infty$, the effective potential approaches the classical double barrier potential and quantum effects in the QHD model are suppressed.

3. $O(\hbar^2)$ QHD model. The original $O(\hbar^2)$ QHD model can be derived from the smooth QHD model in the following way. First expand the smoothed quantum potential (8) in a power series in $\hbar$:

$$(31) \qquad \overline{V} = \frac{1}{3}V + O(\hbar^2).$$

The corresponding stress tensor and energy density are

$$(32) \qquad P_{ij} = -nT\delta_{ij} - \frac{\hbar^2 n}{4mT_0}\frac{\partial^2 \overline{V}}{\partial x_i \partial x_j} \approx -nT\delta_{ij} - \frac{\hbar^2 n}{12mT_0}\frac{\partial^2 V}{\partial x_i \partial x_j}$$

$$(33) \qquad W \approx \frac{3}{2}nT + \frac{1}{2}mnu^2 + \frac{\hbar^2 n}{24mT_0}\nabla^2 V.$$

Next use the classical thermal equilibrium relation

(34)
$$n \propto \exp(-\beta V) \quad (?!)$$

where the chess notation indicates a questionable but perhaps strong move, to obtain:

(35)
$$P_{ij} \approx -nT\delta_{ij} + \frac{\hbar^2 n}{12m} \frac{\partial^2 \ln(n)}{\partial x_i \partial x_j}$$

(36)
$$W \approx \frac{3}{2} nT + \frac{1}{2} mnu^2 - \frac{\hbar^2 n}{24m} \nabla^2 \ln(n).$$

At this point we can explain the conceptual difference between the derivation of the original $O(\hbar^2)$ QHD model, which arises from the $\hbar \to 0$ limit, and that of the smooth QHD model, which arises from asymptotics for small potential differences. Both models are derived from a functional expansion of the solution of the Bloch equation in terms of a quantum potential: i.e., the thermal equilibrium density matrix is written as

(37)
$$\rho(\mathbf{R}, \mathbf{s}) = C(\beta) \exp \left\{ -\frac{m}{2\beta\hbar^2} s^2 - \beta V^Q(\mathbf{R}, \mathbf{s}) \right\}, \quad C(\beta) = \left(\frac{m}{2\pi\beta\hbar^2} \right)^{3/2}$$

where V^Q denotes the *density matrix* quantum potential.

In both cases the density matrix quantum potential is assumed to depend on the density n, the temperature T, and the potential energy V, since these are the variables in which the closure of the moment equations is carried out. The conceptual difference lies in the form of the functional ansatz. In the $O(\hbar^2)$ model V^Q is assumed to be of the form

(38)
$$\begin{aligned} V^Q(n(\rho, \mathbf{R}), T(\rho, \mathbf{R}), V(\mathbf{R}); \mathbf{s}) = {} & V_0^Q(n, T, V; \mathbf{s}) \\ & + \hbar^2 V_1^Q(n, T, V; \mathbf{s}) + O(\hbar^4) \end{aligned}$$

where n and T denote the density and temperature derived from the density matrix ρ. The solution of the Bloch equation is then written asymptotically as

(39)
$$\rho(\mathbf{R}, \mathbf{s}) = \rho_0(V(\mathbf{R}); \mathbf{s}) + \hbar^2 \rho_1(V(\mathbf{R}); \mathbf{s}) + O(\hbar^4)$$

where the zeroth-order term ρ_0 is the classical Maxwellian

(40)
$$\rho_0(V(\mathbf{R}); \mathbf{s}) = C(\beta) \exp \left\{ -\frac{m}{2\beta\hbar^2} s^2 - \beta V(\mathbf{R}) \right\}$$

and

(41)
$$n(\rho_0, \mathbf{x}) = C(\beta) \exp\{-\beta V\}, \quad T(\rho_0, \mathbf{x}) = \frac{1}{\beta} = T_0.$$

Comparing Eq. (40) with the zeroth-order term in V^Q yields

(42) $$V_0^Q(n(\rho_0, \mathbf{R}), T(\rho_0, \mathbf{R}), V(\mathbf{R}); \mathbf{s}) = V(\mathbf{R}).$$

Equation (41) then allows us to write

(43) $$V(\mathbf{R}) = -T_0 \ln \left\{ \frac{n}{C(1/T_0)} \right\}$$

in the first-order term V_1^Q, which now depends on the logarithm of the density.

In contrast, the smooth QHD model assumes a small potential (i.e., replaces V by εV) and a functional dependence

(44) $$V^Q = \varepsilon V^Q(n, T, V; \mathbf{s}).$$

The equilibrium density matrix is then given by an asymptotic solution of the Bloch equation of the form

(45) $$\rho(\mathbf{R}, \mathbf{s}) = \rho_0(V(\mathbf{R}); \mathbf{s}) + \varepsilon \rho_1(V(\mathbf{R}); \mathbf{s}) + O(\varepsilon^2)$$

where the zeroth-order term is now the free Maxwellian:

(46) $$\rho_0(V(\mathbf{R}); \mathbf{s}) = C(\beta) \exp \left\{ -\frac{m}{2\beta\hbar^2} s^2 \right\}$$

and

(47) $$n(\rho_0, \mathbf{x}) = C(\beta), \quad T(\rho_0, \mathbf{x}) = \frac{1}{\beta} = T_0.$$

Now the zeroth-order term in V^Q yields the smooth quantum potential $\tilde{V}(\mathbf{R}, \mathbf{s})$ for the density matrix (see Eqs. (18) and (19)). Not surprisingly the results of the two expansions coincide when both asymptotic regimes hold. Thus the $O(\hbar^2)$ QHD model is obtained from the smooth QHD model for a sufficiently smooth potential V by expanding the smooth potential $\overline{V}$ in powers of $\hbar^2$.

The $O(\hbar^2)$ QHD model was mathematically classified in Ref. [2] as having two dispersive Schrödinger modes, two hyperbolic modes (corresponding to contact discontinuities in the tangential velocity), with a parabolic heat conduction mode plus the elliptic Poisson mode. Two of the hyperbolic modes (which allow shock discontinuities to form) in the classical hydrodynamic model have been transmogrified to Schrödinger modes when the quantum corrections are included. In one spatial dimension, the $O(\hbar^2)$ QHD model has two dispersive Schrödinger modes plus the parabolic heat conduction mode and the elliptic Poisson mode.

Problems with the $O(\hbar^2)$ QHD model include: (i) the Schrödinger modes produce unphysical dispersive oscillations as a shock breaks in a (classical) semiconductor device (see Fig. 4); and (ii) the $O(\hbar^2)$ QHD model produces negative differential resistance (NDR) (and hysteresis) for the resonant tunneling diode at 77 K, but no or little NDR at 300 K (see Fig. 7 and discussion below).

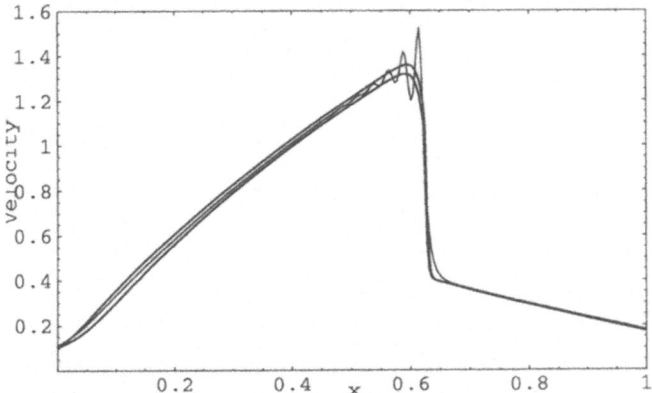

FIG. 4. *Shock wave in a classical semiconductor device illustrating dispersive oscillations in the $O(\hbar^2)$ model (light curve) as the shock breaks. The classical hydrodynamic curve is the upper smooth curve and the smooth QHD curve is the lower smooth curve.*

4. Classification of the smooth QHD equations.

To classify the smooth QHD equations, first we rewrite the equations (2)–(5) (with τ_p, $\tau_w \to \infty$) as (we consider only one spatial dimension for simplicity)

$$(48) \qquad \frac{\partial n}{\partial t} + \frac{\partial}{\partial x}(nu) = 0$$

$$(49) \qquad \frac{\partial u}{\partial t} + u\frac{\partial u}{\partial x} + \frac{1}{mn}\frac{\partial}{\partial x}(nT) + \frac{\hbar^2}{4m^2 n T_0}\frac{\partial}{\partial x}\left(n\frac{\partial^2 \overline{V}}{\partial x^2}\right) + \frac{1}{m}\frac{\partial V}{\partial x} = 0$$

$$(50) \qquad \begin{aligned} & \frac{\partial T}{\partial t} + u\frac{\partial T}{\partial x} + \frac{2}{3}T\frac{\partial u}{\partial x} - \frac{2}{3n}\frac{\partial}{\partial x}\left(\kappa n\frac{\partial T}{\partial x}\right) \\ & \quad - \frac{\hbar^2}{12mn}\frac{\partial}{\partial x}\left(n\frac{\partial^2 u}{\partial x^2}\right) + \frac{\hbar^2 u}{12 m T_0}\frac{\partial^3 \overline{V}}{\partial x^3} + \frac{\hbar^2}{6m T_0}\frac{\partial u}{\partial x}\frac{\partial^2 \overline{V}}{\partial x^2} = 0 \end{aligned}$$

$$(51) \qquad \frac{\partial}{\partial x}\left(\epsilon\frac{\partial V_P}{\partial x}\right) - e^2(N_D - N_A - n) = 0 \ .$$

Next we perturb a solution of the QHD equations with a Fourier mode, linearize the PDEs (48)–(51) with respect to the perturbation, and freeze coefficients. Set

$$(52) \qquad \begin{bmatrix} n \\ u \\ T \\ V \end{bmatrix} = \begin{bmatrix} \overline{n} \\ \overline{u} \\ \overline{T} \\ \overline{V} \end{bmatrix} + e^{-\sigma t + ikx}\begin{bmatrix} \delta n \\ \delta u \\ \delta T \\ \delta V \end{bmatrix}$$

where $[\overline{n},\ \overline{u},\ \overline{T},\ \overline{V}]$ is a (constant) solution of the QHD equations.

We write the linearized QHD equations as

$$(53) \qquad -\mathrm{diag}\{\sigma,\ \sigma,\ \sigma,\ 0\}\,[\delta n,\ \delta u,\ \delta T,\ \delta V] + \mathcal{S}\,[\delta n,\ \delta u,\ \delta T,\ \delta V] = 0$$

where $\mathcal{S}$ is the symbol of the linearized PDE system (48)–(51). Asymptotically as the wavenumber $|k| \to \infty$, $\mathcal{S} - \mathrm{diag}\{\sigma,\ \sigma,\ \sigma,\ 0\}$ has the form

$$(54) \quad \begin{bmatrix} iku - \sigma & ikn & 0 & 0 \\[2mm] ik\frac{T}{mn} & iku - \sigma & i\frac{k}{m} & i\frac{k}{m} - ik^3\frac{\hbar^2}{4m^2 T_0} \\[2mm] 0 & \frac{2}{3}ikT + ik^3\frac{\hbar^2}{12m} & iku + \frac{2}{3}k^2\kappa - \sigma & -ik^3\frac{\hbar^2 u}{12mT_0} \\[2mm] 0 & 0 & 0 & -k^2\epsilon \end{bmatrix}$$

where we have dropped the bar over the constant solution.

The asymptotic eigenvalues σ of the symbol as $|k| \to \infty$ determine the mathematical type of the PDE system. In fact, we need only consider the upper 3×3 block $\mathcal{S}_3$ of the symbol, since the coupling of the transport equations (48)–(50) to Poisson's equation (51) only introduces the elliptic Poisson mode, and does not affect the modes of the transport equations.

To leading order in k, the eigenvalues of the symbol and corresponding modes are

$$(55) \quad \begin{cases} \left(\dfrac{\kappa}{3} \pm \sqrt{\left(\dfrac{\kappa}{3}\right)^2 - \dfrac{\hbar^2}{12m^2}}\ \right) k^2 & \text{Schrödinger/parabolic} \\[6mm] iku & \text{hyperbolic} \\[4mm] -\epsilon k^2 & \text{elliptic.} \end{cases}$$

The "Schrödinger/parabolic" modes are parabolic if $\kappa^2 > 3\hbar^2/(4m^2)$ and are dispersive Schrödinger modes if $\kappa^2 < 3\hbar^2/(4m^2)$. In Si and GaAs semiconductor devices, $\kappa^2 \gg 3\hbar^2/(4m^2)$.

5. Conservative upwind numerical methods. Hyperbolic methods from computational gasdynamics like the ENO, piecewise parabolic, and discontinuous Galerkin methods are well suited for simulating the transient classical and quantum hydrodynamic models. (The parabolic mode is treated as a source term.) Steady-state solutions may be obtained as the asymptotic large t limit.

We have also developed [15, 16, 2] a simple, robust steady-state method that is an order of magnitude faster than the transient solvers in 1D. The 1D steady-state equations are discretized using a conservative upwind method adapted from computational fluid dynamics. Then the discretized equations are solved by a damped Newton method. This steady-state conservative upwind method is used for the simulations in Section 6.

Table 2 indicates the range of methods that have been used to simulate the classical and quantum hydrodynamic models. (A blank indicates only that the numerical method has not been tried with the model in question.)

6. Simulations of the resonant tunneling diode. We present simulations of a GaAs resonant tunneling diode with $Al_{0.3}Ga_{0.7}As$ double barriers at 300 K. The barrier height is equal to 280 meV. The diode consists

TABLE 2

Method	Type	CHD	$O(\hbar^2)$ QHD	smooth QHD
damped Newton	steady-state upwind FD	stable	stable	stable
ENO*	4th-order Godunov FD	stable		
RKDG[†]	2nd-order upwind FE	stable	stable	
CLAWPACK[‡]	2nd-order Godunov FD	stable		

*Fatemi, Gardner, Jerome, Osher, and Rose [10]; Gardner, Jerome, and Shu [11].
[†] Chen, Cockburn, Gardner, and Jerome [12].
[‡] Gardner and Hernandez [13] using LeVeque's code [14].

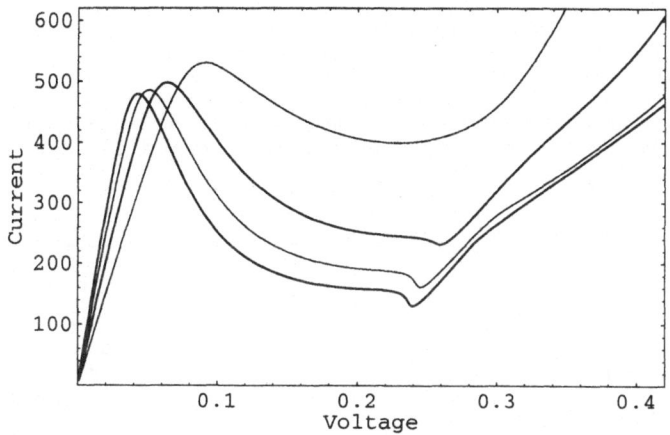

FIG. 5. *Current density in kiloamps/cm² vs. voltage for the smooth QHD model with various amounts of heat conduction for the RTD at 300 K.* $\kappa_0 = 1, 0.8, 0.6,$ *and 0.4 from bottom to top at 0.2 volts. The barrier height is 280 meV.*

of n^+ source (at the left) and drain (at the right) regions with the doping density $N = 10^{18}$ cm^{-3}, and an n channel with $N = 5 \times 10^{15}$ cm^{-3}. The channel is 200 Å long, the barriers are 25 Å wide, and the quantum well between the barriers is 50 Å wide. Note that the device has 50 Å spacers between the barriers and the contacts.

In Fig. 5, the experimental signal of quantum resonance—negative differential resistance, a region of the current-voltage (I-V) curve where the current *decreases* as the applied voltage is increased—is displayed for a range of values κ_0 for the amount of heat conduction in Eq. (9).

In the classical hydrodynamic model, current-voltage curves typically show very little change as the amount of heat conduction is varied (see Fig. 6). The dramatic shift in the I-V curves and the change in the peak to valley current ratios in the smooth QHD simulations indicate that the model—like the original QHD model—is too sensitive to the heat flux term. Comparison with Wigner-Boltzmann simulations should help address this issue.

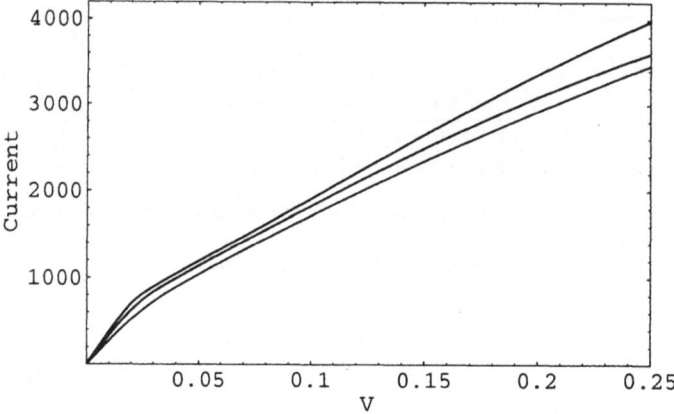

FIG. 6. *Current density in kiloamps/cm^2 vs. voltage for various amounts of heat conduction for the classical $n^+/n/n^+$ diode at 300 K. $\kappa_0 = 2$, 1, and 0.5 from top to bottom.*

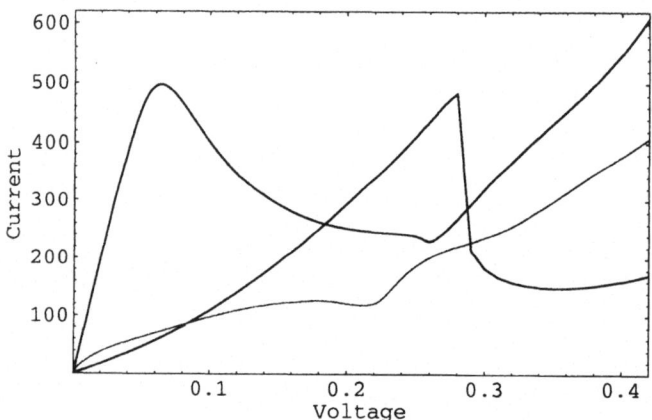

FIG. 7. *Current density in kiloamps/cm^2 vs. voltage comparing the smooth QHD (peaks near 0.1 volts), $O(\hbar^2)$ QHD (lowest curve), and (one band one effective mass) NEMO (peaks near 0.3 volts) models for the RTD at 300 K. $\kappa_0 = 0.6$ for the QHD models. The barrier height is 280 meV.*

Fig. 7 compares the smooth QHD and $O(\hbar^2)$ QHD I-V curves with that predicted by the simulator NEMO [4, 5], which is based on the mixed state Schrödinger equation approach using non-equilibrium Green's functions. The NEMO current-voltage curves have been extensively verified against actual resonant tunneling devices. The $O(\hbar^2)$ current exhibits extremely weak NDR and a shoulder indicating a virtual resonance. Simulations have demonstrated that it is difficult to achieve NDR at 300 K with the $O(\hbar^2)$ QHD model, while NDR is experimentally observed at 300 K and is predicted by the smooth QHD simulations. The smooth QHD I-V curves

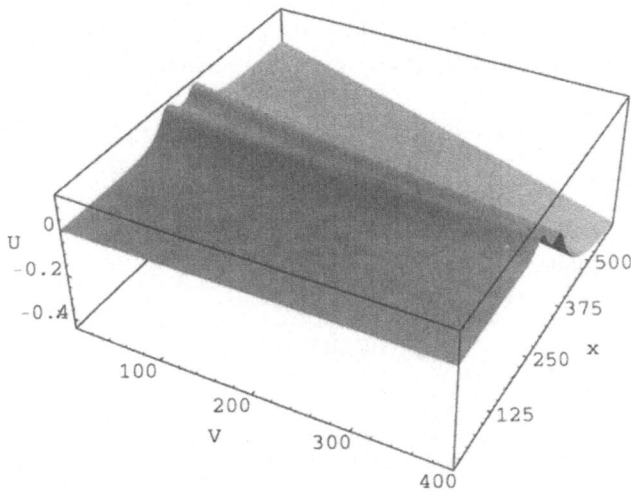

FIG. 8. *Smooth effective potential U in eV for applied voltages between 0 and 400 millivolts for 280 meV double barriers at 300 K. x is in Å.*

do show NDR and the correct peak current, but the current peak is at a voltage that is too low.

Fig. 8 illustrates the smooth effective potential U for the resonant tunneling diode for applied voltages 0–400 millivolts with $\kappa_0 = 0.6$. At zero applied voltage, the 280 meV barrier height is effectively reduced to 120 meV by the smoothing in Eq. (30). (With the addition of the smoothed Poisson potential U_P, the effective U has a maximum of 230 meV in Fig. 8 at an applied voltage of one millivolt.) The effective quantum well is parabolic near its minimum.

The resonant peak of the current-voltage curve occurs as the electrons tunneling through the first barrier come into resonance with the energy levels of the quantum well. The resonant peak in Fig. 5 occurs at 60 millivolts for $\kappa_0 = 0.6$. As the voltage bias increases above 60 millivolts, the resonance effect rapidly decreases because the right barrier height is progressively reduced and electrons tunnel out of the well through the thin portion of the effective parabolic well.

7. Conclusion. The current-voltage curves for the resonant tunneling diode at 300 K computed with the smooth QHD model, while qualitatively correct—the I-V curves do show NDR and the correct peak current, unlike the $O(\hbar^2)$ model I-V curve—peak at a voltage that is too low and display too much sensitivity to the amount of heat conduction.

Simulations of the resonant tunneling diode using the Wigner-Boltzmann-Poisson equations are planned to determine when the smooth QHD model gives solutions and current-voltage curves which agree closely with the kinetic theory results. These simulations should help determine physically accurate models for heat conduction and the relaxation times appearing in the transport equations. More accurate models for heat conduction and the relaxation times may improve the agreement between the smooth QHD model and NEMO.

There should be a technologically important range of parameters (device size, ambient temperature, potential barrier height, applied voltage, semiconductor material, etc.) in which the smooth QHD model gives solutions and current-voltage curves that are very close to those given by the Wigner-Boltzmann-Poisson system.

REFERENCES

[1] H.L. GRUBIN AND J.P. KRESKOVSKY, "Quantum moment balance equations and resonant tunnelling structures," *Solid-State Electronics*, **32**: 1071–1075, 1989.

[2] C.L. GARDNER, "The quantum hydrodynamic model for semiconductor devices," *SIAM Journal on Applied Mathematics*, **54**: 409–427, 1994.

[3] C.L. GARDNER AND C. RINGHOFER, "Smooth quantum potential for the hydrodynamic model," *Physical Review*, E **53**: 157–167, 1996.

[4] http://www.raytheon.com/rsc/ses/spr/spr_nan/nan_nem.htm (NEMO website).

[5] R. LAKE, G. KLIMECK, R.C. BOWEN, AND D. JOVANOVIC, "Single and multiband modeling of quantum electron transport through layered semiconductor devices," *Journal of Applied Physics*, **81**: 7845–7869, 1997.

[6] R.P. FEYNMAN, *Statistical Mechanics: A Set of Lectures.* Reading, Massachusetts: W.A. Benjamin, 1972.

[7] C.L. GARDNER AND C. RINGHOFER, "The Chapman-Enskog Expansion and the Quantum Hydrodynamic Model for Semiconductor Devices," *VLSI Design*, **10**: 415–435, 2000.

[8] C.L. GARDNER, "Resonant tunneling in the quantum hydrodynamic model," *VLSI Design*, **3**: 201–210, 1995.

[9] H.L. GRUBIN, T.R. GOVINDAN, J.P. KRESKOVSKY, AND M.A. STROSCIO, "Transport via the Liouville equation and moments of quantum distribution functions," *Solid-State Electronics*, **36**: 1697–1709, 1993.

[10] E. FATEMI, C.L. GARDNER, J.W. JEROME, S. OSHER, AND D.J. ROSE, "Simulation of a steady-state electron shock wave in a submicron semiconductor device using high-order upwind methods," in *Computational Electronics: Semiconductor Transport and Device Simulation*, pp. 27–32, Boston: Kluwer Academic Publishers, 1991.

[11] C.L. GARDNER, J.W. JEROME, AND C.-W. SHU, "The ENO method for the hydrodynamic model for semiconductor devices," in *High Performance Computing, 1993: Grand Challenges in Computer Simulation*, pp. 96–101, San Diego: The Society for Computer Simulation, 1993.

[12] Z. CHEN, B. COCKBURN, C.L. GARDNER, AND J.W. JEROME, "Quantum hydrodynamic simulation of hysteresis in the resonant tunneling diode," *Journal of Computational Physics*, **117**: 274–280, 1995.

[13] C.L. GARDNER, A. GELB, AND J.R. HERNANDEZ, "A Comparison of Modern Hyperbolic Methods for Semiconductor Device Simulation: NTK Central Scheme vs. CLAWPACK," accepted for publication in *VLSI Design*.

[14] http://www.amath.washington.edu/~claw/ (CLAWPACK website).

[15] C.L. GARDNER, J.W. JEROME, AND D.J. ROSE, "Numerical methods for the hydrodynamic device model: Subsonic flow," *IEEE Transactions on Computer-Aided Design of Integrated Circuits and Systems*, **8**: 501–507, 1989.

[16] C.L. GARDNER, "Numerical simulation of a steady-state electron shock wave in a submicrometer semiconductor device," *IEEE Transactions on Electron Devices*, **38**: 392–398, 1991.

A REVIEW ON SMALL DEBYE LENGTH AND QUASI-NEUTRAL LIMITS IN MACROSCOPIC MODELS FOR CHARGED FLUIDS

INGENUIN GASSER*

Abstract. We consider macroscopic models for charged fluids of drift-diffusion and of hydrodynamic type. In both cases we study the small Debye length limit on different time scales. In certain scalings we end up in the quasi-neutral regime. The limits are performed and the limit problems are identified.

Key words. Hydrodynamic model, Drift-Diffusion model, Debye length, Quasi-neutral limit.

AMS(MOS) subject classifications. 35Q60, 76W05.

1. Introduction. In this paper we consider electrically charged media. Examples are plasmas, semiconductor materials, charged particles in solutions (electrophoresis) etc. (see [30, 25, 4] for applications in plasma physics, semiconductor physics and electrophoresis, respectively).

There are many different ways to model such kind of charged media. In microscopic models the dynamics of single charged particles under the influence of the motion of the other particles is described. This leads in general to many particle models and under additional assumptions to single particle models. In the kinetic picture a simplified description via phase space distribution functions is used. Here we deal with a macroscopic fluid-dynamic description in terms of macroscopic quantities like densities and current densities under influence of the electric (in general electromagnetic) field.

On the macroscopic level many different models are known. The most complete models are hydrodynamic models where the differently charged fluids are described by Euler- or Navier-Stokes type equations (e.g. MHD equations). Here in this paper we will use so called isentropic or isothermal hydrodynamic models obtained from the above models by an additional assumption in the constitutive relations. These equations for the differently charged fluids are coupled via the Poisson equation representing the Maxwell equations in case of no or negligible magnetic forces.

A simplified version of the hydrodynamic models are the drift-diffusion models. Drift-diffusion models are relaxation time approximations of the hydrodynamic models (see Section 3). In these models the only independent macroscopic variables are the charge densities (and the electric field).

---

*Fachbereich Mathematik, Universität Hamburg, Bundesstrasse 55, D-20146 Hamburg, Germany. The author thanks the Institute for Mathematics and its Applications (IMA) for the kind hospitality and the financial support. The work of the author was supported partly by the TMR project *Asymptotic methods in kinetic theory*.

Further simplifications lead to so called (pure) drift models or to (in general nonlinear) diffusion equations.

The scaled Debye length is a (dimensionless) parameter appearing in both hydrodynamic and drift-diffusion models. This parameter expresses the relation between a characteristic non-neutral length and a characteristic length in the fluid determined by the space scale. In many physically relevant scalings this parameter is very small. Therefore, the vanishing Debye length limit is of particular interest. We study this limit on the significant time scales. This gives a clear picture of the relations between the various models. In Section 2 we show that the pure drift models and the (nonlinear) diffusion models are small Debye length approximations of the drift diffusion equations. In Section 3 we will show that these simplified models can be obtained also from the hydrodynamic model using a small Debye length limit combined with a relaxation time limit.

2. The small Debye length limit in drift-diffusion model. We consider a bipolar drift diffusion model of the form

$$(2.1) \qquad n_t = \mu_n div(\nabla n^{\gamma_n} + nE),$$

$$(2.2) \qquad p_t = \mu_p div(\nabla p^{\gamma_p} - pE),$$

$$(2.3) \qquad -\lambda^2 div E = n - p - C$$

with $x \in \Omega \subset \mathbb{R}^d$, Ω bounded with smooth boundary, $t \geq 0$ and $E = \nabla\Phi$. The name of the model originates in the two competing effects, (nonlinear) diffusion and drift caused by the electric field. A standard reference in semiconductor applications is [25]. Electrophoretic applications are given in [4]. The problem has to be supplemented with initial conditions

$$(2.4) \qquad n(t = 0, x) = n_0(x), \qquad p(t = 0, x) = p_0(x)$$

and with boundary conditions. In order to keep things simple we choose homogeneous Neumann boundary conditions for the fluxes

$$(2.5) \qquad (\nabla n^{\gamma_n} + nE) \cdot \nu = 0,$$

$$(2.6) \qquad (\nabla p^{\gamma_p} - pE) \cdot \nu = 0,$$

$$(2.7) \qquad E \cdot \nu = 0 \text{ on } \partial\Omega,$$

where ν is the normal vector on the boundary $\partial\Omega$. We require the related compatibility condition

$$(2.8) \qquad \int_\Omega (n_0(x) - p_0(x) - C(x))dx = 0.$$

The quantities n, p, E, Φ are the densities of the negatively and positively charged fluids, the electric field and the electric potential, respectively. The function $C = C(x)$ describes the background charge. The constants

μ_n and μ_p are the so called mobilities. The constant λ denotes the scaled Debye length. For the isentropic exponents γ_n and γ_p we assume $\gamma_n, \gamma_p \geq 1$. Confusingly, the case $\gamma_n, \gamma_p > 1$ is often referred as "nonlinear" drift diffusion model although the model is nonlinear even in the case $\gamma_n = \gamma_p = 1$.

We are mainly interested in the behaviour of the solutions of the problem (2.1–2.8) in the vanishing Debye length limit $\lambda \to 0$. There are two significant time scales in this problem. One is the original (slow) time scale t and the other one is the fast initial time layer scale $s = \frac{t}{\lambda^2}$. In the next two sections we consider these two problems.

Before passing to these two problems we state the main a priori estimates used in Subsections 2.1 and 2.2.

LEMMA 2.1. Let $n_0, p_0 \in L^q(\Omega)$, nonnegative, satisfying the compatibility condition (2.6) and $C \in L^\infty(\Omega)$. Then, the solutions $n(t), p(t)$ of (2.1–2.5) satisfy

$$(2.9) \qquad \|n(t)\|_1 = \|n_0\|_1, \quad \|p(t)\|_1 = \|p_0\|_1, \qquad q = 1,$$

$$
\begin{aligned}
(2.10) \quad &\|n(t)\|_q^q + \|p(t)\|_q^q + \lambda^2 \int_0^t (\|\nabla n^{\frac{q+\gamma_n-1}{2}}\|_2^2 + \|\nabla p^{\frac{q+\gamma_n-1}{2}}\|_2^2) ds \\
&\leq c e^{K\|C\|_\infty \frac{t}{\lambda^2}} (\|n_0\|_q^q + \|p_0\|_q^q) \qquad 1 < q < \infty
\end{aligned}
$$

with c and K depending on μ_n, μ_p, q.

The proof of this lemma is given in [8, 9]. Note the (bad) λ-dependence in the case $C \neq 0$ and $q > 1$.

Another important quantity is the entropy. It is defined by

$$(2.11) \quad e(t) = \int_\Omega (n \frac{n^{\gamma_n-1}-1}{\gamma_n-1} + p \frac{p^{\gamma_p-1}-1}{\gamma_p-1} + \frac{\lambda^2}{2}|E|^2) dx + e_0,$$

where $(\frac{A^{\gamma-1}-1}{\gamma-1})|_{\gamma=1} := \ln A$ for $A > 0$ and the constant e_0 is chosen such that the entropy $e^\lambda(t)$ is a nonnegative quantity. An easy calculation gives

LEMMA 2.2. Let $t > 0$. Then

$$(2.12) \quad \frac{d}{dt} e(t) = - \int_\Omega \left(\mu_n \frac{|\nabla n^{\gamma_n} + nE|^2}{n} + \mu_p \frac{|\nabla p^{\gamma_p} - pE|^2}{p} \right) dx$$

holds for the solutions of system (2.1–2.7).

We recall that the existence analysis of the bipolar drift diffusion problem was done in [6, 7] for the standard drift diffusion model and in [15] in the "nonlinear" case.

2.1. The initial time layer. First we consider the problem on the small time scale $s = \frac{t}{\lambda^2}$. We rescale

$$(2.13) \qquad s = \frac{t}{\lambda^2},$$

(2.14) $n^\lambda(s, x) = n(\lambda^2 s, x)$ $p^\lambda(s, x) = p(\lambda^2 s, x)$,

(2.15) $F^\lambda(s, x) = \lambda^2 E(\lambda^2 s, x)$ $V^\lambda(s, x) = \lambda^2 \Phi(\lambda^2 s, x)$.

This corresponds to an initial time layer analysis. The system (2.1–2.8) in this scaling reads

(2.16) $$n^\lambda_s = \mu_n div(\lambda^2 \nabla(n^\lambda)^{\gamma_n} + n^\lambda F^\lambda),$$

(2.17) $$p^\lambda_s = \mu_p div(\lambda^2 \nabla(p^\lambda)^{\gamma_p} - p^\lambda F^\lambda),$$

(2.18) $$-div F^\lambda = n^\lambda - p^\lambda - C$$

with initial conditions

(2.19) $$n^\lambda(t = 0, x) = n_0(x), \qquad p^\lambda(t = 0, x) = p_0(x)$$

and homogeneous Neumann boundary conditions for the fluxes

(2.20) $$(\lambda^2 \nabla(n^\lambda)^{\gamma_n} + n^\lambda F^\lambda) \cdot \nu = 0,$$

(2.21) $$(\lambda^2 \nabla(p^\lambda)^{\gamma_n} - p^\lambda F^\lambda) \cdot \nu = 0,$$

(2.22) $$F^\lambda \cdot \nu = 0, \quad \text{on } \partial\Omega.$$

Formally, if $\lambda \to 0$, the following limit problem is obtained for the limit quantities n, p and F of n^λ, p^λ and F^λ, respectively,

(2.23) $$n_s = \mu_n div(Fn),$$
(2.24) $$p_s = -\mu_p div(Fp),$$
(2.25) $$-div F = n - p - C,$$

with the initial conditions

(2.26) $$n(t = 0, x) = n_0(x), \qquad p(t = 0, x) = p_0(x)$$

and homogeneous Neumann boundary conditions

(2.27) $$F \cdot \nu = 0, \quad \text{on } \partial\Omega.$$

Many authors have investigated the initial time layer problem for the standard drift diffusion model (i.e. the case $\gamma_n = \gamma_p = 1$) [1, 26, 29, 31]. All these results are valid in one space dimension with quite strong assumptions on the data. Recently, the multi-dimensional case was solved in [10] for the standard drift diffusion model and in [8, 9] for the "nonlinear model".

THEOREM 2.1.   *Let $C \in L^\infty(\Omega)$ and $S > 0$. Let $n_0, p_0 \in L^q(\Omega)$ with $q > \max(\frac{2d}{d+1}, \frac{\gamma_n+1}{2}, \frac{\gamma_p+1}{2})$, nonnegative and the compatibility condition (2.8) hold. Then, as $\lambda \to 0$ the following convergences hold (after extracting subsequences):*

- $n^\lambda \rightharpoonup n$, $p^\lambda \rightharpoonup p$ *weak-$\star$ in* $L^\infty((0,S); L^q(\Omega))$.
- $F^\lambda \to F$ *strongly in* $C([0,S]; L^p(B))$

$$for\ 1 < p < \begin{cases} \dfrac{dq}{d-q} & for\ q < d, \\ \infty & for\ d \le q. \end{cases}$$

- $F^\lambda n^\lambda \rightharpoonup Fn$, $F^\lambda p^\lambda \rightharpoonup Fp$ *in* $\mathcal{D}'([0,S) \times \Omega)$

and n, p, F *satisfy the formal limit equations (2.23–2.26) in* $\mathcal{D}'([0,S) \times \Omega)$.

The proof of this theorem can be found in [8, 9]. It is based on the a priori estimates in Lemma 2.1. The passage to the limit in the nonlinear terms $n^\lambda F^\lambda$ and $p^\lambda F^\lambda$ is done showing strong convergence of the electric field F^λ.

A final comment on the condition $q > max(\frac{2d}{d+1}, \frac{\gamma_n+1}{2}, \frac{\gamma_n+1}{2})$. The first term in the maximum is related to the nonlinearities $n^\lambda F^\lambda$, $p^\lambda F^\lambda$ and the second part to the nonlinear diffusivities.

2.2. The quasineutral regime. On the original (slow) time scale t the situation is completely different. In order to see the formal limit we assume $C \equiv 0$ and take a special linear combination of the equations for n^λ and p^λ, namely

$$(2.28) \quad \left(\frac{1}{\mu_n} n^\lambda + \frac{1}{\mu_p} p^\lambda\right)_t = \Delta\left((n^\lambda)^{\gamma_n} + (p^\lambda)^{\gamma_n}\right) + div((n^\lambda - p^\lambda)E^\lambda)$$

$$(2.29) \quad (n^\lambda - p^\lambda)_t = \Delta\left(\mu_n(n^\lambda)^{\gamma_n} + \mu_p(p^\lambda)^{\gamma_n}\right) + div((\mu_n n^\lambda + \mu_p p^\lambda)E^\lambda)$$

$$(2.30) \quad -\lambda^2 div E^\lambda = n^\lambda - p^\lambda,$$

(λ again denotes the dependence on λ). Formally, if $\lambda \to 0$, the variable $n^\lambda - p^\lambda$ tends to zero, which expresses the quasineutrality. Suppose that $(n^\lambda - p^\lambda)E^\lambda$ also tends to zero, then we are left with a nonlinear diffusion type equation as limiting problem for the limit w of n^λ and p^λ

$$(2.31) \quad w_t = \frac{\mu_n \mu_p}{\mu_n + \mu_p} \Delta(w^{\gamma_n} + w^{\gamma_p})$$

$$(2.32) \quad w(0,x) = \frac{1}{\mu_n + \mu_p}(\mu_p n_0(x) + \mu_n p_0(x))$$

and the boundary conditions

$$(2.33) \quad \nabla w \cdot \nu = 0 \text{ on } \partial\Omega.$$

There are very few results concerning the quasineutral limit. In the literature there is the result [29] based on asymptotic analysis. The first and only rigorous results were obtained in [8] for the case $C \equiv 0$, in [9, 10, 11] for special cases $C \neq 0$ and in [9] for cases with additional recombination-generation terms.

THEOREM 2.2. *Let $C \equiv 0$ and $T > 0$. Let $n_0 = p_0 \in L^q(\Omega)$, nonnegative with $q = \max(\gamma_n, \gamma_p)$ uniformly in λ. Let $\frac{1}{r} = \frac{1}{2} - \frac{1}{d}$. Then, as $\lambda \to 0$ the following convergences hold (after extracting subsequences):*

- $n^\lambda \to w$ *strongly in* $L^r((0,T) \times \Omega)$, $1 \le r < 2\dfrac{r-1}{r}\gamma_n$,

- $p^\lambda \to w$ *strongly in* $L^r((0,T) \times \Omega)$, $1 \le r < 2\dfrac{r-1}{r}\gamma_p$.

- $(n^\lambda - p^\lambda)E^\lambda \to 0$ *strongly in* $L^{\frac{2(q+1)}{q+2}}((0,T) \times \Omega)$

and w satisfies the nonlinear diffusion equation (2.31–2.32) in $\mathcal{D}'([0,T) \times \Omega)$.

The proof of this theorem can be found in [8, 9]. The main ingredient is the use the entropy and the entropy dissipation terms.

In the case of nonvanishing background charge $C \ne 0$ things are more difficult. The limit problem is more delicate. As an example we consider the case of the standard drift diffusion model $\gamma_n = \gamma_p = 1$ with a constant background charge $C = const$. Then the formal limit problem reads

$$(2.34) \qquad n_t = div\left(\frac{2n - C}{\frac{n}{\mu_p} + \frac{n-C}{\mu_n}} \nabla n \right).$$

Note that for $\mu_n \ne \mu_p$ this is already a nonlinear diffusion equation. More general results on the case $C \ne 0$ can be found in [9, 10] and in [11].

3. The small Debye length limit in hydrodynamic model.

Here we consider the bipolar hydrodynamic model in one space dimension. Hydrodynamic models are more sophisticated models than drift diffusion models. Indeed, we will show below that the drift diffusion model is nothing but a relaxation time limit model of the hydrodynamic model. From a mathematical viewpoint hydrodynamic models are systems of hyperbolic conservation laws with inhomogeneous terms. Therefore, we only consider a one dimensional whole space setting. This kind of models are used in plasma [30] and in semiconductor physics [25].

Denoting by n, p, j, g and E the charge densities, current densities, pressures and electric field, the scaled equations of the hydrodynamic model are given by

$$(3.1) \qquad n_r + j_x = 0$$

$$(3.2) \qquad j_r + \left(\frac{j^2}{n} + n^{\gamma_n} \right)_x = -nE - \frac{j}{\mu_n \tau}$$

$$(3.3) \qquad p_r + g_x = 0$$

$$(3.4) \qquad g_r + \left(\frac{g^2}{p} + p^{\gamma_n} \right)_x = pE - \frac{g}{\mu_p \tau}$$

$$(3.5) \qquad -\lambda^2 E_x = n - p - C$$

where $x \in \mathbb{R}$ and time $r \geq 0$. The time scale r is different from time scale t in Section 2. We prescribe initial conditions

(3.6) $$n(r = 0, x) = n_0(x), \qquad j(r = 0, x) = j_0(x)$$

(3.7) $$p(r = 0, x) = p_0(x), \qquad g(r = 0, x) = g_0(x).$$

The adiabatic exponents $\gamma_n, \gamma_p \geq 1$ have the same meaning as in Section 2. The positive constants τ and λ denote the relaxation time and the Debye length, respectively. The relaxation term describes in a very rough manner the damping effect of a possible neutral background structure. λ denotes again the Debye length.

Let us comment on the existence theory of solutions for the hydro-dynamic model. In fact, for the model mentioned above there is a global (in time) existence theory of weak entropy solutions available. A standard reference concerning the isentropic Euler equations (the equations for a single fluid with no electric field and no relaxation term) is [5], the most complete results can be found in [22, 21]. There, all values of $\gamma_n > 1$ (or $\gamma_p > 1$) are treated. As far as the (inhomogeneous) equations (3.1–3.7) are concerned the unipolar isentropic case (a single charged fluid only) was studied in [23, 24] and recently by [16], the bipolar case in [27]. The corresponding unipolar isothermal case ($\gamma_n = 1$) was solved in [28]. Due to the lack of a global existence theory for all but the isentropic or isothermal Euler equations in one space dimension, the cited multidimensional results [19, 20] are based on the assumption of global existence in time of a weak entropy solutions.

In many applications both the scaled Debye length and the scaled relaxation time are very small. Therefore, there are various possible limits one can study, $\tau \to 0$, $\lambda \to 0$ and combinations of the two limits. First we comment on the two separate limits, $\tau \to 0$ with fixed λ and $\lambda \to 0$ with fixed τ.

We start with a discussion of the so called relaxation limit $\tau \to 0$ (for fixed λ). This limit is based on the rescaling

(3.8) $$t = \tau r,$$

(3.9) $$n^\tau(t, x) = n(\frac{t}{\tau}, x) \qquad p^\tau(t, x) = p(\frac{t}{\tau}, x)$$

(3.10) $$j^\tau(t, x) = \frac{1}{\tau} j(\frac{t}{\tau}, x) \qquad g^\tau(t, x) = \frac{1}{\tau} g(\frac{t}{\tau}, x).$$

Then the rescaled equations read

(3.11) $$n_t^\tau + j_x^\tau = 0$$

(3.12) $$\tau^2 j_t^\tau + \left(\tau^2 \frac{(j^\tau)^2}{n^\tau} + (n^\tau)^{\gamma_n}\right)_x = -n^\tau E^\tau - \frac{j^\tau}{\mu_n}$$

(3.13) $$p_t^\tau + g_x^\tau = 0$$

(3.14) $$\tau^2 g_t^\tau + \left(\tau^2 \frac{(g^\tau)^2}{p^\tau} + (p^\tau)^{\gamma_n}\right)_x = p^\tau E^\tau - \frac{g^\tau}{\mu_p}$$

(3.15) $$-\lambda^2 E_x^\tau = n^\tau - p^\tau - C.$$

In the (formal) limit $\tau \to 0$ the limits n, p, E of n^τ, p^τ, E^τ satisfy the drift diffusion equations

(3.16) $$n_t = \mu_n((n^{\gamma_n})_x + nE)_x$$

(3.17) $$p_t = \mu_p((p^{\gamma_p})_x - pE)_x$$

(3.18) $$-\lambda^2 E_x = n - p - C.$$

This are exactly the drift diffusion equations (2.1–2.3) in one dimension. This (nonlinear) limit process was studied first by [24] in the one dimensional unipolar case and recently by [16]. A corresponding three dimensional result is due to [20]. The one dimensional bipolar case is treated in [27], the three dimensional bipolar case in [19]. The corresponding analysis in the one dimensional isothermal case ($\gamma_n = \gamma_p = 1$) was done in [18].

In (3.16–3.18) one can perform a small Debye length analysis as presented in Section 2 which – depending on the time scale we choose – leads to a bipolar pure drift model (2.23–2.25) or to a nonlinear diffusion model (2.31–2.32).

Now let us comment on the quasineutral limit $\lambda \to 0$ for fixed τ in (3.1–3.7). We consider the case of no background charge $C \equiv 0$. The Poisson equation (3.5) requires the limits of n and p to be equal. We denote it (again) by w. The momentum equations (3.2) and (3.4) remain unchanged. Taking the difference of the two continuity equations (3.1), (3.3) and assuming the current densities to be equal at $x = -\infty$ (or $x = +\infty$) one concludes the same limit v of the current densities j and g. The electric field is eliminated by summing the momentum equations. Therefore we expect the following neutral relaxing hydrodynamic limit problem

(3.19) $$w_r + v_x = 0$$

(3.20) $$v_r + \left(\frac{v^2}{w} + \frac{1}{2}(w^{\gamma_n} + w^{\gamma_p})\right)_x = -\left(\frac{1}{2\mu_n} + \frac{1}{2\mu_p}\right)\frac{v}{\tau}$$

with initial conditions

(3.21) $$w(r = 0, x) = \frac{1}{2}(n_0(x) + p_0(x)),$$

(3.22) $$v(r = 0, x) = \frac{1}{2}(j_0(x) + g_0(x)).$$

The justification of this formal limit is an open challenging problem. The only rigorous results regarding this problem concern sufficiently smooth solutions [2] (i.e up to the breakdown of classical solutions in time) and the case of travelling wave solutions [3]. The difficulties arise in the fact that oscillations in the solutions cannot be controlled in the limit, especially in case of no ($\tau = \infty$) or weak relaxation mechanism.

In (3.19–3.22) one can perform a relaxation limit, i.e. rescaling

$$(3.23) \qquad t = \tau r,$$

$$(3.24) \qquad w^\tau(t, x) = w(\frac{t}{\tau}, x), \qquad v^\tau(t, x) = \frac{1}{\tau} v(\frac{t}{\tau}, x)$$

in (3.19–3.20) we obtain

$$(3.25) \quad w_t^\tau + v_x^\tau = 0$$

$$(3.26) \quad \tau^2 v_t^\tau + \left(\tau^2 \frac{(v^\tau)^2}{w^\tau} + \frac{1}{2} ((w^\tau)^{\gamma_n} + (w^\tau)^{\gamma_n}) \right)_x = -(\frac{1}{2\mu_n} + \frac{1}{2\mu_p}) v^\tau.$$

It is easy to see that in the limit $\tau \to 0$ we obtain the limit problem (2.31–2.32) for the limit w of w^τ (the momentum equation gives in the limit an explicit expression for the current).

In the next two subsections we consider combinations of the two limits, i.e. assuming the relation $\lambda^2 = \tau^{1+\alpha}, -1 < \alpha$ and considering $\tau \to 0$. The difference lies in times scales we choose. We will see that in both cases we are able to recover from the hydrodynamic model limit models obtained already in the drift diffusion case by a small Debye length analysis.

The analysis of the next two subsections makes extensive use of entropies. There are the standard entropies or (electro-)mechanical energies

$$(3.27) \quad e(r) = \int_{\mathbb{R}} \left(\frac{j^2}{2n} + \frac{g^2}{2p} + \frac{1}{\gamma_n - 1} n^{\gamma_n} + \frac{1}{\gamma_p - 1} p^{\gamma_p} \right) dx,$$

$$(3.28) \quad \tilde{e}(r) = \int_{\mathbb{R}} \left(\frac{j^2}{2n} + \frac{g^2}{2p} + \frac{1}{\gamma_n - 1} n^{\gamma_n} + \frac{1}{\gamma_p - 1} p^{\gamma_p} + \frac{\lambda^2}{2} |E|^2 \right) dx$$

where (3.28) satisfies the following lemma.

LEMMA 3.1. *Let $r > 0$. Then*

$$(3.29) \qquad \frac{d}{dr} \tilde{e}(r) = - \int_{\mathbb{R}} \left(\frac{1}{\mu_n} \frac{j^2}{n} + \frac{1}{\mu_n} \frac{g^2}{p} \right) dx$$

holds for the solutions of system (3.1–3.7).

It comes out that it is convenient to use the "modified" entropies

$$(3.30) \quad h(r) = \int_{\mathbb{R}} \left(\frac{(j + nE)^2}{2n} + \frac{(g - pE)^2}{2p} + \frac{1}{\gamma_n - 1} n^{\gamma_n} + \frac{1}{\gamma_p - 1} p^{\gamma_p} \right) dx,$$

$$(3.31) \quad \tilde{h}(r) = \int_{\mathbb{R}} \left(\frac{(j + nE)^2}{2n} + \frac{(g - pE)^2}{2p} + \frac{1}{\gamma_n - 1} n^{\gamma_n} + \frac{1}{\gamma_p - 1} p^{\gamma_p} + \frac{\lambda^2}{2} |E|^2 \right) dx.$$

Also, so called higher order entropies like

$$(3.32) \quad h_k(r) = \sum_{i=0}^{k} c_i \int_{\mathbb{R}} \Big((j + nE)^{2i} \, n^{(\gamma_n - 1)(k - i) - 2i + 1} + (g - pE)^{2i} \, p^{(\gamma_p - 1)(k - i) - 2i + 1} \Big) dx,$$

are employed. The constants $c_i, i = 0, \ldots, k$ are known (up to multiplicative factors) and related only to the homogeneous problem, i.e. the isentropic gasdynamic equations. There is a well defined procedure to calculate all the constants $c_i, i = 0, \ldots, k$ based on the explicit construction of the entropies (see [22] or [16]).

3.1. The fast problem. Assuming $\lambda^2 = \tau^{1+\alpha}$ $(-1 < \alpha < 1)$ and the scaling

$$(3.33) \qquad s = \tau^{-\alpha} r,$$

$$(3.34) \qquad n^\tau(s, x) = n(\tau^\alpha s, x), \qquad j^\tau(s, x) = \tau^\alpha j(\tau^\alpha s, x)$$

$$(3.35) \qquad p^\tau(s, x) = p(\tau^\alpha s, x), \qquad g^\tau(s, x) = \tau^\alpha g(\tau^\alpha s, x)$$

$$(3.36) \qquad F^\tau(s, x) = \tau^{1+\alpha} E(\tau^\alpha s, x)$$

in (3.1–3.5) gives

$$(3.37) \qquad n_s^\tau + j_x^\tau = 0$$

$$(3.38) \qquad \tau^{1-\alpha} j_s^\tau + \left(\tau^{1-\alpha} \frac{(j^\tau)^2}{n^\tau} + \tau^{1+\alpha} (n^\tau)^{\gamma_n} \right)_x = -n^\tau F^\tau - \frac{j^\tau}{\mu_n}$$

$$(3.39) \qquad p_s^\tau + g_x^\tau = 0$$

$$(3.40) \qquad \tau^{1-\alpha} g_s^\tau + \left(\tau^{1-\alpha} \frac{(g^\tau)^2}{p^\tau} + \tau^{1+\alpha} (p^\tau)^{\gamma_p} \right)_x = p^\tau F^\tau - \frac{g^\tau}{\mu_p}$$

$$(3.41) \qquad -F_x^\tau = n^\tau - p^\tau - C.$$

Note that the scaling $\alpha = -1$ corresponds exactly to the relaxation time scaling (3.8–3.10) (λ fixed). In the case $\alpha > 0$ (3.33) is a short time scaling. The Poisson equations is not singularly perturbed and the two charged fluids persist in the limit. The limits n, p, F of n^τ, p^τ, F^τ satisfy formally the bipolar drift model (2.23–2.25).

A result concerning this limit reads as follows [12].

THEOREM 3.1. *Let n^τ, p^τ, j^τ g^τ and F^τ be an entropy solution of (3.37–3.41). Let $C \in L^1(\mathbb{R}) \cap L^\infty(\mathbb{R})$, $S > 0$ and $\gamma_n = \gamma_p$. Let $\gamma > 2$ and $-1 < \alpha < \frac{\gamma-5}{3(\gamma-1)}$. Then, as $\tau \to 0$ the following convergences (after extracting subsequences) hold:*

- *$n^\tau \rightharpoonup n$, $p^\tau \rightharpoonup p$ weakly in $L^\gamma((0, S) \times \mathbb{R})$,*
- *$F^\tau \to F$ strongly in $C([0, S]; L_{loc}^s(\mathbb{R}))$ with $1 < s < \infty$,*
- *$J^\tau \rightharpoonup J = nF$, $G^\tau \rightharpoonup G = pF$ weakly in $L_{loc}^q([0, S) \times \mathbb{R})$ with $q = 2\frac{\gamma}{\gamma+1}$,*

and the limits n, p and F satisfy the system (2.23–2.26) in $\mathcal{D}'([0, S) \times \mathbb{R})$.

To our knowledge this is the only result in this direction. The proof and details on "entropy solutions" can also be found in [12]. A key point in the proof is to use the "modified" entropies (3.30–3.31) instead of the standard entropies (3.27–3.28). The cases $\gamma \leq 2$ and $\gamma_n \neq \gamma_p$ are still open problems (we believe of technical nature).

3.2. The quasineutral regime. Here we study (again) the combined limit $\lambda^2 = \tau^{1+\alpha}$, $-1 < \alpha$ and $\tau \to 0$ in the equations (3.1–3.5). We consider the long time rescaling

$$(3.42) \qquad t = \tau r,$$

$$(3.43) \qquad j^\tau(t,x) = \frac{1}{\tau} j(\frac{t}{\tau}, x)$$

$$(3.44) \qquad g^\tau(t,x) = \frac{1}{\tau} g(\frac{t}{\tau}, x).$$

Note that the new time scale is a long time scale and that the the electric field is not rescaled. Then the rescaled equations (3.1–3.5) for $C \equiv 0$ read

$$(3.45) \qquad n_t^\tau + j_x^\tau = 0$$

$$(3.46) \qquad \tau^2 j_t^\tau + \left(\tau^2 \frac{(j^\tau)^2}{n^\tau} + (n^\tau)^{\gamma_n}\right)_x = -n^\tau E^\tau - \frac{j^\tau}{\mu_n}$$

$$(3.47) \qquad p_t^\tau + g_x^\tau = 0$$

$$(3.48) \qquad \tau^2 j_t^\tau + \left(\tau^2 \frac{(g^\tau)^2}{p^\tau} + (p^\tau)^{\gamma_p}\right)_x = p^\tau E^\tau - \frac{g^\tau}{\mu_p}$$

$$(3.49) \qquad -\tau^{1+\alpha} E_x^\tau = n^\tau - p^\tau$$

with initial data

$$(3.50) \qquad n^\tau(t=0,x) = n_0^\tau(x), \qquad j^\tau(t=0,x) = \frac{1}{\tau} j_0^\tau(x)$$

$$(3.51) \qquad p^\tau(t=0,x) = p_0^\tau(x), \qquad g^\tau(t=0,x) = \frac{1}{\tau} g_0^\tau(x).$$

As $\tau \to 0$ the same ideas as in the quasineutral limit (with fixed τ) lead to the limit problem. The limits of n^τ and p^τ (denoted by w) satisfy the limit problem (2.31–2.32).

The following result concerns this limit [14].

THEOREM 3.2. *Let $T > 0$ and assume $\gamma_n = \gamma_p$. Let $n^\tau, p^\tau, j^\tau, g^\tau, E^\tau$ be a weak entropy solution of (3.45–3.49). Assume the initial data are such that the entropies $e(0)$, $\tilde{e}(0)$, $h(0)$, $h_2(0)$, $\tilde{h}(0)$ are uniformly bounded in τ.*
Then there exists a time scale with $-1 < \alpha \le \min(-\frac{2}{3}, \frac{\gamma-5}{\gamma+1})$ such that as $\tau \to 0$ the following convergences (after extracting subsequences) hold:

- *$n^\tau \to w$, $p^\tau \to w$ strongly in $L^p((0,S) \times \mathbb{R})$, $1 \le p < 2\gamma - 1$*
- *$(n^\tau - p^\tau)E^\tau \to 0$ strongly in $L^s((0,S) \times \mathbb{R})$, $s = \dfrac{4\gamma}{2\gamma + 1}$*

and the limit w of n^τ, p^τ satisfies the nonlinear diffusion equation (2.31–2.32) in $\mathcal{D}'([0,T] \times \mathbb{R})$.

The proof given in [14] involves the standard entropies $e(r)$, $\tilde{e}(r)$, the "modified" entropies $h(r)$, $\tilde{h}(r)$ and the higher order entropy $h_2(r)$. The strong convergences of the densities are obtained using generalized versions of the "div-curl" lemma [13]. The cases $\gamma \le 2$ and $\gamma_n \ne \gamma_p$ are still open problems (we believe again of technical nature).

4. Summary. The various limits performed after various scalings in hydrodynamic an drift-diffusion models give a quite complete picture of the relations between the models. From both hydrodynamic and drift-diffusion level there is a scaling and a limit procedure to end up in a bipolar drift model or in a nonlinear neutral diffusion model. Also, this is a nonstandard way to obtain nonlinear diffusion equations.

REFERENCES

[1] F. BREZZI AND P.A. MARKOWICH, *A convection-diffusion problem with small diffusion coefficient arising in semiconductor physics*, Bollettino UMI **7**(2B): 903–930 (1988).

[2] S. CORDIER AND E. GRENIER, *Quasineutral limit of Euler-Poisson systems arising from plasma physics*, Preprint LAN, Universite' Paris 6, R96030 (1996).

[3] S. CORDIER, P. DEGOND, P.A. MARKOWICH, AND C. SCHMEISER, *Quasineutral limit of travelling waves for the Euler-Poisson model*, in Mathematical and numerical aspects of wave propagation, Gary (ed.) 1995, pp. 724–733.

[4] Z. DEYL, *Electrophoresis: A survey of techniques and applications*, Elsevier, Amsterdam, 1979.

[5] R. DiPERNA, *Convergence of the viscosity method for isentropic gas dynamics*, Commun. Math. Phys. **91**: 1–30 (1983).

[6] H. GAJEWSKI, *On existence, uniqueness and asymptotic behaviour of solutions of the basic equations for carrier transport in semiconductors*, ZAMM **65**: 101–108 (1985).

[7] H. GAJEWSKI, *On the uniqueness of solutions to the drift-diffusion model of semiconductor devices*, Math. Models Methods Appl. Sci. **4**: 121–133 (1994).

[8] I. GASSER, *The initial time layer problem and the quasineutral limit in a nonlinear drift diffusion model for semiconductors*, Nonlin. Diff. Eq. Appl. NoDEA **8**(3): 237–249 (2001).

[9] I. GASSER, *On the Relations between different macroscopic models for semiconductors*, Habilitation thesis, Fachbereich Mathematik, Universität Hamburg, 2000.

[10] I. GASSER, C.D. LEVERMORE, P.A. MARKOWICH, AND C. SCHMEISER, *The initial time layer problem and the quasineutral limit in the drift diffusion model*, Eur. J. Appl. Math **12**(4): 497–512 (2001).

[11] I. GASSER, L. HSIAO, P.A. MARKOWICH, AND S. WANG, *Quasineutral limit in a nonlinear drift diffusion model for semiconductors*, to appear in J. Math. Anal. Appl. (2002).

[12] I. GASSER AND P. MARCATI, *The Combined Relaxation and Vanishing Debye Length Limit in the Hydrodynamic Model for Semiconductors*, Math. Meth. in the Appl. Sci. M^2AS, **24**(2): 81–92 (2001).

[13] I. GASSER AND P. MARCATI, *On a generalization of the Div-Curl lemma*, preprint (2000).

[14] I. GASSER AND P. MARCATI, *A Quasineutral Limit in a Hydrodynamic Model for Charged Fluids*, to appear in Monatshefte der Mathematik (2002).

[15] A. JÜNGEL, *Qualitative behavior of solutions of a degenerate nonlinear drift-diffusion model for semiconductors*, Math. Models Methods Appl. Sci. **5**(4): 497–518 (1995).

[16] A. JÜNGEL AND Y.J. PENG, *A hierarchy of hydrodynamic models for plasmas. Part II: zero-relaxation-time limits*, Comm. Partial Differential Equations **24**(5, 6): 1007–1033 (1999).

[17] A. JÜNGEL AND Y.J. PENG, *A hierarchy of hydrodynamic models for plasmas. Quasineutral limits in the drift diffusion equations*, submitted (2000).

[18] S. JUNCA AND M. RASCLE, *Relaxation du système d'Euler-Poisson isotherme vers les équations de dérive-diffusion*, to appear in Quart. Appl. Math. (2000).

[19] C. LATTANZIO, *On the 3-D bipolar isentropic Euler-Poisson model for semiconductors and the drift-diffusion limit*, to appear in Math. Models Methods Appl. Sci. (2000).

[20] C. LATTANZIO AND P. MARCATI, *The relaxation to the drift-diffusion system for the 3-D isentropic Euler-Poisson model for semiconductors*, Discrete Contin. Dynam. Systems **5**: 449–455 (1999).

[21] P.L. LIONS, B. PERTHAME, AND P.E. SOUGANIDIS, *Existence and stability of entropy solutions for the hyperbolic systems of isentropic gas dynamics in Eulerian and Lagrangian coordinates*, Commun. Pure Appl. Math. **49**(6): 599–638 (1996).

[22] P.L. LIONS, B. PERTHAME, AND E. TADMOR, *Kinetic formulation of the isentropic gas dynamics and p-systems*, Commun. Math. Phys. **163** (2): 415–431 (1994).

[23] P. MARCATI AND R. NATALINI, *Weak solutions to a hydrodynamic model for semiconductors: the Cauchy problem*, Proc. Royal Soc. Edinburgh **28**: 115–131 (1995).

[24] P. MARCATI AND R. NATALINI *Weak solutions to a hydrodynamic model for semiconductors and relaxation to the drift-diffusion equation*, Arch. Rational Mech. Anal. **129**: 129–145 (1995).

[25] P.A. MARKOWICH, C. RINGHOFER, AND C. SCHMEISER, *Semiconductors equations*, Springer Verlag, Wien, New York, 1990.

[26] P.A. MARKOWICH AND P. SZMOLYAN, *A system of convection-diffusion equations with small diffusion coefficient arising in semiconductor physics*, JDE **81**: 234–254 (1989).

[27] R. NATALINI, *The bipolar hydrodynamic model for semiconductors and the drift-diffusion equations*, J. of Math. Anal. and Appl. **198**: 262–281 (1996).

[28] F. POUPAUD, M. RASLCE, AND J.P. VILA, *Global solutions to the isothermal Euler-Poisson system with arbitrarily large data*, J. Differ. Equations **123** (1): 93–121 (1995).

[29] C. RINGHOFER, *An asymptotic analysis of a transient p − n−junction model*, SIAM J. Appl. Math. **47**: 624–642 (1879).

[30] A. SITENKO AND V. MALNEV, *Plasma physics theory*, Chapman & Hall, London 1995.

[31] P. SZMOLYAN, *Asymptotic methods for transient semiconductor equations*, COMPEL **8**: 113–122 (1989).

GLOBAL SOLUTION OF THE CAUCHY PROBLEM FOR THE RELATIVISTIC VLASOV–POISSON EQUATION WITH CYLINDRICALLY SYMMETRIC DATA[*]

ROBERT T. GLASSEY[†] AND JACK SCHAEFFER[‡]

Key words. Vlasov–Poisson equation, Plasmas, Global Existence, Initial value problems.

AMS(MOS) subject classifications. 35L60, 35Q99, 82C21, 82C22, 82D10.

1. Introduction. The dynamics of a collisionless plasma are described by the Vlasov–Maxwell system. If very large velocities are possible, relativistic corrections are necessary. When magnetic effects are ignored this formally becomes the relativistic Vlasov–Poisson equation. In this lecture we will sketch the proof of global existence for the Cauchy Problem for symmetric data; the details will appear in [GSC6]. The initial datum for the phase space density $f_0(x,v)$ is assumed to be sufficiently smooth, nonnegative and cylindrically symmetric. We will show that if the (two–dimensional) angular momentum is bounded away from zero on the support of $f_0(x,v)$, then a C^1 solution to the Cauchy problem exists for all times.

The relativistic Vlasov–Maxwell system (RVM) in three dimensions is given by

$$(1.1) \qquad \partial_t f_\alpha + \hat{v}_\alpha \cdot \nabla_x f_\alpha + e_\alpha (E + c^{-1} \hat{v}_\alpha \times B) \cdot \nabla_v f_\alpha = 0$$

$$(1.2) \qquad \partial_t E = c \nabla \times B - j, \qquad \nabla \cdot E = \rho$$

$$(1.3) \qquad \partial_t B = -c \nabla \times E, \qquad \nabla \cdot B = 0.$$

Here $x \in \mathbb{R}^3$ is position, $v \in \mathbb{R}^3$ is momentum, and c is the speed of light. f_α gives the number density in phase space of particles of species α (with mass m_α and charge e_α) and the velocity of these particles is given by

$$(1.4) \qquad \hat{v}_\alpha = (m_\alpha^2 + c^{-2}|v|^2)^{-1/2} v.$$

The charge and current densities are given by

$$(1.5) \qquad \rho(t,x) = 4\pi \int \sum_\alpha e_\alpha f_\alpha \, dv$$

$$(1.6) \qquad j(t,x) = 4\pi \int \sum_\alpha e_\alpha f_\alpha \hat{v}_\alpha \, dv$$

[*]Research supported in part by NSF DMS 9321383 and NSF DMS 9731956.
[†]Department of Mathematics, Indiana University, Bloomington, IN 47405–5701; glassey@indiana.edu.
[‡]Department of Mathematics, Carnegie Mellon University, Pittsburgh, PA 15213; js5m+@andrew.cmu.edu.

respectively. Initial data for f_α, E, and B are prescribed at $t = 0$. Here we will simplify the argument by taking the case of one species. Hence the subscript α will be omitted and we will take $e_\alpha = 1$, $m_\alpha = 1$ and $c = 1$. We will also drop the factors of 4π in ρ, j.

It is known ([S1], cf. also [AU, D]) that as $c \to \infty$, under appropriate regularity assumptions, the solution to the Cauchy problem for (RVM) converges to the solution of the Vlasov–Poisson system

$$(VP) \qquad \qquad \partial_t f + v \cdot \nabla_x f + E \cdot \nabla_v f = 0$$

where $E = \nabla U$, $\Delta U = \rho$. In this lecture we consider a hybrid model, the relativistic Vlasov–Poisson system

$$(RVP) \qquad \qquad \partial_t f + \hat{v} \cdot \nabla_x f + E \cdot \nabla_v f = 0$$

where again $E = \nabla U$, $\Delta U = \rho \equiv \int_{\mathbb{R}^3} f(t, x, v)\, dv$ and now (with $m = c = 1$)

$$(1.7) \qquad \qquad \hat{v} = \frac{v}{\sqrt{1 + |v|^2}}.$$

Global smooth existence has been established for (VP) ([LP, PF, S2]) but its relativistic counterpart (RVP) remains largely open (see however [GSC1]). The classical kinetic energy expression contains a *second* moment of f in v but the relativistic problem has essentially only *one* v moment. A consequence of this is that the density ρ lies in $L^{5/3}$ for (VP) but only in $L^{4/3}$ for (RVP). Every known proof for (VP) for general data cited above fails for (RVP).

Our initial value $0 \leq f_0(x, v) \in C_0^1(\mathbb{R}^6)$ is assumed to be cylindrically symmetric. This means that we can write

$$f_0(x, v) = f_0(r, u, \alpha, x_3, v_3)$$

where $r = \sqrt{x_1^2 + x_2^2}$, $u = \sqrt{v_1^2 + v_2^2}$ and $x_1 v_1 + x_2 v_2 = ru \cos \alpha$. No size restrictions are imposed on f_0.

For $T > 0$ arbitrary, the maximum momentum support is measured by

$$Q = Q(T) = \sup\{|v| : \text{there exists} \quad x, s \in [0, T] \quad \text{with} \quad f(s, x, v) \neq 0\} + 3.$$

We will also use a "planar support function"

$$P = P(T) = \sup\{|u| : \text{there exists } v_3, x, s \in [0, T] \text{ with } f(s, x, v) \neq 0\} + 3.$$

The estimates will be made on arbitrary time intervals $[0, T]$.

2. Cylindrical symmetry and a sufficient condition for existence. The Cauchy problem to be solved is

$$(RVP) \qquad \partial_t f + \hat{v} \cdot \nabla_x f + E \cdot \nabla_v f = 0 \quad (x,\, v \in \mathbb{R}^3,\, t > 0), \quad f(0,x,v) = f_0(x,v)$$

where $E = \nabla U$, $\quad \Delta U = \rho \equiv \int_{\mathbb{R}^3} f(t,x,v)\, dv$ and

$$(2.1) \qquad \hat{v} = \frac{v}{\sqrt{1 + |v|^2}}.$$

From [Z] and [B], a sufficient condition for global existence is this: Let $T > 0$ be arbitrary. If the function $Q(t)$ is bounded above by some continuous function on $[0,T]$, there exists a unique solution on $[0,\infty) \times \mathbb{R}^6$ to the initial-value problem for (RVP) of class C^1. Therefore the problem is reduced to finding an a priori bound on Q.

As consequences of the symmetry assumption, we have

$$\rho = \rho(t,r,x_3) = 2 \int_0^\infty \int_{-\infty}^\infty \int_0^\pi f(t,r,u,\alpha,x_3,v_3)\, d\alpha\, dv_3\, u\, du,$$

$$E(t,x) = \frac{(x_1,x_2,0)}{r} E_r + (0,0,1) E_3$$

where E_r, E_3 are functions of t,r,x_3. In particular, ρ has no angular dependence. In these variables (RVP) takes the explicit form

$$f_t + \frac{u \cos \alpha}{\sqrt{1+|v|^2}} f_r + E_r \cos \alpha f_u - \sin \alpha \left(\frac{u}{r\sqrt{1+|v|^2}} + \frac{E_r}{u} \right) f_\alpha + \hat{v}_3 f_{x_3} + E_3 f_{v_3} = 0.$$

We see that the *characteristic ordinary differential equations are*

$$(2.2) \qquad
\begin{aligned}
\dot{r} &= \frac{u \cos \alpha}{\sqrt{1 + |v|^2}} \\
\dot{u} &= E_r \cos \alpha \\
\dot{\alpha} &= -\sin \alpha \left(\frac{u}{r\sqrt{1+|v|^2}} + \frac{E_r}{u} \right) \\
\dot{x}_3 &= \hat{v}_3 \\
\dot{v}_3 &= E_3.
\end{aligned}$$

A simple calculation shows that the *angular momentum*

$$(2.3) \qquad\qquad ru \sin \alpha = \text{const.}$$

along the characteristics.

From $\Delta U = \rho$ we have after a standard calculation

$$4\pi |E(t,x)| \leq \int_{\mathbb{R}^3} \frac{\rho(t,y)}{|x-y|^2} dy$$

(2.4)
$$= \int_0^\infty \int \rho(t,r',z') \int_0^{2\pi} \frac{r' d\theta\, dz'\, dr'}{r^2+(r')^2-2rr'\cos\theta+(x_3-z')^2}$$

$$= 2\pi \int_0^\infty \int \frac{\rho(t,r',z')r'\, dr'\, dz'}{D_- D_+}$$

where

(2.5)
$$D_\pm^2 = (r \pm r')^2 + (x_3 - z')^2.$$

We define

(2.6)
$$v_0 = \sqrt{1 + |v|^2} = \sqrt{1 + u^2 + v_3^2}$$

and the Kinetic Energy density to be

(2.7)
$$K = K(t,x) = \int_{\mathbb{R}^3} v_0 f(t,x,v)\, dv.$$

The invariant energy is well-known to be

(2.8)
$$\int_{\mathbb{R}^3} K\, dx + \frac{1}{2} \int_{\mathbb{R}^3} |E(t,x)|^2\, dx = \text{const.}$$

A Strengthened Space Time Estimate. Let $0 \leq f \in C_0^1(\mathbb{R}^6)$ be a (not necessarily symmetric) solution of (RVP) with finite energy. Let $\omega = \frac{x}{|x|}$ for $x \in \mathbb{R}^3$. Then the following estimate holds:

(2.9)
$$\int_0^\infty \int_{\mathbb{R}^6} \frac{|v \times \omega|^2}{|x| v_0} f(t,x,v)\, dv\, dx\, dt + \int_0^\infty \int_{\mathbb{R}^3} \frac{|\omega \cdot E(t,x)|^2}{|x|} dx\, dt \leq \text{const.}$$

This estimate, an analog of the "Royal Estimate" of Morawetz ([MO]) appears in [GS2] and is applied there to the more general Vlasov–Maxwell system. Under the hypothesis of cylindrical symmetry for (RVP), a stronger version is valid: Let $0 \leq f \in C_0^1(\mathbb{R}^6)$ be a cylindrically symmetric solution of (RVP) with finite energy. Then

(2.10)
$$\int_0^\infty \int_{\mathbb{R}^6} \frac{u^2 \sin^2 \alpha}{r v_0} f(t,x,v)\, dv\, dx\, dt + \frac{1}{2} \int_0^\infty \int_{\mathbb{R}^3} \frac{|E(t,x)|^2}{r} dx\, dt \leq \text{const.}$$

We use only the first term in (2.10) but point out that the second integral includes every component of the E field and possesses the stronger singularity r^{-1}. The proof is achieved by multiplying the cylindrical form of (RVP) by $\frac{x_1 v_1 + x_2 v_2}{r}$ and by integrating by parts.

We now turn to the estimation of P in terms of Q. Our main result is:

$$P \le cQ^{\frac{2}{3}} \ln Q$$

for some constant c. There are several ingredients in the proof. The first is an observation on the support of a cylindrically symmetric solution f of (RVP). We note that

$$\{v : f(t,x,v) \ne 0\} \subseteq \{v : |u \cos \alpha| \le P, u \sin \alpha \le \min(P, cr^{-1}), |v_3| \le Q\}$$

by the conservation of angular momentum. Let us define

$$(2.11) \qquad\qquad M(r) = \min\left(P, cr^{-1}\right).$$

Then

$$\rho = \int f \, dv \le cPQM(r).$$

Furthermore, for any $R > 0$,

$$\rho \le \int_{\substack{|u \cos \alpha| \le P, |v_3| \le R \\ u \sin \alpha \le M(r)}} f \, dv + R^{-1} \int_{|v_3| > R} v_0 f \, dv \le cPMR + R^{-1}K$$

where K is the kinetic energy density from (2.7). We choose $R = (KM^{-1}P^{-1})^{\frac{1}{2}}$ and obtain

$$(2.12) \qquad\qquad \rho \le c(MPK)^{\frac{1}{2}}.$$

The second ingredient is to transfer this decay in r to a corresponding statement about the decay of the field E. The major estimate is this: There exists a constant c depending only on the data such that we have for all x, t

$$|E(t,x)| \le g(r) \equiv \begin{cases} cP^{\frac{4}{3}}Q^{\frac{1}{3}}, & \text{if } r < (PQ)^{-\frac{2}{3}} \\ cP \ln^{\frac{1}{2}}(Q) \cdot r^{-\frac{1}{2}} & \text{if } (PQ)^{-\frac{2}{3}} < r < P^{-1} \\ cP^{\frac{1}{2}} \ln^{\frac{1}{2}}(Q) \cdot r^{-1} & \text{if } r > P^{-1} \end{cases}$$

For fixed time, the solution f to (RVP) has compact support in the space variable. This follows from our assumption that $f_0(x, v)$ has compact support and from the fact that f is constant on the characteristics, the solutions $X = X(s,t,x,v)$, $V = V(s,t,x,v)$ of

$$\dot{x} = \hat{v}, \quad \dot{v} = E$$

with initial values $X = x$, $V = v$ when $s = t$. Thus

$$f(t,x,v) = f_0(X(0,t,x,v), V(0,t,x,v))$$

and

$$|x - X(0, t, x, v)| \leq \int_0^t |\hat{V}|\, ds \leq t.$$

Put $z = x_3$ and define the set S by

(2.13) $$S = \{(r', z') : D_- < R\}.$$

From (2.4) and (2.12) we have

(2.14)
$$\begin{aligned}
4\pi|E(t, x)| \ &\leq 2\pi \int_0^\infty \int \frac{\rho(t, r', z')r'\, dr'\, dz'}{D_- D_+} \\
&\leq c \iint_S \frac{PQM(r')r'\, dr'\, dz'}{D_- D_+} \\
&\quad + c \iint_{S^c} \frac{(PM(r')K(t, r', z'))^{\frac{1}{2}}r'\, dr'\, dz'}{D_- D_+}.
\end{aligned}$$

We notice that

$$\frac{r'M(r')}{D_+} \leq \frac{r' \min\left(P, \frac{c}{r'}\right)}{r + r'}$$

is bounded above by

$$\frac{Pr'}{r + r'} \leq P$$

as well as by

$$\frac{c}{r + r'} \leq \frac{c}{r}.$$

Therefore we have

(2.15) $$\frac{r'M(r')}{D_+} \leq \min\left(P, \frac{c}{r}\right) = M(r).$$

Using this in both terms of (2.14) we get

(2.16)
$$\begin{aligned}
|E| &\leq cPQM(r) \iint_S \frac{dz'\, dr'}{D_-} + c \iint_{S^c} \frac{P^{\frac{1}{2}}\left(\frac{r'M(r')}{D_+}\right)^{\frac{1}{2}}(r'K)^{\frac{1}{2}}\, dz'\, dr'}{D_-(D_+)^{\frac{1}{2}}} \\
&\leq cPQM(r) \int_0^R \frac{2\pi\lambda\, d\lambda}{\lambda} + c(PM)^{\frac{1}{2}}\left(\iint_{\substack{S^c \\ (r')^2 + (z')^2 < C}} \frac{dz'\, dr'}{D_-^2 D_+}\right)^{\frac{1}{2}}.
\end{aligned}$$

The first term in (2.16) is bounded by $cPQMR$. We will estimate the second term there in two ways. First, on the support of f,

(2.17)
$$\iint_{\substack{S^c \\ (r')^2+(z')^2<C}} \frac{dz'\, dr'}{D_-^2\, D_+} \leq \frac{1}{r} \iint_{\substack{S^c \\ (r')^2+(z')^2<C}} \frac{dz'\, dr'}{D_-^2}$$

$$\leq \frac{1}{r} \iint_{R^2<\alpha^2+\beta^2<C+2(r^2+z^2)} \frac{d\alpha\, d\beta}{\alpha^2 + \beta^2}$$

$$\leq \frac{1}{r} \int_R^C \frac{2\pi\lambda\, d\lambda}{\lambda^2}$$

$$= \frac{C}{r} \ln \frac{C}{R}.$$

However, we also have

(2.18)
$$\iint_{\substack{S^c \\ (r')^2+(z')^2<C}} \frac{dz'\, dr'}{D_-^2\, D_+} \leq \iint_{S^c} \frac{dz'\, dr'}{D_-^3} \leq \int_R^\infty \frac{2\pi\lambda\, d\lambda}{\lambda^3} = cR^{-1}.$$

Using (2.18) in (2.16) we get

$$|E| \leq cPQM(r)R + c(PM)^{\frac{1}{2}}(cR^{-1})^{\frac{1}{2}}$$

and taking $R = Q^{-\frac{2}{3}}(PM)^{-\frac{1}{3}}$ we find

(2.19)
$$|E| \leq c(MP)^{\frac{2}{3}}Q^{\frac{1}{3}} \leq cP^{\frac{4}{3}}Q^{\frac{1}{3}}.$$

Similarly, using (2.17) in (2.16) we get

$$|E| \leq cPQM(r)R + c(PM)^{\frac{1}{2}} \left(\frac{c}{r} \ln \frac{c}{R}\right)^{\frac{1}{2}}.$$

Using the choice $R = Q^{-1}(PMr)^{-\frac{1}{2}}$ we obtain

(2.20)
$$|E| \leq c \left(\frac{PM}{r} \ln Q\right)^{\frac{1}{2}}.$$

which is the major estimate in the form of $g(r)$.

Next we consider estimates on the characteristics in a special but representative case. The characteristic ordinary differential equations in r, u, α are the first three equations written in (2.2). Recall that $v_0 = \sqrt{1 + |v|^2}$ from (2.6). We *assume* that $\dot{r} \geq 0$ on $[t_1, t_2]$ and that $f \neq 0$ so $r \leq C$. (The general case is treated in [GSC6]). Using the ODE's we get

$$\left|\frac{d}{dt}u^2\right| = |2uE_r \cos\alpha| = 2|E_r|v_0|\dot{r}| \leq cQg(r)|\dot{r}|$$

so by the definition of the function $g(r)$ we have

$$
\begin{aligned}
|u^2(t_2)-u^2(t_1)| &\le cQ \int_{t_1}^{t_2} g(r)|\dot r|\, dt \\
&= cQ \left| \int_{t_1}^{t_2} g(r)\dot r\, dt \right| \\
&= cQ \left| \int_{r(t_1)}^{r(t_2)} g(r)\, dr \right| \\
&\le cQ \left(\int_0^{P^{-1}} (\ln Q)^{\frac12} Pr^{-\frac12} dr + \int_{P^{-1}}^{c} (\ln Q)^{\frac12} P^{\frac12} r^{-1} dr \right) \\
&= cQ \left(2(\ln Q)^{\frac12} P(P^{-1})^{\frac12} + (\ln Q)^{\frac12} P^{\frac12} \ln(cP) \right) \\
&\le cQ(\ln Q)^{\frac12} P^{\frac12} \ln P
\end{aligned}
$$

(2.21)

When we use the definition of P as the supremum of u, the above estimate for P in terms of Q follows.

At this point we may ask: are the current estimates sufficient to bound Q? Unfortunately the answer is no. Let us assume further that the angular momentum is bounded away from zero on the support of the solution: $0 < c < ru\sin\alpha < rP$ on the support of f. Substituting this and the bound for P into the second inequality defining $g(r)$, we get

$$
|E| \le cP(\ln Q)^{\frac12} \cdot r^{-\frac12} \le cP^{\frac32}(\ln Q)^{\frac12} < cQ(\ln Q)^2
$$

on the support of the solution. This estimate is not sufficient to bound Q, and it remains to remove one logarithmic factor here.

3. Use of the spacetime estimate. Therefore our estimates need to be strengthened. It is at this juncture that we use the spacetime estimate. Define

$$
\mathcal{L}(t) = \iint f(t,x,v) \frac{u^2 \sin^2 \alpha}{rv_0}\, dv\, dx,
\tag{3.1}
$$

$$
\ell(t,x) = \int \frac{f(t,x,v)}{v_0}\, dv.
\tag{3.2}
$$

We know that

$$
\int_0^{\infty} \mathcal{L}(t)\, dt < \infty.
\tag{3.3}
$$

Now we make the assumption that

(3.4) $ru\sin\alpha \ge c > 0$ for some $c > 0$ on the support of $f_0(x,v)$.

Then for a constant $c > 0$

$$
\mathcal{L}(t) \ge c \iint \frac{f}{r^3 v_0}\, dv\, dx = c \int \frac{\ell(t,x)}{r^3}\, dx.
\tag{3.5}
$$

Next, for every $R \geq 1$ we have

$$\rho(t, x) \leq \int_{v_0 < R} \frac{R}{v_0} f \, dv + \int_{v_0 > R} \frac{v_0}{R} f \, dv \leq R\ell + R^{-1}K.$$

Taking $R = (K/\ell)^{\frac{1}{2}}$ we obtain

$$(3.6) \qquad\qquad \rho \leq 2(K\ell)^{\frac{1}{2}}.$$

The field is now estimated above in terms of $\mathcal{L}(t)$. Recall that $D_\pm = \sqrt{(r \pm r')^2 + (z - z')^2}$ and let $S^c = \{(r', z') : D_- > R\}$. Beginning with (2.4) we use the angular momentum bound on the integral over S and (3.6) on the complement to get

$$(3.7) \quad \begin{aligned} |E(t,x)| \ &\leq c \int_0^\infty \int \frac{\rho(t, r', z') r' \, dr' \, dz'}{D_- D_+} \\ &\leq c \iint_S \frac{(PQ(r')^{-1}) r' \, dz' \, dr'}{D_- r} \\ &\quad + c \iint_{S^c} \frac{(r')^{\frac{3}{2}} (K(r')^{-3}\ell)^{\frac{1}{2}} r' \, dz' \, dr'}{D_- r'} \\ &\leq cPQRr^{-1} + c \iint_{S^c} \frac{(r')^{\frac{1}{2}} (K(r')^{-3}\ell)^{\frac{1}{2}} r' \, dz' \, dr'}{D_-} \\ &\leq cPQRr^{-1} \\ &\quad + c \sup_{S^c} \left(\frac{(r')^{\frac{1}{2}}}{D_-} \right) \iint_{S^c} (r'K)^{\frac{1}{2}} \left[(\ell(r')^{-3})^{\frac{1}{2}} (r')^{\frac{1}{2}} \right] dr' dz'. \end{aligned}$$

For the expression $\frac{r'}{(D_-)^2}$ on S^c we have the inequalities

$$\frac{r'}{(D_-)^2} \leq \frac{|r - r'| + r}{(D_-)^2} \leq \frac{1}{D_-} + \frac{r}{(D_-)^2} \leq \frac{1}{R} + \frac{r}{R^2} = \frac{r + R}{R^2}.$$

Thus from above

$$(3.8) \quad \begin{aligned} |E| \ &\leq cPQRr^{-1} + c\frac{\sqrt{r + R}}{R} \iint_{S^c} (r'K)^{\frac{1}{2}} \left[(\ell(r')^{-3})^{\frac{1}{2}} (r')^{\frac{1}{2}} \right] dr' dz' \\ &\leq cPQRr^{-1} + c\frac{\sqrt{r + R}}{R} \mathcal{L}^{\frac{1}{2}} \end{aligned}$$

where we have used the Schwarz inequality, the energy bound (2.8) and the relation (3.5). Now we take

$$R = (PQ)^{-\frac{1}{2}} r^{\frac{3}{4}} \mathcal{L}^{\frac{1}{4}}.$$

By virtue of the angular momentum assumption, we have $r > cP^{-1}$ on the support of f. Thus by the energy bound

$$\mathcal{L} \leq \iint \frac{fu^2}{rv_0} \, dv \, dx \leq cP \iint f v_0 \, dv \, dx \leq cP$$

and hence (recall that $Q > 1$)

$$R \leq c(PQ)^{-\frac{1}{2}} r^{\frac{3}{4}} P^{\frac{1}{4}} \leq cP^{-\frac{1}{4}} r^{\frac{3}{4}} \leq cr.$$

Using this in (3.8) we then get

$$
\begin{aligned}
|E(t,x)| \ & \leq cPQr^{-1}(PQ)^{-\frac{1}{2}} r^{\frac{3}{4}} \mathcal{L}^{\frac{1}{4}} + c \frac{\sqrt{r}\mathcal{L}^{\frac{1}{2}}}{(PQ)^{-\frac{1}{2}} r^{\frac{3}{4}} \mathcal{L}^{\frac{1}{4}}} \\
& = c(PQ)^{\frac{1}{2}} r^{-\frac{1}{4}} \mathcal{L}^{\frac{1}{4}} \\
& \leq cP^{\frac{3}{4}} Q^{\frac{1}{2}} \mathcal{L}^{\frac{1}{4}}
\end{aligned}
$$
(3.9)

on the support of f. Finally, we use the bound on P in terms of Q to obtain

(3.10)
$$|E(t,x)| \leq cQ(\ln Q)^{\frac{3}{4}} \mathcal{L}^{\frac{1}{4}}.$$

We can now finish the proof. We take an arbitrary characteristic $(X(s), V(s))$ on which $f(s, X(s), V(s)) \neq 0$. From $\dot{V} = E$ we get

$$
\begin{aligned}
|V(t)| \ & \leq c + \int_0^t |E(s, X(s))| \, ds \\
& \leq c + c \int_0^t Q(s)(\ln Q(s))^{\frac{3}{4}} \mathcal{L}^{\frac{1}{4}}(s) \, ds \\
& \leq c + c \left(\int_0^t Q(s)^{\frac{4}{3}} \ln Q(s) \, ds \right)^{\frac{3}{4}} \left(\int_0^t \mathcal{L}(s) \, ds \right)^{\frac{1}{4}} \\
& \leq c + c \left(\int_0^t Q(s)^{\frac{4}{3}} \ln Q(s) \, ds \right)^{\frac{3}{4}}.
\end{aligned}
$$
(3.11)

Thus

$$Q(t) \leq c + c \left(\int_0^t Q(s)^{\frac{4}{3}} \ln Q(s) \, ds \right)^{\frac{3}{4}}$$

from which it follows that

$$Q(t)^{\frac{4}{3}} \leq c_1 \exp\left(\exp(ct)\right)$$

and the proof is complete.

REFERENCES

[AN]   ANDRÉASSON, H. *Controlling the Propagation of the Support for the Relativistic Vlasov Equation with a Selfconsistent Lorentz Invariant Field,* In. Univ. Math. J. **45**: 617–642, 1996.

[AU]   ASANO, K. AND UKAI, S. *On the Vlasov–Poisson Limit of the Vlasov–Maxwell Equation,* Patterns and Waves, pp. 369–383, 1986.

[BDN]   BARDOS, C., DEGOND, P., AND NGOAN, H. T. *Existence Globale des solutions des équations de Vlasov–Poisson relativistes en dimension 3,* C. R. Acad. Sci. Paris, t. **310**, Série I, No. 6, pp. 265–268, 1985.

[B]   BATT, J. *Global Symmetric Solutions of the Initial–Value Problem of Stellar dynamics,* J. Diff. Eqns. **25**: 342–364, 1977.

[BR]   BATT, J. AND REIN, G. *Global Classical Solutions of the Periodic Vlasov–Poisson System in Three Dimensions,* C.R. Acad. Sci. Paris, t. **313**, Serie I: 411–416, 1991.

[D]   DEGOND, P. *Local Existence of Solutions of the Vlasov–Maxwell Equations and Convergence to the Vlasov–Poisson Equations for Infinite Light Velocity,* Internal Report **117**, Centre de Mathématiques Appliquées, Ecole Polytechnique, Paris, Dec. 1984.

[GJP]   GASSER, I., JABIN, P-E, AND PERTHAME, B. *Regularity and Propagation of Moments in some nonlinear Vlasov Systems,* Proc. Roy. Soc. Edinburgh, Sect. A **130**: 1259–1273, 2000.

[GMP]   GASSER, I., MARKOWICH, P., AND PERTHAME, B. *Dispersion and Moments Lemma revisited,* J. Differential Equations **156**: 254–281, 1999.

[G]   GLASSEY, R. *The Cauchy Problem in Kinetic Theory,* S.I.A.M., Philadelphia, 1996.

[GSC1] GLASSEY, R. AND SCHAEFFER, J. *On Symmetric Solutions of the Relativistic Vlasov–Poisson System,* Comm. Math. Phys. **101**: 459–473, 1985.

[GSC2] GLASSEY R. AND SCHAEFFER, J. *Global Existence of the Relativistic Vlasov–Maxwell System with Nearly Neutral Initial Data,* Comm. Math. Phys. **119**: 353–384, 1988.

[GSC3] GLASSEY R. AND SCHAEFFER, J. *The Relativistic Vlasov Maxwell System in Two Space Dimensions: Part I,* Arch. Rat. Mech. Anal. **141**: 331–354, 1998.

[GSC4] GLASSEY R. AND SCHAEFFER, J. *The Relativistic Vlasov Maxwell System in Two Space Dimensions: Part II,* Arch. Rat. Mech. Anal. **141**: 355–374, 1998.

[GSC5] GLASSEY R. AND SCHAEFFER, J. *The "two and one–half Dimensional" Relativistic Vlasov Maxwell System,* Comm. Math. Phys. **185**: 257–284, 1997.

[GSC6] GLASSEY R. AND SCHAEFFER, J. *On Global Symmetric Solutions to the Relativistic Vlasov–Poisson Equation in Three Space Dimensions,* Math. Meth. Appl. Sci. **24**: 143–157, 2001.

[GS2]   GLASSEY, R. AND STRAUSS, W. *Singularity Formation in a Collisionless Plasma Could Occur Only at High Velocities,* Arch. Rat. Mech. Anal. **92**: 59–90, 1986.

[GS2]   GLASSEY, R. AND STRAUSS, W. *Remarks on Collisionless Plasmas,* Contemporary Mathematics, **28**: 269–279, 1984.

[GS3]   GLASSEY, R. AND STRAUSS, W. *Absence of Shocks in an Initially Dilute Collisionless Plasma,* Comm. Math. Phys. **113**: 191–208, 1987.

[H1]   HORST, E. *On the Classical Solutions of the Initial Value Problem for the Unmodified Nonlinear Vlasov–Equation, Parts I and II,* Math. Meth. Appl. Sci. **3**: 229–248, 1981; and **4**: 19–32, 1982.

[H2]   HORST, E. *On the Asymptotic Growth of the Solutions of the Vlasov–Poisson System,* Math. Meth. Appl. Sci. **16**: 75–85, 1993.

[LDP]   LIONS, P.L. AND DIPERNA, R. *Global Solutions of Vlasov–Maxwell Systems,* Comm. Pure Appl. Math. **42**: 729–757, 1989.

[LP]   LIONS, P.-L., PERTHAME, B. *Propagation of Moments and Regularity for the Three Dimensional Vlasov–Poisson System*, Invent. Math. **105**: 415–430, 1991.

[MO]   MORAWETZ, C. *Time Decay for the Nonlinear Klein–Gordon Equation*, Proc. Royal Soc. A **306**: 291–296, 1968.

[PER] PERTHAME, B. *Time Decay, Propagation of Low Moments and Dispersive Effects for Kinetic Equations*, Comm. P.D.E. **21**: 659–686, 1996.

[PF]   PFAFFELMOSER, K. *Global Classical Solution of the Vlasov–Poisson System in Three Dimensions for General Initial Data*, J. Diff. Eqns., **95**(2): 281–303, 1992.

[S1]   SCHAEFFER, J. *The Classical Limit of the Relativistic Vlasov–Maxwell System*, Comm. Math. Phys. **104**: 403–421, 1986.

[S2]   SCHAEFFER, J. *Global Existence of Smooth Solutions to the Vlasov Poisson System in Three Dimensions*, Comm. in Part. Diff. Eqns. **16**(8 & 9): 1313–1335, 1991.

[W1]   WOLLMAN, S. *Global-in-Time Solutions of the Two Dimensional Vlasov–Poisson System*, Comm. Pure Appl. Math. **33**: 173–197, 1980.

[Z]   ZIEMER, W. *Global Solution of the Initial Value Problem of Stellar Dynamics Using Relativistic Coefficients*, Master's Thesis, Carnegie Mellon University, November, 1986.

MESOSCOPIC SCALE MODELING FOR CHEMICAL VAPOR DEPOSITION IN SEMICONDUCTOR MANUFACTURING

MATTHIAS K. GOBBERT* AND CHRISTIAN RINGHOFER†

Abstract. Low pressure chemical vapor deposition is a process, by which a thin layer of solid material is deposited onto the microstructured surface of a silicon wafer. Reactor scale models are used to model the flow of reacting chemicals on the scale of the chemical reactor. Feature scale models are designed to predict the evolution of the deposited film inside an individual feature. For a length scale intermediate to both classical models, the concept of a mesoscopic scale model is introduced. Depending on the typical length scale of the mesoscopic domain, either a continuum model or a kinetic model is appropriate to describe the flow mathematically. In both regimes, a homogenization technique is applied to the boundary condition at the microstructured surface to derive an equivalent boundary condition that is amenable to effective numerical simulations. In the near-continuum regime, this equivalent mesoscopic scale model is used in stand-alone to analyze cluster-to-cluster effects as well as to interface with both classical models to obtain an integrated process simulator. In the transition regime, a numerical example is given that validates the analysis.

Key words. Boltzmann equation, homogenization, multiscale modeling, chemical vapor deposition.

1. Introduction. Chemical vapor deposition (CVD) is one of the fundamental production step in the manufacturing of semiconductor chips. It consists of pumping chemical gases through a reactor chamber and across the surface of a (typically) silicon wafer. By carefully choosing the chemicals and operating conditions of the reactor, a chemical reaction at the surface is induced that results in the formation of a solid product. This product is designed to stick to the surface, thus forming layer upon layer of new material over the time of the deposition process.

In order to reduce the amount of extraneous particles that would contaminate the substrate, the process is typically run at very low pressure. A typical value for the total pressure in low pressure chemical vapor deposition (LPCVD) is 1 Torr, where 760 Torr equals atmospheric pressure.

Test runs of the chemical reactor are costly both due to the expensive chemicals and the cost of cleaning up the hazardous exhaust gases.

Communicating author; Department of Mathematics and Statistics, University of Maryland, Baltimore County, 1000 Hilltop Circle, Baltimore, MD 21250 (gobbert@math.umbc.edu). This research was supported in part by the Institute for Mathematics and its Applications with funds provided by the National Science Foundation. The first author would also like to acknowledge partial support from the National Science Foundation under grant DMS-9805547.

†Department of Mathematics, Arizona State University, Tempe, AZ 85287-1804 (ringhofer@asu.edu). This research was supported in part by the Institute for Mathematics and its Applications with funds provided by the National Science Foundation. The second author would also like to acknowledge partial support from the National Science Foundation under grant DMS-9706792.

They are also wasteful due to the distructive testing. Therefore, computer simulations are being used to reduce the number of prototype runs.

The first models for chemical vapor deposition were reactor scale models that are designed to simulate the flow of the gaseous chemicals throughout the entire reactor chamber. At the typically low pressures used, the mean free path λ of the gas particles between collisions is on the order of 0.03 cm. For the typical domain size of $L = 30$ cm, the Knudsen number Kn, defined as the ratio of λ over L, is about Kn $= 0.001$. Therefore, this regime is governed by continuum flow, and fluid equations based on the Navier-Stokes equations appropriately describe the gas flow [15].

The second class of models are feature scale models. They are designed to predict the film growth in a single feature. Their typical length scale is $L < 3$ μm, hence the Knudsen number Kn is larger than 100. The flow has to be modeled strictly by kinetic equations, and collisions with the feature walls dominate over particle-particle collisions [2].

Traditionally, reactor scale and feature scale models have been used independently with success on their respective scale [2, 17]. Some studies have combined the models by using the reactor scale model to supply input fluxes at the feature mouth to the feature scale model [4, 13]. Such a combination necessitates the assumption of uniform distribution of features across the whole wafer.

However, to increase the efficiency of the deposition production step, the trend is towards using larger wafers. At the same time, the typical dimension of each semiconductor component has fallen (far) below 1 μm. Both of these developments imply that the assumptions used to combine the classical models directly are not valid any more due to the vast difference in length scale between them.

This vast difference in length scale is one of the reasons, why the authors and co-workers have introduced the concept of a mesoscopic scale model in the past [9, 12, 16]. Figure 1 shows a sketch of the cross-section of a rotationally symmetric single wafer reactor and indications of the typical domains associated with the reactor scale model (RSM), several instances of mesoscopic scale models, and arrows pointing to the locations of several representative feature scale models (FSM).

The length scale intended to be covered by the mesoscopic scale model ranges from 0.01 cm to 1.0 cm. When compared to the mean free path of the gas particles in the pressure regime under consideration, we notice that the Knudsen number ranges from 0.01 to 1 in order of magnitude. When the Knudsen number is 'small' (in this range), say, between 0.01 and 0.1, it is justified to approximate the flow using a continuum model. But for the case of a 'large' Knudsen number (in the mesoscopic range) of about 1, the system lies in the transition regime when both collisions and transport are equally important, and the Boltzmann equation must be used to describe the gas flow.

In both regimes of the mesoscopic scale model, there are still from hundreds to thousands of microscopic features along the wafer surface,

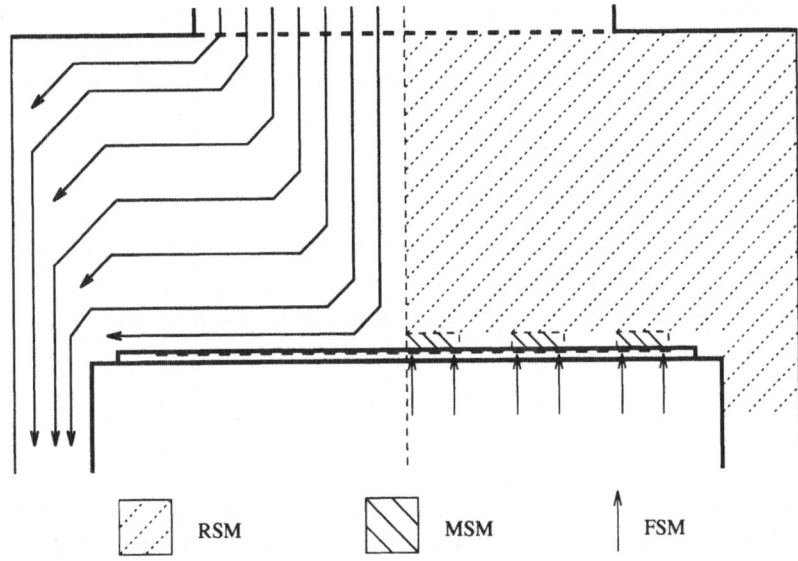

RSM   MSM   FSM

FIG. 1. *Schematic of a typical rotationally symmetric single wafer reactor with flow pattern and typical domains of the models.*

and the numerical resolution of the mesoscopic scale model is still not possible for realistic surface structures. However, the typical wafer surface exhibits structure in that the features are not randomly placed, but rather are organized in clusters of (by design) identical features. This fact leads us to propose a mathematical surface model that depends on a slowly changing (macroscopic) and a fast changing (microscopic) variable. Since the features throughout each cluster are identical, we assume periodicity in the microscopic variable, and perform a boundary homogenization to obtain an equivalent problem on a flat surface. This equivalent problem is then amenable to efficient numerical computations.

The remainder of this paper is organized as follows: We summarize the analytical results for both regimes of the mesoscopic scale model and present numerical results obtained using the approach. As indicated above, the mesoscopic scale model can be used as an interface between the classical models. Additionally, it is available to analyze the effect of clustering effects (microloading) in stand-alone mode. Both of these features will be discussed in Section 2 for the case of 'small' Knudsen numbers. Section 3 covers then the new theoretical results for the case of a 'large' Knudsen number and presents a first numerical validation.

2. Modeling of the 'small' Knudsen number regime.

2.1. Theory. For comparably 'large' mesoscopic domains and/or relatively 'high' pressures, the Knudsen number for the mescopic scale model is on the order of 0.1. In this regime, the gas flow is described by continuum equations. Since we want to model the flow very close to the wafer surface, we assume additionally that the flow is dominated by diffusion. In dimensionless form, the model is then given by the reaction-diffusion equation

$$(2.1) \qquad \frac{\partial c}{\partial t} = -\text{div} F + R_g, \quad F = -D \nabla c,$$

where c is the molar concentration and F the flux of one gaseous species. D is the diffusivity matrix for gaseous species in the mixture and R_g the gas-phase reaction term. On both sides of the domain, no-flux conditions are used

$$-e_1^T F = 0 \quad \text{at } x = 0,$$
$$e_1^T F = 0 \quad \text{at } x = 1,$$

where $e_1 = (1,0)^T$ denotes the first unit vector corresponding to the x-direction. The boundary condition at the gas-phase interface is given by the known function $c^{top}(x)$, which represents a (trial) solution of the reactor scale model, as

$$c = c^{top}(x) \quad \text{at } y = 1.$$

Along the wafer surface, the flux is given as a function of the species generation rates on the surface, namely

$$\nu^T F = S(c, x, y) \quad \text{at } (x, y) \in \Gamma_\varepsilon,$$

where ν is the unit outward normal vector at $(x, y)^T \in \Gamma_\varepsilon$.

The definition of the problem is completed by specifying the function for the wafer surface Γ_ε. It is necessarily rough for the application problem at hand. However, the typical wafer surface exhibits structure in that the features are not randomly placed, but are organized in clusters of (by design) identical features. This fact is used to propose a model for the wafer surface in dimensionless form

$$(2.2) \qquad x = s + \varepsilon \alpha(s, \tfrac{s}{\varepsilon}), \quad y = \varepsilon \beta(s, \tfrac{s}{\varepsilon}), \quad 0 \le s \le 1,$$

where the functions α and β are known functions of the slowly varying (macroscopic) variable s and the fast changing (microscopic) variable $\sigma s / \varepsilon$. Since the features are assumed identical locally, we assume that

$$\alpha(s, \sigma) = \alpha(s, \sigma + 1), \quad \beta(s, \sigma) = \beta(s, \sigma + 1) \quad \text{for all } (s, \sigma).$$

Here, ε is the dimensionless ratio of the typical feature size and the domain size and satisfies $0 < \varepsilon \ll 1$.

Using ε as homogenization parameter, we make the asymptotic ansatz

$$c(x, y, t) = \tilde{c}(x, y, t) + \hat{c}(\tfrac{x}{\varepsilon}, \tfrac{y}{\varepsilon}, x, t) + \mathcal{O}(\varepsilon),$$

where $\tilde{c}$ represents the solution in the bulk of the domain and $\hat{c}$ is the zeroth-order correction inside the boundary layer. In [8, 11], an asymptotic analysis has been performed to show that $\tilde{c}$ satisfies the same equation and boundary conditions as the original problem, except that the condition at the wafer surface is replaced by

$$-e_2^T F = \tilde{\sigma} \, S(c, x, 0) \quad \text{at } y = 0$$

with

$$\tilde{\sigma} = \int_0^1 \sqrt{\left(1 + \frac{\partial \alpha}{\partial \sigma}\right)^2 + \left(\frac{\partial \beta}{\partial \sigma}\right)^2} \, d\sigma$$

and $e_2 = (0,1)^T$. The quantity $\tilde{\sigma}$ is asymptotically equal to the surface area of one period of σ. Physically, it is a measure of the additional surface area in a structured area compared to a flat area. The results here were stated in two dimensions for simplicity of presentation; they hold in three dimensions in the same manner [11].

2.2. The mesoscopic scale model in stand-alone mode. The mesoscopic scale model has been used to analyze cluster-to-cluster effects in thermally induced deposition of silicon dioxide from tetraethoxysilane (TEOS) on silicon wafers with oxygen as an inert gas. A single-species model is used for the surface reactions [1]; this simple chemistry is chosen, because it involves a single species (TEOS) and suffices to demonstrate the method. The reaction rate expression for the sole solid species SiO_2 is modeled as

$$R_{SiO_2} = k_0 \exp\left(-\frac{E_a}{RT}\right) \frac{(P_{TEOS})^{0.5}}{1 + k_3 (P_{TEOS})^{0.5}}$$

with coefficients $k_0 = 37.55$ mol / (s cm^2), $k_3 = 0.25$ 1 / $\sqrt{\text{torr}}$, and $E_a = 46.5$ kcal / mol, where E_a denotes the activation energy of the reaction. R denotes the universal gas constant, and T is the ambient temperature. For this single-component demonstration, the net flux of TEOS to the surface is given by

$$S(c, x, t) = R_{SiO_2}(P_{TEOS}),$$

where the partial pressure of TEOS is computed from the molar concentration of TEOS via $P_{TEOS} = (c_{TEOS}/c_{total})P_{total}$. It is assumed that there are no gas-phase reactions, that is $R_g = 0$ in (2.1). See [12] for additional details on the model.

For demonstration purposes, a mesoscopic domain length of $X = 8$ mm, encompassing four feature clusters of length 1 mm each, was chosen. The domain was chosen to extend into the gas-phase to a value of $Y = 0.8$ mm. The individual features are infinite trenches of feature aspect

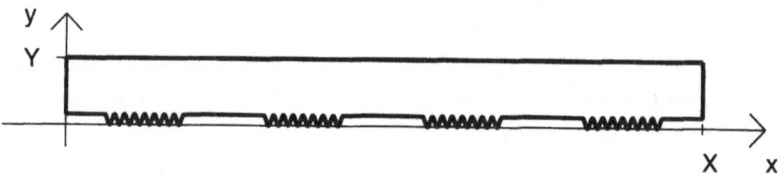

FIG. 2. *Sketch of a possible domain of the mesoscopic scale model with associated coordinate system.*

ratio 2 with initial width 1 μm and a distance of 1 μm between features; hence, the small scale period within the feature clusters is $x_p = 2$ μm. At this spacing, there are 500 features per cluster. The dimensionless small scale parameter ε is chosen as the ratio $\varepsilon = x_p/X = 0.25 \cdot 10^{-3}$.

For the purposes of this demonstration, this structure is approximated by an explicitly given, smooth function. The surface function is chosen as

$$y = \tilde{h}(x) = \begin{cases} a + a \cos\left(2\pi \frac{x}{x_p}\right) & \text{if } x \text{ is inside a feature cluster,} \\ 2a & \text{if } x \text{ is inside a flat area} \end{cases}$$

with constant amplitude $a = Ax_p/4 = 1$ μm, where $A = 2$ is the feature aspect ratio of the trenches. Figure 2 shows an example of this function with a larger x_p chosen for better visibility. This results in four clusters with 500 features each; refer to this case as Case 1.

As an example for greater feature density, 1000 features per cluster can be approximated by replacing 2π by 4π in the argument of the cosine. Their small scale period is then $x_p = 1$ μm. Maintaining the amplitude $a = 1$ μm, their feature aspect ratio is then 4. Case 2 uses two clusters of 1000 features and two clusters of 500 features, such that the geometry of the right half of the domain agrees with the geometry of Case 1.

The following results demonstrate the capability of the mesoscopic scale model to study feature-to-feature effects on the scale of several feature clusters, i.e., how varying feature density inside clusters affects the concentration levels and the species fluxes at the wafer surface. The main operating conditions include the total pressure is 5 torr and the temperature is 1000 K, while the partial pressure of TEOS at the top of the domain is fixed at 0.2 torr.

Figure 3 shows the concentration profiles throughout the two-dimensional domain, where the solid lines correspond to Case 1 and the dashed lines to Case 2. The concentration levels are lower closer to the wafer surface than at the gas-phase interface at the top. They are also lower inside the feature clusters than in the flat areas in between them; this is explained by the increase in surface reactions due to the increase in surface area available for reactions inside the feature clusters. It can be observed that the concentration levels in the denser clusters are lower than in the

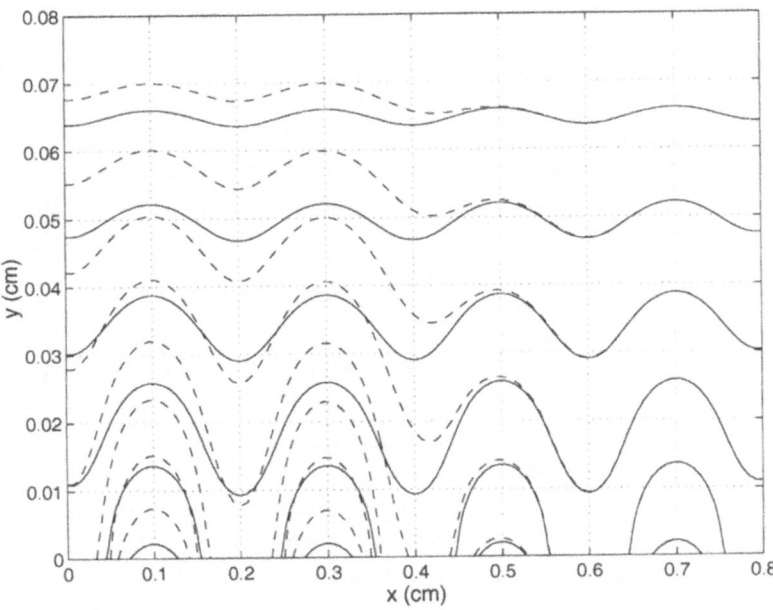

FIG. 3. *Comparison of concentration profiles. The solid lines depict the results for four feature clusters with 500 features each. The dashed lines mark the results for two clusters with 1000 features and two with 500 features.*

others. However, the concentration levels in the right-hand side of the domain (encompassing the clusters with 500 features) are equal in both cases.

The effect of the higher feature density in the two left feature clusters increases the depletion of reactants in those areas due to an increase in available surface area. This shows that the method predicts the concentration levels and species fluxes and accounts for the variations from one feature cluster to the next on the macroscopic scale. The method is also capable of representing the variations inside a feature cluster. The scale of these variations is clearly still far above the microscopic scale of an individual feature, but is one order of magnitude below that of the macroscopic scale of the overall die; note that predictions on the scale of several hundred features are not reasonably obtainable using a feature scale model.

2.3. A basic multiscale simulator. As example of the combination of the mesoscopic scale model with the classical models for a three-scale integrated simulator, we used thermally induced deposition of silicon dioxide (SiO_2) from tetraethyloxysilane (TEOS), with argon (Ar) as the carrier gas with a more complicated stoichiometry than above. The kinetic model [14] involves six gaseous reacting species and one inert carrier gas, which participate in four gas-phase reactions and eight surface reactions. We focus attention on triethyloxysilane, a reactive intermediate formed from TEOS. It is a major contributor to film deposition in the chemistry model [14].

For the wafer surface, we consider three cases, all of which consider a flat 200 mm wafer with at most one die of non-flat area for demonstration purposes. The first case consists of a blanket wafer without any features. The second case considers a die between wafer radius 46.7 mm and 50.0 mm. All features in the die are modeled as infinite trenches and taken as 1 µmin depth and 1 µmin width, hence with aspect ratio 1, and with flat areas of 2 µmseparating the features, resulting in a pitch of 3 µm. These two cases are used to validate the three-scale approach, since they agree with geometries that a two-scale simulator is capable of representing. The third surface case poses a more realistic geometry, in which the die does not consist of uniform features but rather contains three clusters of features of width 0.4 mm each. The individual features inside each cluster are the same as those in the second case.

For the purposes of the presentation here, we restrict our attention to obtaining the pseudo steady-state solution for the deposition model, which might then be used as the initial solution for a transient deposition simulation. Due to the feedback from the other models, each model's solution must satisfy the governing equations on its own scale and be consistent with the ones obtained on the other scales, and hence an iterative solution is necessary.

Figure 4 shows the geometry of the three-scale model as well as the meshes used on each scale. The arrows indicate the positions of the feature scale models. The simulations on the reactor scale and the mesoscopic scale were performed using the fluid-dynamics package FIDAP 7.6 [7], while the Ballistic Transport and Reaction Model [3] implemented in the software package EVOLVE 4.1a [6] was used for the feature scale. See [9] for additional details on the chemistry and the simulations.

Figure 5 contains plots of the mass fraction of triethyloxysilane along the wafer surface vs. wafer radius. The three curves are obtained by using the three-scale deposition model for each of these cases. It is verified in [9] that the curves of the two-scale model for the first and second surface case agree with the corresponding ones of the three-scale model. Notice first that all three curves agree throughout the flat regions of the wafer. Secondly, since triethyloxysilane is depleted due to surface reactions, the mass fraction decreases with increasing total surface area available for deposition. That is why the curve for the third case lies in between the curves for the other cases throughout the area inside the die. The species is depleted more by the surface reactions, because more surface area is available for deposition.

Figure 6 shows the same quantities as Figure 5, but now on the length scale of the mesoscopic scale model. As expected, the level of the mass fraction decreases again with an increase in total surface area, hence the curve for the second case is lower than the one for the first case. The

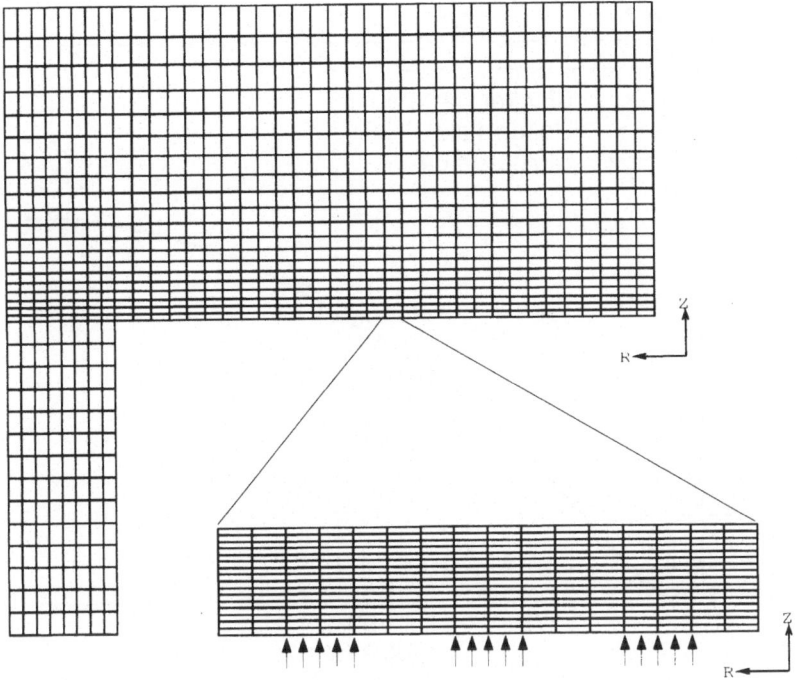

FIG. 4. *Schematic of the setup for the three-scale deposition model and the meshes of the reactor scale and mesoscopic scale models. The arrows indicate the positions of the feature scale models.*

curves for the flat wafer and the uniformly patterned die do not exhibit any significant variations, which is correct for their uniform surface model on the mesoscopic scale. The variations expected for the surface that involves three clusters in the die can only be captured by the mesoscopic scale model. It provides information in this case, which is not accessible to the other models, because they cannot resolve the details of the surface structure inside the die.

These results demonstrate that the three-scale deposition model provides information on a new length scale, namely on the scale of feature clusters inside a die. For instance, the information about microloading, seen in Figure 6 for the case of three clusters of features inside the die, cannot be obtained from the two-scale model.

3. Modeling of the 'large' Knudsen number regime.

3.1. Theory. Consider now the lower end of the range of domain sizes of the mesoscopic scale model and/or more realistic lower pressures. In this case, the mean free path of the particles is inevitably on the same order of magnitude as the domain size. Hence, the Knudsen number is of order unity, and the process is inside the transition regime.

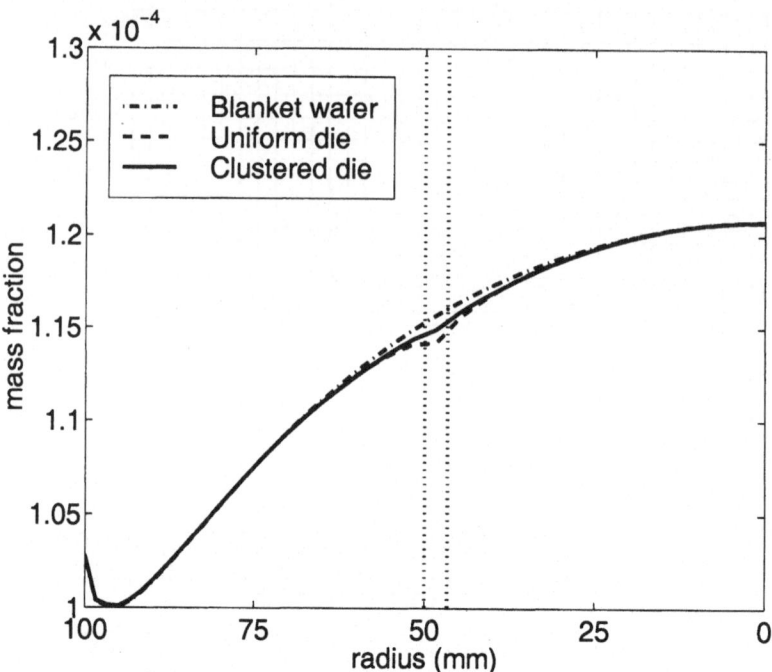

FIG. 5. *Mass fraction of triethyloxysilane vs. wafer radius across the wafer. The dotted lines mark the location of the die on the wafer surface.*

The flow of the gas particles in this regime is described by the Boltzmann equation for gas dynamics. While all chemical species are rarefied, the background gas is still an order of magnitude denser than the reacting species. Therefore, together with the assumption that the background gas is spatially uniform and at equilibrium, only collisions between a reacting species and the background gas need be considered. This results in the linear Boltzmann equation for each reacting species of the form

$$(3.1) \qquad \frac{\partial f}{\partial t} + v_1 \frac{\partial f}{\partial x} + v_2 \frac{\partial f}{\partial y} = Q(f)$$

with collision operator

$$Q(f)(x,v,t) = \int \sigma(v,v') \left[\frac{f(x,v',t)}{M(v')} - \frac{f(x,v,t)}{M(v)} \right] dv';$$

see [5, Chapter IV] for a detailed derivation. Here, $\sigma(v,v') = \sigma(v',v)$ denotes the scattering cross-section, which describes the probability that a particle with velocity v' before a collision scatters to a velocity v after the collision. We assume that the reacting species are introduced into the reactor chamber at the beginning of the processing step, i.e., that $f = 0$ at $t = 0$.

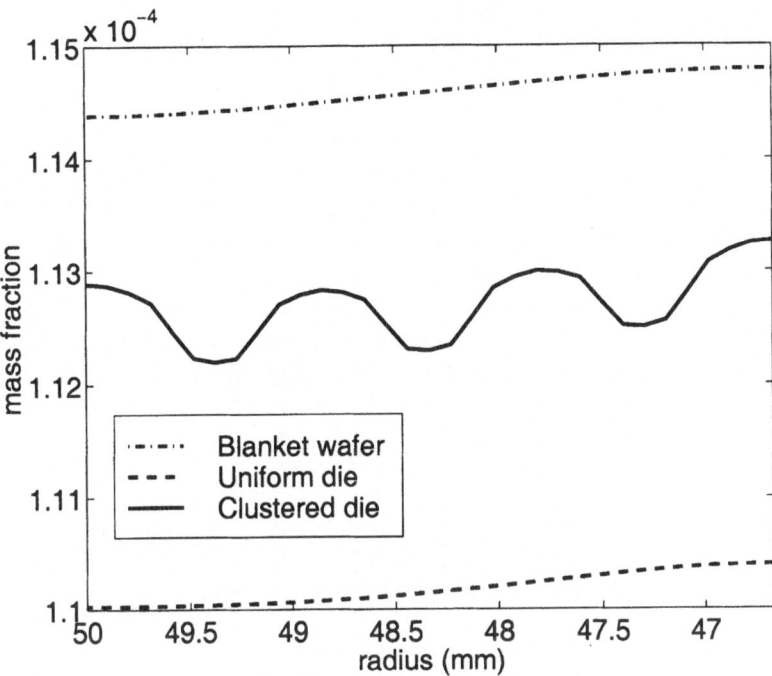

FIG. 6. *Mass fraction of triethyloxysilane vs. wafer radius across the die.*

The rough wafer surface Γ_ε is modeled for now as given by a function

(3.2) $$y = \tilde{h}(x) = \varepsilon\, h(x, \tfrac{x}{\varepsilon}).$$

with the small parameter $0 < \varepsilon \ll 1$. In dimensionless form, $h(x, \xi)$ is periodic in ξ with period 1, i.e.,

$$h(x, \xi) = h(x, \xi + 1) \quad \text{for all } (x, \xi).$$

At this surface, we apply the reaction condition for all inflowing particles

(3.3) $$f(x, y, v, t) = M(v) \int_{n \cdot w > 0} a(x, w)\, f(x, y, w, t)\, dw \quad \text{for all } n \cdot v < 0$$

for $(x, y)^T \in \Gamma_\varepsilon$, where $a(x, w)$ denotes a given function. It is again clear that the model in its present form with a microstructured surface is not numerically tractable due to the high cost of resolving the domain close to the rough surface Γ_ε. The fundamental goal is therefore again to obtain a model with a reduced boundary condition on a flat surface $y = 0$ that gives equivalent results for $f(x, y, v, t)$ in the bulk of the gaseous domain away from the surface in an asymptotic sense using the expansion parameter ε:

(3.4) $$f(x, y, v, t) = M(v) \int_{w_2 < 0} \tilde{a}(x, v, w)\, f(x, y, w, t)\, dw \quad \text{for all } v_2 > 0$$

for $y = 0$, where $\tilde{a}(x, v, w)$ needs to be determined.

To this end, we make the ansatz

$$f(x, y, v, t) = \tilde{f}(x, y, v, t) + \hat{f}(\tfrac{x}{\varepsilon}, \tfrac{y}{\varepsilon}, \tfrac{t}{\varepsilon}, x, v, t) + \mathcal{O}(\varepsilon),$$

where $\tilde{f}$ denotes the bulk variable, for which we wish to derive a numerically tractable model, and $\hat{f}$ is the small-scale correction, which is assumed to be periodic in ξ in the same way as the surface function. Due to the hyperbolic nature of the Boltzmann equation, we have to assume that $\hat{f}$ decays weakly with ε at any fixed distance from the surface, that is, we assume that small scale fluctuations in the inner solution $\hat{f}$ average out to zero at any fixed finite distance above the surface as $\varepsilon \to 0$; formally, we require that

$$(3.5) \qquad \lim_{\varepsilon \to 0} \int \hat{f}(\tfrac{x}{\varepsilon}, \tfrac{y}{\varepsilon}, \tfrac{t}{\varepsilon}, x, v, t)\psi(x)\, dx = 0$$

for all test functions $\psi(x)$.

The asymptotic ansatz leads to the inner problem for the layer term $\hat{f}(\xi, \eta, \tau, x, v_1, v_2, t)$ as a function of the layer variables ξ, η, τ

$$\frac{\partial \hat{f}}{\partial \tau} + v_1 \frac{\partial \hat{f}}{\partial \xi} + v_2 \frac{\partial \hat{f}}{\partial \eta} = 0$$

for any fixed x, v_1, v_2, t. This problem is solved by the method of characteristics, which are followed back to the boundary or to the initial condition, whichever comes first.

Introduce $\rho(\xi, \tau, x, t)$ as short-hand for the integral on the right-hand side of the boundary condition (3.3). It turns out that in the bulk of the domain, that is, for all $\eta > \max_{(x,\xi)} h(x, \xi)$

$$\hat{f}(\xi, \eta, \tau, x, t, v) = H(v_2)H(\tau - \tfrac{\eta}{v_2} - \phi_3)$$

$$\times \left[\rho(\xi - \tfrac{v_1}{v_2}\eta - v_1\phi_3, \tau - \tfrac{\eta}{v_2} - \phi_3, x, t)M(v) - \tilde{f}(x, 0, t, v)\right]$$

with

$$\phi_3(\xi, \eta, x, v) = \min\left\{z \in \mathbb{R} : -v_2 z = h(x, \xi - \tfrac{v_1}{v_2}\eta - v_1 z)\right\},$$

where $H(z)$ denotes the Heaviside function. The damping in the weak sense given by (3.5) is then satisfied if and only if

$$\tilde{f}(x, 0, v, t) = M(v) \int_0^1 \rho(\xi - v_1\phi_4, \infty, x, t)\, d\xi \quad \text{for all } v_2 > 0$$

holds with

$$\phi_4(\xi, x, v) = \min\{z \in \mathbb{R} : -v_2 z = h(x, \xi - v_1 z)\}.$$

This is the desired boundary condition for $\tilde{f}$, except that an explicit formula for ρ needs to be developed still.

This can be done by introducing an integral kernel representation for ρ with integral kernel $K^\infty(x, w, \xi)$, that is,

$$\rho(\xi, \infty, x, t) = \int K^\infty(x, v, \xi)\, \tilde{f}(x, 0, v, t)\, dv$$

This representation for ρ has the advantage that the kernel does not depend on time t, hence it can be pre-computed once at the beginning. To compute it, we can derive the system of equation

$$K^\infty(x, v, \xi) = H(\sigma(v))a(x, v)\chi(\xi, x, v)$$
$$+ \int H(\sigma(w))a(x, w)K^\infty(x, v, \xi - w_1\phi_1(w))\, (1 - \chi(\xi, x, w))\, M(w)\, dw$$

with

$$\chi(\xi, x, v) = \begin{cases} 0 & \text{if } \phi_1(\xi, x, v) < \infty, \\ 1 & \text{if } \phi_1(\xi, x, v) = \infty. \end{cases}$$

and

$$\phi_1(\xi, x, v) = \phi_0(\xi, h(x, \xi), x, v).$$

Therefore, we need to solve the numerical problem given by the Boltzmann equation for the bulk term $\tilde{f}(x, y, v, t)$

$$\frac{\partial \tilde{f}}{\partial t} + v_1 \frac{\partial \tilde{f}}{\partial x} + v_2 \frac{\partial \tilde{f}}{\partial y} = Q(\tilde{f})$$

with boundary condition for inflowing particles on the *flat* surface $y = 0$

$$\tilde{f}(x, 0, v, t) = M(v) \int_{w_2 < 0} \tilde{a}(x, v, w)\, \tilde{f}(x, y, w, t)\, dw \quad \text{for } v_2 > 0$$

with

$$\tilde{a}(x, v, w) = \int_0^1 K^\infty(x, w, \xi - v_1\phi_4(v))\, d\xi$$

and

$$\phi_4(\xi, x, v) = \min\left\{ z \in \mathbb{R} : -v_2 z = h(x, \xi - v_1 z) \right\}.$$

This problem is tractable numerically, since it is posed on a domain with a flat reacting surface and the values of $\tilde{a}(x, v, w)$ can be precomputed for all times. See [10] for more details of the analysis.

3.2. Numerical validation. As a validation problem for the analytic result, we have considered the linear Boltzmann equation with a relaxation time approximation using $\sigma(v, v') = (1/\tau)M(v)M(v')$ with relaxation time τ in the collision operator. The wafer surface Γ_ε is modeled by

$$y = \tilde{h}(x) = \frac{\varepsilon}{4}\left(1 + \cos\left(2\pi\frac{x}{\varepsilon}\right)\right).$$

At this reacting surface, let R denotes the sticking factor (the probability that a particle sticks to the surface). Then the inflow into the gaseous domain is equal to $(1 - R)$ times the outflow from the gaseous domain, namely

$$(3.6) \int_{n\cdot v<0} |n \cdot v| \, f(x, y, v, t) \, dv = (1 - R) \int_{n\cdot v>0} |n \cdot v| \, f(x, y, v, t) \, dv$$

where $n = n(x, y)$ denotes the unit outward normal vector at position $(x, y) \in \Gamma_\varepsilon$. This can be re-written in the form

$$f(x, y, v, t) = M(v) \int_{n\cdot w>0} a(x, w) f(x, y, w, t) \, dw$$

with

$$a(x, w) = \frac{1 - R}{c} |n \cdot w|, \quad c = \int_{n\cdot v<0} |n \cdot v| \, M(v) \, dv.$$

The problem is completed by choosing Maxwellian inflow at the top and periodic boundary conditions at both sides. This chosen setup of the problem is representative of the application under consideration. We choose $\tau = 1$ and $R = 0.5$ as values.

The small parameter $0 < \varepsilon \ll 1$ is chosen large enough to enable a full resolution of the original problem. The Figures 7, 8, and 9 show the solutions f for the original problem on the microstructured surface with $\varepsilon = 1/4$, $1/8$, and $1/16$, respectively, as well as the homogenized solution in the bulk of the domain $\tilde{f}$.

For a numerical comparison, consider the results in Table 1. It shows

TABLE 1
Error in density for $y > 0.5$ and total mass.

	$\max_{y>0.5} |\tilde{\rho} - \rho|$	$|\tilde{M} - M|$
$\varepsilon = \frac{1}{4}$	0.014665	0.021205
$\varepsilon = \frac{1}{8}$	0.006289	0.010215
$\varepsilon = \frac{1}{16}$	0.003449	0.006929

the errors in the values of the physical density $\rho(x, y, t) = \int f(x, y, v, t) \, dv$, when comparing only across the upper half of the domain, i.e., for $y > 0.5$, and the error in the total mass $M = \int\int \rho(x, y, t) \, dx \, dy$ across the whole domain.

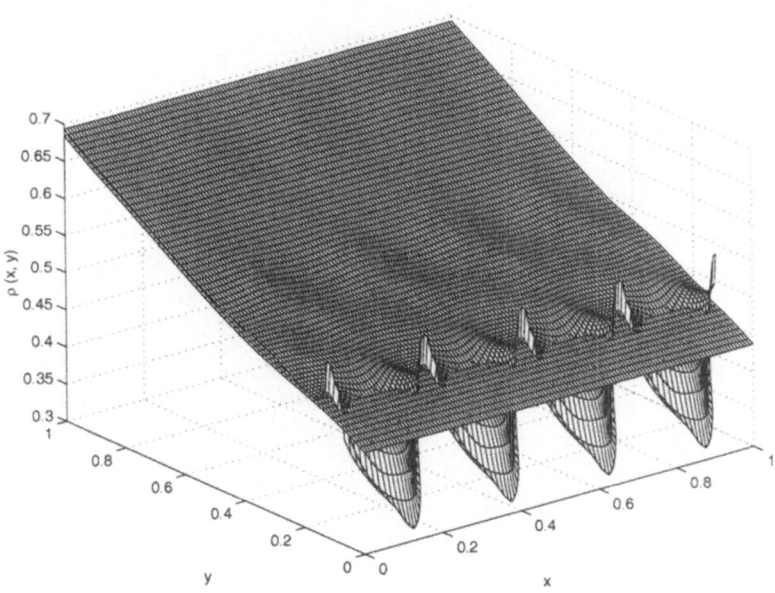

FIG. 7. *Comparison of densities for the homogenized and the structured domain with $\varepsilon = 1/4$.*

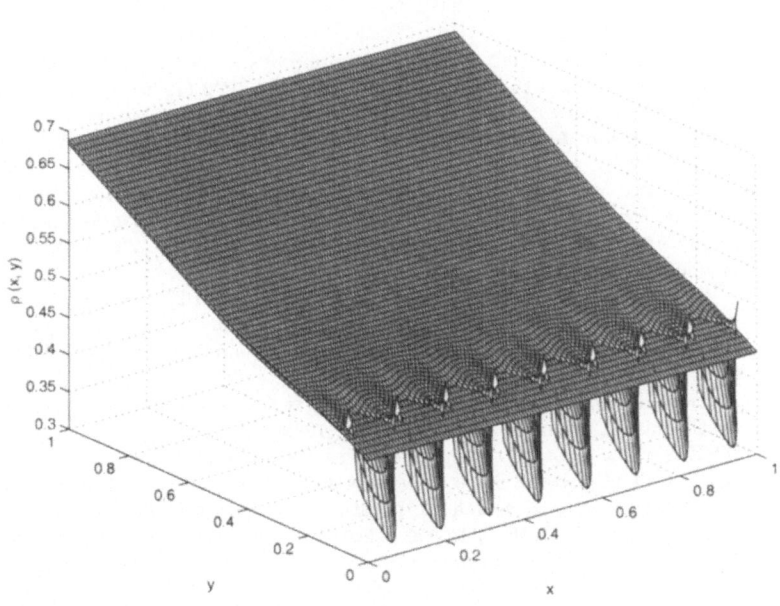

FIG. 8. *Comparison of densities for the homogenized and the structured domain with $\varepsilon = 1/8$.*

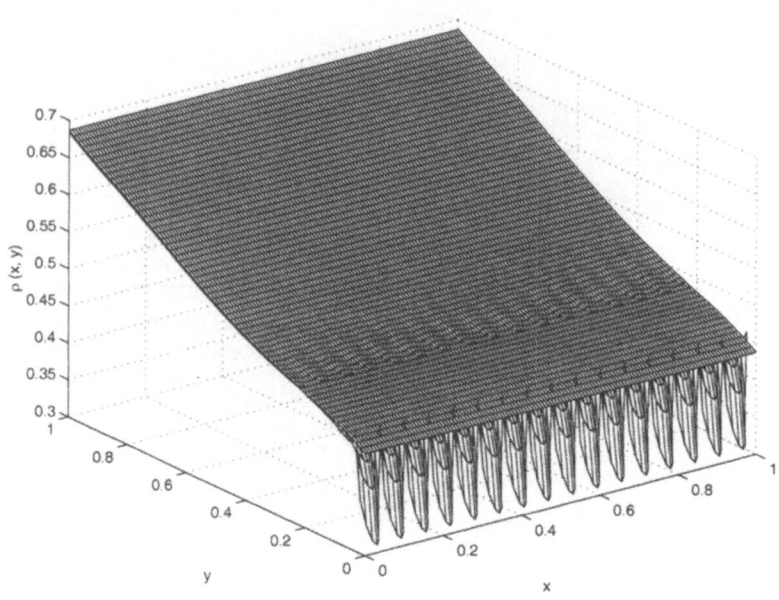

FIG. 9. *Comparison of densities for the homogenized and the structured domain with $\varepsilon = 1/16$.*

REFERENCES

[1] A.C. ADAMS AND C.D. CAPIO, *The decomposition of silicon dioxide films at reduced pressure*, Journal of the Electrochemical Society, **126** (1979), pp. 1042–1046.

[2] T.S. CALE, T.H. GANDY, AND G.B. RAUPP, *A fundamental feature scale model for low pressure deposition processes*, Vacuum Science Technology, A **9** (1991), pp. 524–529.

[3] T.S. CALE AND V. MAHADEV, *Feature scale transport and reaction during low pressure deposition processes*, in Modeling of Film Deposition for Microelectronic Applications, S. Rossnagel and A. Ulman, eds., Vol. **22** of Thin Films, Academic Press, 1996, p. 175.

[4] T.S. CALE, J.-H. PARK, T.H. GANDY, G.B. RAUPP, AND M.K. JAIN, *Step coverage predictions using combined reactor scale and feature scale models for blanket tungsten LPCVD*, Chemical Engineering Communications, **119** (1993), pp. 197–220.

[5] C. CERCIGNANI, *The Boltzmann Equation and Its Applications*, Vol. **67** of Applied Mathematical Sciences, Springer-Verlag, 1988.

[6] EVOLVE is a low pressure transport and reaction simulator developed by Timothy S. Cale at Arizona State University and Motorola, Inc. with funding from the Semiconductor Research Corporation, the National Science Foundation, and Motorola. EVOLVE 4.1a was released in February 1996.

[7] FIDAP 7.6, Fluid Dynamics International, 500 Davis St., Ste. 600, Evanston, IL 60201 (1996).

[8] M.K. GOBBERT, T.S. CALE, AND C.A. RINGHOFER, *The combination of equipment scale and feature scale models for chemical vapor deposition via a homogenization technique*, VLSI Design, **6** (1998), pp. 399–403.

[9] M.K. GOBBERT, T.P. MERCHANT, L.J. BORUCKI, AND T.S. CALE, *A multiscale simulator for low pressure chemical vapor deposition*, Journal of the Electrochemical Society, **144** (1997), pp. 3945–3951.

[10] M.K. GOBBERT AND C.A. RINGHOFER, *A homogenization technique for the Boltzmann equation for low pressure chemical vapor deposition.* In preparation.

[11] ———, *An asymptotic analysis for a model of chemical vapor deposition on a microstructured surface*, SIAM Journal on Applied Mathematics, **58** (1998), pp. 737–752.

[12] M.K. GOBBERT, C.A. RINGHOFER, AND T.S. CALE, *Mesoscopic scale modeling of microloading during low pressure chemical vapor deposition*, Journal of the Electrochemical Society, **143** (1996), pp. 2624–2631.

[13] A. HASPER, J. HOLLEMAN, J. MIDDELHOEK, C.R. KLEIJN, AND C.J. HOOGENDOORN, *Modeling and optimization of the step coverage of tungsten LPCVD in trenches and contact holes*, Journal of the Electrochemical Society, **138** (1991), pp. 1728–1738.

[14] P. HO. presented at the Schumacher Forum for Future Trends, San Diego, CA, February 1995, and private communication.

[15] C.R. KLEIJN AND C. WERNER, *Modeling of Chemical Vapor Deposition of Tungsten Films*, Vol. **2** of Progress in Numerical Simulation for Microelectronics, Birkhäuser Verlag, Basel, 1993.

[16] T.P. MERCHANT, M.K. GOBBERT, T.S. CALE, AND L.J. BORUCKI, *Multiple scale integrated modeling of deposition processes*, Thin Solid Films, **365** (2000), pp. 368–375.

[17] D.W. STUDINER, J.T. HILLMAN, R. ARORA, AND R.F. FOSTER, *LPCVD titanium nitride in a rotating disk reactor: Process modelling and results*, in Advanced Metallization for ULSI Applications 1992, T.S. Cale and F. Pintchovski, eds., Materials Research Corporation, 1993, pp. 211–217.

ASYMPTOTIC LIMITS
IN MACROSCOPIC PLASMA MODELS

ANSGAR JÜNGEL*

Abstract. A model hierarchy of macroscopic equations for plasmas consisting of electrons and ions is presented. The model equations are derived from the transient Euler-Poisson system in the zero-relaxation-time, zero-electron-mass and quasineutral limits. These asymptotic limits are performed using entropy estimates and compactness arguments. The resulting limits equations are Euler systems with a nonlinear Poisson equation and nonlinear drift-diffusion equations.

Key words. Zero-relaxation-time limit, zero-electron-mass limit, quasineutral limit, entropy functional, weak solutions, compensated compactness, compactness by convexity, plasmas.

AMS(MOS) subject classifications. Primary 35K55, 35L60, 35B25, 82D10.

1. Introduction. For the description of physical phenomena in plasmas, fluid dynamical models like the hydrodynamic (or Euler-Poisson) equations are widely used. The numerical solution of the hydrodynamic models requires a lot of computing power and special algorithms. In some situations, however, the model equations can be approximated by simpler models in the sense that a small parameter appearing in the equations—for instance, the (scaled) momentum relaxation time, the electron mass, or the Debye length—is set equal to zero. Some of these formal limits are known by physicists, see, e.g., [2, 4, 13]. Then the natural question arises if the solution of the full model converges, as the parameter tends to zero, to a solution of the limit model. In this review we present recent results how these asymptotic limits can be made rigorously.

More specifically, we consider an unmagnetized plasma consisting of electrons with (scaled) mass m_e and charge $q_e = -1$ and of a single species of ions with mass m_i and charge $q_i = +1$. Denoting by $n_e = n_e(x,t)$, $j_e = j_e(x,t)$ (n_i, j_i, respectively) the scaled particle density and current density of the electrons (ions, respectively) and by $\phi = \phi(x,t)$ the electrostatic potential, these variables satisfy the following scaled Euler-Poisson system **(HD-EI)**:

$$(1.1) \qquad \partial_t n_\alpha + \operatorname{div} j_\alpha = 0,$$

*Fachbereich Mathematik und Statistik, Universität Konstanz, Fach D193, 78457 Konstanz, Germany (juengel@fmi.uni-konstanz.de). The author acknowledges partial support from the Gerhard-Hess Program of the Deutsche Forschungsgemeinschaft, grant number JU 359/3-1, the TMR Project "Asymptotic methods in kinetic theory", grant number ERB-FMBX-CT97-0157, and from the German-French DAAD-Procope Program.

$$(1.2) \quad m_\alpha \partial_t j_\alpha + m_\alpha \operatorname{div}\left(\frac{j_\alpha \otimes j_\alpha}{n_\alpha}\right) + \nabla p_\alpha(n_\alpha) = -q_\alpha n_\alpha \nabla \phi - m_\alpha \frac{j_\alpha}{\tau_\alpha},$$

$$(1.3) \qquad\qquad\qquad\qquad -\lambda^2 \Delta \phi = n_i - n_e,$$

where $\alpha = e, i$ and $(x, t) \in \mathbb{R}^d \times (0, \infty)$. Here, $j_\alpha \otimes j_\alpha$ denotes the tensor product with components $j_{\alpha k} j_{\alpha l}$ for $k, l = 1, \ldots, d$. The pressure functions are of the form

$$(1.4) \qquad\qquad p_\alpha(n_\alpha) = a_\alpha n_\alpha^{\gamma_\alpha}, \quad n_\alpha \geq 0,$$

where $a_\alpha > 0$ and $\gamma_\alpha \geq 1$ are constants. The plasma is called *isothermal* if $\gamma_\alpha = 1$ and *adiabatic* if $\gamma_\alpha > 1$. The system (1.1)–(1.3) is complemented by initial conditions for n_α and j_α and by boundary conditions for ϕ ($\alpha = e, i$):

$$(1.5) \qquad n_\alpha(x, 0) = n_{\alpha 0}(x), \quad j_\alpha(x, 0) = j_{\alpha 0}(x), \quad x \in \mathbb{R}^d,$$

$$(1.6) \qquad\qquad\qquad \lim_{|x| \to \infty} \phi(x, t) = 0, \quad \text{a.e. } t > 0.$$

The homogeneous boundary condition for ϕ means that the plasma is in equilibrium at infinity.

The physical parameters are the (scaled) momentum relaxation time constants $\tau_e, \tau_i > 0$ of the electrons and ions, respectively, and the Debye length $\lambda > 0$. More precisely, these parameters are defined by

$$(1.7) \qquad \tau_\alpha = \frac{\tau_\alpha^*}{\tau_0}, \quad \lambda = \sqrt{\frac{\varepsilon_0 k_B T_0}{q^2 L^2 N}}, \quad m_\alpha = \frac{m_\alpha^* v_0^2}{k_B T_0},$$

where τ_α^* is the unscaled relaxation time, $\tau_0 = L/v_0$ a typical time scale, L a typical length, v_0 a typical velocity, ε_0 the permittivity constant, k_B the Planck constant, T_0 the ambient temperature, q the elementary charge, N a typical density and m_α^* the unscaled particle mass. We wish to perform rigorously the limits $\tau_\alpha \to 0$ for long time and small current density (i.e. we rescale $t \to t/\tau^2$ and $j_\alpha = \tau j_\alpha$ with $\tau = \tau_e = \tau_i \to 0$ and refer to this limit as the *zero-relaxation-time limit*) and the limits $m_e \to 0$ (*zero-electron-mass limit*) and $\lambda \to 0$ (*quasineutral limit*).

We now explain these asymptotic limits into more detail. Usually, the ions are heavy compared to the electrons, i.e. $m_i^* \gg m_e^*$. Therefore, if v_0^2 is equal to $k_B T_0 / m_i^*$, we obtain (see (1.7))

$$m_i = 1, \quad m_e = \frac{m_e^*}{m_i^*} \ll 1.$$

Letting formally $m_e \to 0$ in Eq. (1.2) for $\alpha = e$, we get

$$0 = \nabla p_e(n_e) - n_e \nabla \phi = n_e \nabla(h_e(n_e) - \phi),$$

where h_α ($\alpha = e, i$) is the enthalpy function defined by $h_\alpha'(s) = p_\alpha'(s)/s$ ($s > 0$), $h_\alpha(1) = 0$. Hence, if $n_e > 0$, we conclude $h_e(n_e) = \phi$ or, introducing the function $f_e = h_e^{-1}$, $n_e = f_e(\phi)$. The integration constant can

$$(\text{QH-EI}) \xleftarrow{\lambda \to 0} (\text{HD-EI}) \xrightarrow{m_e \to 0} (\text{HD-I}) \xrightarrow{\lambda \to 0} (\text{QH-I})$$

$$'\tau_e = \tau_i \to 0' \Big\downarrow \qquad\qquad \Big\downarrow '\tau_e = \tau_i \to 0' \qquad \Big\downarrow '\tau_i \to 0' \qquad\qquad \Big\downarrow '\tau_i \to 0'$$

$$(\text{QD-EI}) \xleftarrow{\lambda \to 0} (\text{DD-EI}) \xrightarrow{m_e \to 0} (\text{DD-I}) \xrightarrow{\lambda \to 0} (\text{QD-I})$$

FIG. 1. *A hierarchy of macroscopic plasma models. The limits* $'\tau_\alpha \to 0'$ *denote the zero-relaxation-time limits as explained in the text.*

be set equal to zero by choosing a reference point for the potential. Hence, the system (HD-EI) reduces in the zero-electron-mass limit to the model **(HD-I)** (see Fig. 1):

$$(1.8) \qquad\qquad\qquad \partial_t n_i + \operatorname{div} j_i = 0,$$

$$(1.9) \qquad m_i \partial_t j_i + m_i \operatorname{div}\left(\frac{j_i \otimes j_i}{n_i}\right) + \nabla p_i(n_i) = -n_i \nabla\phi - m_i \frac{J_i}{\tau_i},$$

$$(1.10) \qquad\qquad\qquad -\lambda^2 \Delta\phi = n_i - f_e(\phi).$$

Another set of equations is derived in the zero-relaxation-time limit in the models (HD-EI) and (HD-I). For this, introduce a scaling of time $s = t\tau$ (with $\tau = \tau_e = \tau_i$) and define

$$(1.11) \qquad N_\alpha(x, s) = n_\alpha\left(x, \frac{s}{\tau}\right), \quad J_\alpha(x, s) = \frac{1}{\tau}j_\alpha\left(x, \frac{s}{\tau}\right),$$

$$(1.12) \qquad\qquad\qquad \Phi(x, s) = \phi\left(x, \frac{s}{\tau}\right).$$

Setting again $t = s$, the model (HD-EI) becomes

$$(1.13) \quad \partial_t N_\alpha + \operatorname{div} J_\alpha = 0,$$

$$(1.14) \quad \tau^2 m_\alpha \partial_t J_\alpha + \tau^2 m_\alpha \operatorname{div}\left(\frac{J_\alpha \otimes J_\alpha}{N_\alpha}\right) + \nabla p_\alpha(N_\alpha) = -q_\alpha N_\alpha \nabla\Phi - m_\alpha J_\alpha,$$

$$(1.15) \quad -\lambda^2 \Delta\Phi = N_i - N_e.$$

Letting formally $\tau \to 0$, we obtain the drift-diffusion-model **(DD-EI)**:

$$(1.16) \qquad m_\alpha \partial_t N_\alpha - \operatorname{div}(\nabla p_\alpha(N_\alpha) + q_\alpha N_\alpha \nabla\Phi) = 0,$$

$$(1.17) \qquad\qquad\qquad -\lambda^2 \Delta\Phi = N_i - N_e,$$

where $\alpha = e, i$. Using the diffusion scaling (1.11)–(1.12) in the model (HD-I) and letting $\tau \to 0$, we obtain the model **(DD-I)**:

$$(1.18) \qquad m_i \partial_t N_i - \operatorname{div}(\nabla p_i(N_i) + N_i \nabla\Phi) = 0,$$

$$(1.19) \qquad\qquad\qquad -\lambda^2 \Delta\Phi = N_i - f_e(\Phi).$$

This model can also be derived from (DD-EI) in the limit $m_e \to 0$. Indeed, from (1.16) for $\alpha = e$ we obtain

$$0 = \mathrm{div}\left(\nabla p_e(N_e) - N_e \nabla \Phi\right) = \mathrm{div}\left(N_e \nabla (h_e(N_e) - \Phi)\right).$$

Usually, drift-diffusion equations are studied in bounded domains with mixed Dirichlet-Neumann boundary conditions (see Section 3). Hence, if $h_e(N_e) - \Phi = 0$ on a part of the domain boundary and $N_e > 0$ in the domain, we conclude $h_e(N_e) = \Phi$ or $N_e = f_e(\Phi)$ in the domain, which gives (1.19).

The quasineutral limit $\lambda \to 0$ in the model (HD-EI) implies $n \overset{\mathrm{def}}{=} n_e = n_i$. Hence adding Eqs. (1.2) for $\alpha = e, i$, simplifying $\tau = \tau_e = \tau_i$ and introducing the center-of-mass current density $j = (m_e j_e + m_i j_i)/(m_e + m_i)$, we obtain the model (QH-EI):

$$(1.20) \qquad \partial_t n + \mathrm{div}\, j = 0, \quad \mathrm{div}(j_e - j_i) = 0,$$

$$(1.21) \qquad \begin{aligned} \partial_t j + \mathrm{div}\left(\frac{j \otimes j}{n}\right) &+ \frac{m_e m_i}{(m_e + m_i)^2} \mathrm{div}\left(\frac{(j_e - j_i) \otimes (j_e - j_i)}{n}\right) \\ &+ \frac{1}{m_e + m_i} \nabla(p_e(n) + p_i(n)) = -\frac{j}{\tau}. \end{aligned}$$

Notice that the electric field $-\nabla \phi$ is eliminated in the equations. Under some assumptions, the above system can be simplified. Indeed, consider the equations in one space dimension in a bounded domain Ω with periodic boundary conditions and with no relaxation term, $\tau = +\infty$, and assume that the mean initial current densities for the electrons and ions are equal, $\int_\Omega j_e(x,0)dx = \int_\Omega j_i(x,0)dx$. Then from the second equation in (1.20) follows that $j_e - j_i$ is constant in space and from Eq. (1.21) we conclude after integration over Ω, that $j_\alpha(t)$ equals $\int_\Omega j_\alpha(x,0)dx/\mathrm{meas}(\Omega)$ and therefore, $j_e - j_i = 0$.

Performing the zero-relaxation-time limit in the above equations gives the diffusion model (QD-EI):

$$(1.22) \qquad \partial_t N - \frac{1}{m_e + m_i} \Delta(p_e(N) + p_i(N)) = 0,$$

where $N \overset{\mathrm{def}}{=} N_e = N_i$. In the case of an adiabatic plasma (i.e. $p_\alpha(n) = a_\alpha n^{\gamma_\alpha}$ with $\gamma_\alpha > 1$ and $\alpha = e, i$), this equation is of degenerate type. This model is well known in plasma physics [4, p. 160]. Let us assume that the electrons and ions are described by the same density-pressure relation but with different diffusivities, which account for the different masses:

$$p_\alpha(s) = D_\alpha p(s), \quad \alpha = e, i.$$

Then Eq. (1.22) becomes

$$\partial_t N - D \Delta p(N) = 0 \quad \text{with} \quad D = \frac{D_e + D_i}{m_e + m_i}.$$

This means that the new diffusivity of the quasi-neutral plasma is given by the so-called *ambipolar* diffusion coefficient D [4, p. 160].

The model (QD-EI) can also be obtained directly from (DD-EI) after letting $\lambda \to 0$ and adding Eqs. (1.16) for $\alpha = e$ and $\alpha = i$.

The limit $\lambda \to 0$ in the system (HD-I) leads to $n_i = f_e(\phi)$ or $\phi = h_e(n_i)$ and the model **(QH-I)**

$$(1.23) \qquad \partial_t n_i + \operatorname{div} j_i = 0,$$

$$(1.24) \qquad m_i \partial_t j_i + m_i \operatorname{div}\left(\frac{j_i \otimes j_i}{n_i}\right) + \nabla(p_i(n_i) + p_e(n_i)) = -m_i \frac{j_i}{\tau_i},$$

since $n_i \nabla \phi = n_i \nabla h_e(n_i) = \nabla p_e(n_i)$. These equations can be also found in the physical literature [2]. Performing the zero-relaxation-time limit in (1.23)–(1.24) we obtain the diffusion model **(QD-I)**:

$$(1.25) \qquad \partial_t N_i - \frac{1}{m_i} \Delta(p_i(N_i) + p_e(N_i)) = 0.$$

Clearly, this model follows from (QD-EI) after letting $m_e \to 0$. Moreover, the quasineutral limit in (DD-I) also gives Eq. (1.25).

The (formal) limits of the above models are summarized in Fig. 1. Let us mention that there are other plasma models, e.g. magnetohydrodynamic and kinetic models [13] and nonlinear Poisson models [3], which can be derived from (DD-EI) or (DD-I) in the limit $m_e \to 0$ (see [28]).

In Section 2 we present mathematical results which make rigorously some of the above limits in the whole-space hydrodynamic models. Section 3 is devoted to the proofs of the asymptotic limits in the drift-diffusion equations in bounded domains. Finally, in Section 4, some open problems are mentioned.

2. Asymptotic limits in the hydrodynamic equations. In this section we prove rigorously the zero-relaxation-time, zero-electron-mass and quasineutral limits in the models (HD-EI), (HD-I), respectively.

2.1. Zero-relaxation-time limits. We consider the zero-relaxation-time limits in the hydrodynamic models (HD-EI) and (HD-I) in the one-dimensional case $d = 1$. We study only this case since an existence theory is available only for $d = 1$. Our main goal is to give rigorous proofs of the limits (HD-EI)→(DD-EI) and (HD-I)→(DD-I). Under the assumption (1.4) for $\gamma_\alpha > 1$ and

$$(2.1) \qquad 0 \leq n_{\alpha_0}, \ \frac{j_{\alpha_0}}{n_{\alpha_0}} \in L^\infty(\mathbb{R}) \quad \text{and} \quad n_{\alpha_0}(x) = 0 \quad \text{for } |x| \geq L,$$

where $\alpha = e, i$ and $L > 0$ is a given constant, there exists a global weak entropy solution $(N_\tau, J_\tau, \Phi_\tau)$ of the rescaled problem (HD-I)

$$(2.2) \qquad \partial_t N_\tau + \partial_x J_\tau = 0,$$

$$(2.3) \qquad \tau^2 m_i \left(\partial_t J_\tau + \partial_x \left(\frac{J_\tau^2}{N_\tau} \right) \right) + \partial_x p_i(N_\tau) = -N_\tau \partial_x \Phi_\tau - J_\tau,$$

$$(2.4) \qquad - \partial_x^2 \Phi_\tau = N_\tau - f_e(\Phi_\tau),$$

for $x \in \mathbb{R}$, $t > 0$, with the initial and boundary conditions

$$(2.5) \qquad N_\tau(\cdot, 0) = n_{i0}, \quad J_\tau(\cdot, 0) = \tau^{-1} j_{i0} \quad \text{in } \mathbb{R},$$

$$(2.6) \qquad \lim_{|x| \to \infty} \Phi_\tau(x, t) = 0, \quad \text{a.e. } t > 0.$$

This solution satisfies

$$0 \le N_\tau, \; \frac{J_\tau}{N_\tau} \in L^\infty(\mathbb{R} \times \mathbb{R}^+), \quad 0 \le \Phi_\tau \in L^\infty(0, T, W^{2,\infty}(\mathbb{R})),$$

where $T > 0$ [12, 38]. For the existence of weak entropy solutions $(N_{\alpha\tau}, J_{\alpha\tau}, \Phi_\tau)$ to (1.13)–(1.15) with initial and boundary conditions corresponding to (2.5)–(2.6), we refer to [7, 32, 39]. Then it holds:

THEOREM 2.1 ((HD-I)→(DD-I)). *Let the assumptions (1.4) for $\gamma_\alpha >$ 1 ($\alpha = e, i$) and (2.1) hold. Then, as $\tau \to 0$, maybe passing to subsequences, $(N_\tau, J_\tau, \Phi_\tau)$ converges to (N, J, Φ), which is a solution of (1.18)–(1.19), in the following sense:*

$N_\tau \to N$ *strongly in $L^p_{\text{loc}}(\mathbb{R} \times (0, T))$ for any $p \in (1, \infty)$,*

$J_\tau \rightharpoonup J$ *weakly in $L^2(\mathbb{R} \times (0, T))$,*

$\tau^2 J_\tau^2 / N_\tau \to 0$ *strongly in $L^2(\mathbb{R} \times (0, T))$,*

$\Phi_\tau \to \Phi$ *strongly in $L^q_{\text{loc}}(\mathbb{R} \times (0, T))$ for any $q \in [\gamma_e/(\gamma_e - 1), \infty)$.*

The solution (N, J, Φ) satisfies

$$N \in L^\infty(0, T; L^p(\mathbb{R})), \; J \in L^2(\mathbb{R} \times (0, T)), \; \Phi \in L^\infty(0, T; W^{2,r}(\mathbb{R})),$$

where $p \in (1, \infty)$ and $r \in [\max(\gamma_e/(\gamma_e - 1), 2), \infty)$.

THEOREM 2.2 ((HD-EI)→(DD-EI)). *Let the assumptions of Theorem 2.1 hold. Then, as $\tau \to 0$, maybe passing to subsequences, $(N_{\alpha\tau}, J_{\alpha\tau}, \Phi_\tau)$ converges to $(N_\alpha, J_\alpha, \Phi)$ which is a solution of (1.16)–(1.17), in the sense of Theorem 2.1, where N_τ, J_τ, N, J are replaced by $N_{\alpha\tau}, J_{\alpha\tau}, N_\alpha, J_\alpha$, respectively.*

The proofs of the above theorems are based on high-energy (or entropy) estimates. More precisely, choosing special convex entropies in the sense of Lions-Perthame-Tadmor [30] and deriving the corresponding entropy inequalities, it is possible to prove that, for instance, $(N_\tau)_\tau$ is bounded in $L^\infty(0, T; L^p(\mathbb{R}))$ for any $p < \infty$ and $(J_\tau)_\tau$ is bounded in $L^2(\mathbb{R} \times (0, T))$. In order to get strong convergence of subsequences of $(N_\tau)_\tau$ and $(\Phi_\tau)_\tau$ we are using the div-curl lemma [42] and the Poisson equation. We refer to [26, 28] for the details and the proofs. Notice that not necessarily the whole sequence $(N_\tau, J_\tau, \Phi_\tau)$ converges since there is no general uniqueness result for the limiting problem (DD-I) or (DD-EI). (The uniqueness question of these models is addressed to in, e.g. [14, 24].)

The relaxation limit for *isothermal* plasmas has been carried out by Junca and Rascle in [20, 21]. They consider a plasma consisting only of ions. The proof of their result is based on an entropy inequality and the de la Vallée-Poisson lemma. Notice that here, L^1 weak convergence of the particle density is enough to pass to the limit in the linear pressure term.

The first mathematical relaxation limit for adiabatic plasmas has been shown by Marcati and Natalini [31, 36] assuming uniform L^∞ estimates. Let us mention some related results. In [5] the relaxation limit has been performed for adiabatic pressure and small smooth initial data. The same authors have also obtained a similar result for the so-called energy hydrodynamic model which includes an equation for the energy [6]. This result has been generalized by Ali et al. [1] for initial data that is a perturbation of the stationary solution of the thermal equilibrium state. A relaxation-time limit for a boundary-value problem has been performed in [19]. The steady-state problem has been addressed to in [33]. At our knowledge, no mathematical results are available for the relaxation limit (QH-EI)→(QD-EI). The limit (QH-I)→(QD-I), however, can be performed using the techniques of the proof of Theorem 2.1. In fact, the proof becomes easier since the electric term does not appear.

2.2. Zero-electron-mass limits. The limit $m_e \to 0$ in the hydrodynamic model is only solved in the one-dimensional case under additional assumptions [18].

THEOREM 2.3 ((HD-EI)→(HD-I)). *Let the assumptions (1.4) and (2.1) hold and set $\delta = m_e$. Let $(n_{\alpha\delta}, j_{\alpha\delta}, E_\delta)$ with $E_\delta = -\partial_x \phi_\delta$ be a weak entropy solution to (1.1)–(1.6). Assume furthermore that the sequences $(n_{e\delta})_\delta$ and $(j_{e\delta}^2/n_{e\delta})_\delta$ are bounded in $L^{2\gamma_e}_{loc}(\mathbb{R} \times (0, T))$ and $L^2_{loc}(\mathbb{R} \times (0, T))$, respectively. Then there is a subsequence of $(n_{\alpha\delta}, j_{\alpha\delta}, E_\delta)$, not relabeled, converging, as $\delta \to 0$, to (n_α, j_α, E) in the following sense:*

$$n_{i\delta} \to n_i, \ j_{i\delta} \to j_i \qquad \text{strongly in } L^p_{loc}(\mathbb{R} \times (0, T)) \text{ for any } p \in [1, \infty),$$

$$n_{e\delta} \to n_e \qquad \text{strongly in } L^p_{loc}(\mathbb{R} \times (0, T)) \text{ for any } p \in [1, \gamma_e + 1],$$

$$j_{e\delta} \rightharpoonup j_e \qquad \text{weakly in } L^2(\mathbb{R} \times (0, T))$$

$$E_\delta \rightharpoonup E \qquad \text{weakly* in } L^\infty(\mathbb{R} \times (0, T)),$$

and (n_α, j_α, E) is a weak solution of (1.8)–(1.9) for $\alpha = i$ and

$$(2.7) \qquad \lambda^2 \partial_x E = n_i - n_e, \quad \partial_x p_e(n_e) = -n_e E \quad \text{in } \mathbb{R} \times (0, T).$$

Notice that if $n_e > 0$ in $\mathbb{R} \times (0, T)$ then we can eliminate n_e in Eq. (2.7) and obtain the nonlinear Poisson equation (1.10) with $E = -\partial_x \phi$. For the proof of Theorem 2.3, we use the div-curl lemma and the monotonicity of p_e to conclude the strong convergence of $(n_{e\delta})_\delta$. The weak convergence of $(j_{e\delta})_\delta$ is a consequence of our assumptions. Finally the strong convergence of the sequences $(n_{i\delta})_\delta$ and $(j_{i\delta})_\delta$ follows from the H_{loc}^{-1} compactness technique of [29].

2.3. Quasineutral limits. The quasineutral limit has been proved for strong solutions of (HD-I) or (HD-EI) locally in time [10, 11]. First we consider the limit in (HD-I).

Let the assumption (1.4) for $\gamma_e, \gamma_i \geq 1$ hold and let $m_i = 1$, $\tau_i = \infty$. Define the ion mean velocity by $u_i = j_i/n_i$ and let $(n_{i\lambda}, u_{i\lambda}, \phi_\lambda)$ be a solution of

$$(2.8) \qquad \partial_t n_{i\lambda} + \partial_x(n_{i\lambda} u_{i\lambda}) = 0,$$

$$(2.9) \qquad \partial_t u_{i\lambda} + u_{i\lambda} \partial_x u_{i\lambda} + \frac{1}{n_{i\lambda}} \partial_x p_i(n_{i\lambda}) = -\partial_x \phi_\lambda,$$

$$(2.10) \qquad -\lambda^2 \partial_x^2 \phi_\lambda = n_{i\lambda} - f_e(\phi_\lambda) \quad \text{in } \mathbb{R} \times (0, \infty),$$

$$(2.11) \qquad n_{i\lambda}(\cdot, 0) = n_{i0}, \ u_{i\lambda}(\cdot, 0) = u_{i0} \quad \text{in } \mathbb{R}.$$

Assume that the plasma is uniform and electrically neutral at infinity, i.e. let ρ be a smooth, strictly positive function, constant outside of $[-1, 1]$, and tending to $\rho_\pm$ as $x \to \pm\infty$, let $\lim_{|x| \to \infty} \phi(x, t) = 0$, a.e. $t > 0$ and let n_{i0}, u_{i0} satisfy

$$n_{i0} - \rho, \ u_{i0} \in H^s(\mathbb{R}), \quad n_{i0} \geq \underline{n} > 0 \quad \text{for some } s > \frac{3}{2} \text{ and } \underline{n} > 0.$$

If $\lambda = 0$ we obtain the following system:

$$(2.12) \qquad \partial_t n_i + \partial_x(n_i u_i) = 0,$$

$$(2.13) \qquad \partial_t u_i + u_i \partial_x u_i + \frac{1}{n_i}(p_e'(n_i) + p_i'(n_i))\partial_x n_i = 0.$$

THEOREM 2.4 ((HD-I)→(QH-I)). *Let the above assumptions hold and let (n_i, u_i) be a strong solution to (2.12)–(2.13) with initial conditions corresponding to (2.11), on the time interval $[0, T]$, $T > 0$. Then there exist solutions $(n_{i\lambda}, u_{i\lambda})$ to (2.8)–(2.11) existing at least on $[0, T]$. Moreover, for any $T' < T$ there exist constants $c > 0$, $s' < s$ such that*

$$\|n_{i\lambda} - n_i\|_{L^\infty(0,T';H^{s'}(\mathbb{R}))} + \|u_{i\lambda} - u_i\|_{L^\infty(0,T';H^{s'}(\mathbb{R}))} \leq c\lambda,$$

where $c > 0$ is independent of λ.

The existence of strong local-in-time solutions to the limit problem (2.12)–(2.13) follows from classical results (see [11]). In order to prove Theorem 2.4, the system (2.12)–(2.13) is written equivalently as a system for the functions $\bar{n}$ and $\bar{u}$ which are the first-order expansions of the solutions in powers of λ, i.e. $n_{i\lambda} = n_i + \lambda \bar{n}$ and $u_{i\lambda} = u_i + \lambda \bar{u}$. Then it is shown in [11], using pseudodifferential techniques and energy estimates, that $(\bar{n}, \bar{u})$ are bounded in $H^{s'}(\mathbb{R})$ uniformly in λ.

Let us consider now a plasma described by (HD-EI) in the one-dimensional torus $\mathbb{T}$ and assume the condition (1.4) for $\gamma_e = \gamma_i = 1$. Furthermore, let $m_e = m_i = 1$ and $\tau_e = \tau_i = \infty$. Define similarly as above the mean velocities $u_\alpha = j_\alpha/n_\alpha$, $\alpha = e, i$, and let $(n_{\alpha\lambda}, u_{\alpha\lambda}, \phi_\lambda)$ be a solution of $(\alpha = e, i)$

$$(2.14) \qquad \partial_t n_{\alpha\lambda} + \partial_x(n_{\alpha\lambda} u_{\alpha\lambda}) = 0,$$

$$(2.15) \qquad \partial_t(n_{\alpha\lambda} u_{\alpha\lambda}) + \partial_x(n_{\alpha\lambda} u_{\alpha\lambda}^2 + a_\alpha n_{\alpha\lambda}) = -q_\alpha n_{\alpha\lambda} \partial_x \phi_\lambda,$$

$$(2.16) \qquad -\lambda^2 \partial_x^2 \phi_\lambda = n_{i\lambda} - n_{e\lambda} \quad \text{in } \mathbb{T} \times (0, \infty),$$

$$(2.17) \qquad n_{\alpha\lambda}(\cdot, 0) = n_{\alpha 0}, \; u_{\alpha\lambda}(\cdot, 0) = u_{\alpha 0} \quad \text{in } \mathbb{T}.$$

We assume that

$$(2.18) \qquad\qquad n_{e0} = n_{i0}, \quad u_{e0} = u_{i0} \quad \text{in } \mathbb{T}.$$

In order to derive the limit problem, let $\lambda = 0$ in Eqs. (2.14)–(2.17) and set $n \overset{\text{def}}{=} n_e = n_i$. Then, by Eq. (2.14), $j \overset{\text{def}}{=} n_i u_i - n_e u_e = n(u_i - u_e)$ is constant independently of $x \in \mathbb{T}$. Integration of Eq. (2.15) for $\lambda = 0$ in the form

$$\partial_t u_\alpha + \partial_x(\tfrac{1}{2} u_\alpha^2 + a_\alpha \log n_\alpha) = -q_\alpha \partial_x \phi$$

over $\mathbb{T}$ yields

$$\int_{\mathbb{T}} u_\alpha(x, t)dx = \int_{\mathbb{T}} u_{\alpha 0}(x)dx.$$

Hence, by assumption (2.18),

$$j \int_{\mathbb{T}} \frac{dx}{n} = \int_{\mathbb{T}} (u_i - u_e)dx = \int_{\mathbb{T}} (u_{i0} - u_{e0})dx = 0,$$

which gives $j = 0$. Therefore, adding Eqs. (2.15) for $\alpha = e, i$ to eliminate ϕ and setting $u \overset{\text{def}}{=} (m_e u_e + m_i u_i)/(m_e + m_i)$, we obtain

$$(2.19) \qquad\qquad\qquad \partial_t n + \partial_x(nu) = 0,$$

$$(2.20) \qquad\qquad \partial_t(nu) + \partial_x(nu^2 + (a_e + a_i)n) = 0.$$

In [10] the following theorem is shown by using pseudodifferential techniques:

THEOREM 2.5 ((HD-EI)→(QH-EI)). *Let $n_{\alpha 0}, u_{\alpha 0} \in H^s(\mathbb{T})$ for sufficiently large $s > 0$ and assume (2.18). Then there exists $T > 0$ such that,* *for $\lambda > 0$ small enough, there are solutions $(n_{\alpha\lambda}, u_{\alpha\lambda})$ to (2.14)–(2.17) in $L^\infty(0, T; H^{s'}(\mathbb{T}))$ with $s' < s$ such that, as $\lambda \to 0$,*

$$n_{\alpha\lambda} \to n \quad in \ L^\infty(0, T; H^{s'}(\mathbb{T})),$$
$$u_{\alpha\lambda} \to u \quad in \ L^\infty(0, T; H^{s'}(\mathbb{T})),$$

and (n, u) solves Eqs. (2.19)–(2.20).

There are only a few further results in the mathematical literature. Gasser and Marcati [17] proved an interesting result for a *combined* relaxation and quasineutral limit in (HD-EI) for weak entropy solutions. Indeed, assuming the diffusion scaling $t \to t/\lambda$, $j_i \to \lambda j_i$ and the relation $\lambda = \tau_i^\beta$ with $0 < \beta < 2(\gamma_i - 2)/3(\gamma_i - 1)$ and $\gamma_e = \gamma_i > 2$, they have shown that the solution $(n_{\alpha\lambda}, j_{\alpha\lambda}, \lambda^2 \partial_x \phi_\lambda)$ of (HD-EI) converges in appropriate Lebesgue spaces to a solution of a pure diffusion model for the limit particle density. This model can be written equivalently as an inhomogenous Burgers equation for the limit function of $(\lambda^2 \partial_x \phi_\lambda)_\lambda$.

Furthermore, traveling wave solutions and jump relations in the quasineutral limit have been studied by Cordier et al. [8, 9]. Slemrod has proved the limit $\lambda \to 0$ in the one-dimensional *steady-state* hydrodynamic equations [41].

3. Asymptotic limits in drift-diffusion models.

The limits $m_e \to 0$ and $\lambda \to 0$ in the drift-diffusion models are studied in bounded domains. For this, we assume that assumption (1.4) holds and that

(3.1) $\Omega \subset \mathbb{R}^d$ $(d \geq 1)$ is a bounded domain with Lipschitzian boundary $\partial\Omega = \Gamma_D \cup \Gamma_N$, $\Gamma_D \cap \Gamma_N = \emptyset$, $\mathrm{meas}_{d-1}(\Gamma_D) > 0$, and Γ_N is open in $\partial\Omega$, $T > 0$.

We consider the model (DD-EI) subject to the initial and boundary conditions

(3.2) $N_\alpha = N_{\alpha D}, \qquad \Phi = \Phi_D \qquad \qquad$ on $\Gamma_D \times (0, T)$,

(3.3) $\nabla p_\alpha(N_\alpha) \cdot \nu = \nabla \Phi \cdot \nu = 0 \qquad$ on $\Gamma_N \times (0, T)$,

(3.4) $N_\alpha(\cdot, 0) = N_{\alpha 0} \qquad \qquad \qquad$ in Ω, $\alpha = e, i$,

where ν is the exterior unit normal vector to $\partial\Omega$. Analogous initial and boundary conditions are imposed for the model (DD-I). We assume furthermore that

(3.5) $\begin{aligned} &0 \leq N_{\alpha D} \in C^0([0, T]; L^\infty(\Omega)) \cap H^1(\Omega \times (0, T)), \\ &0 \leq N_{\alpha 0} \in L^\infty(\Omega), \quad \alpha = e, i, \\ &\Phi_D \in L^\infty(\Omega \times (0, T)) \cap H^1(0, T; H^1(\Omega)). \end{aligned}$

Under these assumptions, there exists a global weak solution (N_α, Φ) to (1.16)–(1.17), (3.2)–(3.4), and a global weak solution (N_i, Φ) to (1.18)–(1.19), (3.2)–(3.4) for $\alpha = i$ [22, 23].

3.1. Zero-electron-mass limits. In order to perform the limit m_e $\to 0$ rigorously, we need additional assumptions. Assume that the initial and boundary data are compatible with the zero-electron-mass limit, i.e.

$$(3.6) \qquad N_{eD} = f_e(\Phi_D) \quad \text{in } \Omega \times (0, T), \quad N_{e0} = N_{i0} \quad \text{in } \Omega.$$

THEOREM 3.1 ((DD-EI)→(DD-I)). *Let the assumptions (1.4) and (3.1), (3.5)–(3.6) hold and assume that $N_{\alpha D} \geq \underline{N} > 0$ on $\Gamma_D \times (0, T)$ and $N_{\alpha 0} \geq \underline{N} > 0$ in Ω, for some $\underline{N} > 0$, $\alpha = e, i$. Let $\delta = m_e$ and let $(N_{\alpha \delta}, \Phi_\delta)$ be a weak solution to (1.16)–(1.17), (3.2)–(3.4). Then there exists a subsequence, not relabeled, converging, as $\delta \to 0$, to a weak solution (N_i, Φ) to (1.18)–(1.19), (3.2)–(3.4) for $\alpha = i$, in the following sense:*

$$\begin{aligned} N_{e\delta} &\to f_e(\Phi) && \text{strongly in } L^p(\Omega \times (0,T)) \text{ for any } p \geq 1, \\ N_{i\delta} &\to N_i && \text{strongly in } L^p(\Omega \times (0,T)) \text{ for any } p \geq 1, \\ N_{i\delta} &\rightharpoonup N_i && \text{weakly in } L^2(0,T;H^1(\Omega)), \\ \Phi_\delta &\to \Phi && \text{strongly in } L^2(0,T;H^1(\Omega)). \end{aligned}$$

The result of this theorem remains valid for more general (monotone) pressure functions (see [25]).

The proof is based on entropy estimates and compactness arguments. Indeed, define the *entropy* (or *free energy*) of the model (DD-EI):

$$\eta(t) = \sum_{\alpha = e}^{i} \int_\Omega \left\{ \int_{N_{\alpha D}(t)}^{N_{\alpha \delta}(t)} \left(f_e^{-1}(s) - f_e^{-1}(N_{\alpha D}(t)) \right) ds \right\} dx$$

$$+ \frac{\lambda^2}{2} \int_\Omega |\nabla(\Phi_\delta - \Phi_D)|^2 dx.$$

Then the following entropy inequality holds for a.e. $t > 0$:

$$(3.7) \qquad \eta(t) + \delta^{-1} \int_0^t \int_\Omega N_{e\delta} |\nabla(f_e^{-1}(N_{e\delta}) - \Phi_\delta)|^2 dx\, ds \leq \eta(0) + C,$$

and $C > 0$ depends on $N_{i\delta}$ and Φ_D but not on δ (or λ). Here, we have used the conditions (3.6). By the minimum principle, $N_{e\delta} \geq \underline{N} > 0$ in $\Omega \times (0, T)$, for some $\underline{N} > 0$. Therefore, $f_e^{-1}(N_{e\delta}) - \Phi_\delta \to 0$ strongly in $L^2(0, T; H^1(\Omega))$ and $N_{e\delta} - f_e(\Phi_\delta) \to 0$ strongly in $L^2(\Omega \times (0, T))$. Using the maximum principle and the Poisson equation, it can be shown that $N_{e\delta}$ and Φ_δ converge *weakly* in some Lebesgue spaces to N_e and Φ, respectively. In order to identify N_e and $f_e(\Phi)$ we need the *strong* convergence of one of the sequences $(N_{e\delta})_\delta$ or $(\Phi_\delta)_\delta$. However, since we do not have an appropriate uniform bound for the time derivative $\partial_t N_{e\delta}$ in some space, we cannot conclude the strong compactness for $(N_{e\delta})_\delta$, like for $(N_{i\delta})_\delta$, by an application of Aubin's lemma [40]. Instead the strong compactness of $(\Phi_\delta)_\delta$ is shown by using the monotone Poisson equation (1.19) and an argument which is related to a compactness-by-convexity result [25].

3.2. Quasineutral limits. Recently, the limit $\lambda \to 0$ in the transient drift-diffusion equations has been studied by several authors in the plasma or semiconductor context. We present here the results of [27, 37]. The following theorem concerns the quasineutral limit in the model (DD-EI).

THEOREM 3.2 ((DD-EI)→(QD-EI)). *Let the hypotheses (1.4), (3.1) and (3.5) hold. Furthermore, let $N_{\alpha D} \geq \underline{N} > 0$ on $\Gamma_D \times (0, T)$ and $N_{\alpha 0} \geq \underline{N} > 0$ in Ω, for some $\underline{N} > 0$, $\alpha = e, i$, and let $N_{e0} = N_{i0}$ in Ω. Let $(N_{\alpha\lambda}, \Phi_\lambda)$ be a weak solution to (1.16)–(1.17), (3.2)–(3.4) and let $\omega \subset \Omega$ satisfy $\overline{\omega} \subset \Omega$. Then the sequence $(N_{\alpha\lambda}, \Phi_\lambda)$ converges, as $\lambda \to 0$, in the following sense:*

$$N_{\alpha\lambda} \to N \qquad\qquad \text{strongly in } L^p(\omega \times (0, T)) \text{ for any } p \geq 1,$$

$$N_{\alpha\lambda} \rightharpoonup N, \ \Phi_\lambda \rightharpoonup \Phi \qquad \text{weakly in } L^2(0, T; H^1(\omega)),$$

and N solves Eq. (1.22) and the initial condition $N(\cdot, 0) = N_{e0} = N_{i0}$ in the sense of $H^{-1}(\omega)$. The limit Φ is a solution of

$$\mathrm{div}(N\nabla\Phi) = \frac{1}{m_e + m_i} \Delta(m_e p_i(N) - m_i p_e(N)).$$

Moreover, it holds, as $\lambda \to 0$,

$$(3.8) \qquad \begin{aligned} \|N_{e\lambda} - N_{i\lambda}\|_{L^2(\Omega \times (0,T))} &= \mathcal{O}(\lambda^{1/2}), \\ \|N_{e\lambda} - N_{i\lambda}\|_{L^2(\omega \times (0,T))} &= \mathcal{O}(\lambda). \end{aligned}$$

If in addition the condition $N_{eD} = N_{iD}$ on $\Gamma_D \times (0, T)$ holds, the above convergence results are valid for $\omega = \Omega$ and N satisfies the boundary condition $N = N_{eD}$ on $\Gamma_D \times (0, T)$.

For the quasineutral limit in the model (DD-I) we need additional assumptions since we need strong convergence of Φ_λ and it turns out that we have to estimate $\nabla\Phi_\lambda \cdot \nu$ on Γ_D:

$$(3.9) \qquad \partial\Omega \in C^{1,1}, \ \overline{\Gamma}_D \cap \overline{\Gamma}_N = \emptyset, \text{ and } \Phi_D \in L^2(0, T; H^2(\Omega)).$$

This assumption ensures that $\Phi_\lambda \in L^2(0, T; H^2(\Omega))$ and hence, $\nabla\Phi_\lambda \cdot \nu \in L^2(\Gamma_D \times (0, T))$ [43].

THEOREM 3.3 ((DD-I)→(QD-I)). *Let the assumptions of Theorem 3.2 hold for $\alpha = i$ and let (3.9) hold. Let $(N_{i\lambda}, \Phi_\lambda)$ be a weak solution to (1.16)–(1.17), (3.2)–(3.4) for $\alpha = i$. Then $(N_{i\lambda}, \Phi_\lambda)$ converges, as $\lambda \to 0$, in the following sense:*

$$N_{i\lambda} \to N, \ \Phi_\lambda \to \Phi \quad \text{strongly in } L^2(\Omega \times (0, T)),$$

and (N, Φ) solves (1.25) and $\Phi = f_e^{-1}(N)$ in $\Omega \times (0, T)$, and $N(\cdot, 0) = N_{i0}$ in the sense of $H^{-1}(\omega)$. Moreover, it holds, as $\lambda \to 0$,

$$\begin{aligned} \|N_{i\lambda} - f_e(\Phi_\lambda)\|_{L^2(\Omega \times (0,T))} &= \mathcal{O}(\lambda^{1/2}), & \|\nabla\Phi_\lambda\|_{L^2(\Omega \times (0,T))} &= \mathcal{O}(\lambda^{-1/2}), \\ \|N_{i\lambda} - f_e(\Phi_\lambda)\|_{L^2(\omega \times (0,T))} &= \mathcal{O}(\lambda), & \|\nabla\Phi_\lambda\|_{L^2(\omega \times (0,T))} &= \mathcal{O}(1). \end{aligned}$$

The proofs of the above theorems are based on entropy estimates. Consider the model (DD-EI). Then an entropy inequality similar to (3.7) and an application of the maximum principles yields uniform bounds for the sequences $(f_\alpha^{-1}(N_{\alpha\lambda}) + q_\alpha \Phi_\lambda)_\lambda$ and $(\lambda \Phi_\lambda)_\lambda$ in $L^2(0, T; H^1(\Omega))$ and $L^\infty(0, T; H^1(\Omega))$, respectively. With these bounds, the Poisson equation and Aubin's lemma [40] we obtain the results of Theorem 3.2 [27]. (The estimate (3.8) is proved in [37].) The limit $\lambda \to 0$ can also be performed without the positivity assumption on the initial and boundary data if we assume that $p_e = p_i$ [27].

Concerning the proof of Theorem 3.3, entropy estimates yield uniform bounds for $f_i^{-1}(N_{i\lambda}) + \Phi_\lambda$ and $\lambda \Phi_\lambda$ in $L^2(0, T; H^1(\Omega))$ and $L^\infty(0, T; H^1(\Omega))$, respectively. Then, the strong convergence of Φ_λ (and $N_{i\lambda}$) in $L^2(\Omega \times (0, T))$ is shown by deriving a uniform estimate for $\lambda \nabla \Phi_\lambda \cdot \nu$ on Γ_D, using an idea of [3], and by employing compensated compactness and compactness-by-convexity tools [27]. Under the additional assumptions $p_i(s) = s$ and $N_{iD} = f_e(\Phi_D)$ the positivity condition on N_{eD} and N_{eD} can be removed.

The above theorems show that boundary layers may occur if no compatibility condition on the boundary data is imposed. One may ask which values the limit N takes on the boundary. In view of the uniform estimates for the quasi-Fermi potentials $f_\alpha^{-1}(N_{\alpha\lambda}) + q_\alpha \Phi_\lambda$ in $L^2(0, T; H^1(\Omega))$ it is not very difficult to show that, under the assumptions of Theorem 3.2,

$$f_e^{-1}(N) + f_i^{-1}(N) = f_e^{-1}(N_{eD}) + f_i^{-1}(N_{iD}) \quad \text{on } \Gamma_D \times (0, T).$$

For instance, if $f_e(s) = f_i(s) = \exp(s)$ (this corresponds to isothermal plasmas) then

$$N = \sqrt{N_{eD} N_{iD}} \quad \text{on } \Gamma_D \times (0, T).$$

The quasineutral limit in macroscopic plasma models has been first studied by Brézis et al. in [3]. In this paper the limit $\lambda \to 0$ is shown for the nonlinear Poisson equation, the ion density being fixed. Gasser et al. [15, 16] considered this limit in the semiconductor drift-diffusion equations (i.e. the Poisson equation contains a given function which models fixed background charges) with pure Neumann boundary conditions. Finally, for the one-dimensional stationary equations, asymptotic expansions of the solutions in powers of λ are derived in [34] and a boundary-layer analysis for semiconductor pn-junctions has been performed in [35].

4. Open problems. We present some open problems concerning the discussed asymptotic limits in the hydrodynamic and drift-diffusion plasma models:

- Prove the relaxation-time limit in the magnetohydrodynamic equations (see [13, p. 234]).

- Prove, if possible, the combined zero-relaxation-time and zero-electron-mass limits in the hydrodynamic equations (HD-EI).
- Prove the zero-electron-mass limit in the hydrodynamic equations (HD-EI)→(HD-I) witout the uniform boundedness assumption of Theorem 2.3.
- Perform a boundary-layer analysis for the model (DD-EI) in the zero-electron-mass limit without the first assumption in (3.6).
- Prove the quasineutral limit in the hydrodynamic models for weak entropy solutions.
- Prove the quasineutral limit in the hydrodynamic models in bounded domains with appropriate (not periodic) boundary conditions.
- Prove the quasineutral limit (HD-EI)→(QH-EI) without assuming (2.18).
- Prove the three asymptotic limits in the energy hydrodynamic equations.

REFERENCES

[1] G. ALI, D. BINI, AND S. RIONERO. Global existence and relaxation limit for smooth solutions to the Euler-Poisson model for semiconductors. *SIAM J. Math. Anal.*, **32**: 572–587, 2000.

[2] L. ARTSIMOWITSCH AND R. SAGDEJEW. *Plasmaphysik für Physiker*. Teubner, Stuttgart, 1983.

[3] H. BRÉZIS, F. GOLSE, AND R. SENTIS. Analyse asymptotique de l'équation de Poisson couplée à la relation de Boltzmann. Quasi-neutralité des plasmas. *C. R. Acad. Sci. Paris*, **321**: 953–959, 1995.

[4] F. CHEN. *Introduction to Plasma Physics and Controlled Fusion*, Vol. 1. Plenum Press, New York, 1984.

[5] G.-Q. CHEN, J. JEROME, AND B. ZHANG. Particle hydrodynamic models in biology and microelectronics: singular relaxation limits. *Nonlin. Anal.*, **30**: 233–244, 1997.

[6] G.-Q. CHEN, J. JEROME, AND B. ZHANG. Existence and the singular relaxation limit for the inviscid hydrodynamic energy model. In J. Jerome, editor, *Modelling and Computation for Application in Mathematics, Science, and Engineering*, Oxford, 1998. Clarendon Press.

[7] S. CORDIER. Global solutions to the isothermal Euler-Poisson plasma model. *Appl. Math. Lett.*, **8**: 19–24, 1995.

[8] S. CORDIER, P. DEGOND, P. MARKOWICH, AND C. SCHMEISER. Quasineutral limit of travelling waves for the Euler-Poisson model. In G. Cohen, editor, *Mathematical and Numerical Aspects of Wave Propagation. Proceedings of the Third International Conference, Mandelieu-La Napoule, France*, pp. 724–733, Philadelphia, 1995. SIAM.

[9] S. CORDIER, P. DEGOND, P. MARKOWICH, AND C. SCHMEISER. Travelling wave analysis and jump relations for Euler-Poisson model in the quasineutral limit. *Asymptotic Anal.*, **11**: 209–224, 1995.

[10] S. CORDIER AND E. GRENIER. Quasineutral limit of two species Euler-Poisson systems. *Proceedings of the Workshop "Recent Progress in the Mathematical Theory on Vlasov-Maxwell Equations" (Paris)*, pp. 95–122, 1997.

[11] S. CORDIER AND E. GRENIER. Quasineutral limit of an Euler-Poisson system arising from plasma physics. *Comm. P.D.E.*, **25**: 1099–1113, 2000.

[12] S. CORDIER AND Y.-J. PENG. Système Euler-Poisson non linéaire – existence globale de solutions faibles entropiques. *Mod. Math. Anal. Num.*, **32**: 1–23, 1998.

[13] J. DELCROIX. *Plasma physics.* John Wiley, London, 1965.

[14] J.I. DÍAZ, G. GALIANO, AND A. JÜNGEL. On a quasilinear degenerate system arising in semiconductor theory. Part I: existence and uniqueness of solutions. *Nonlin. Anal. RWA*, **2**: 305–336, 2001.

[15] I. GASSER. The initial time layer problem and the quasi-neutral limit in a nonlinear drift-diffusion model for semiconductors. *Nonlin. Diff. Eqs. Appl.*, **8**: 237–249, 2001.

[16] I. GASSER, D. LEVERMORE, P. MARKOWICH, AND C. SCHMEISER. The initial time layer problem and the quasineutral limit in the drift-diffusion model for semiconductors. *Europ. J. Appl. Math.*, **12**: 497–512, 2001.

[17] I. GASSER AND P. MARCATI. The combined relaxation and vanishing Debye length limit in the hydrodynamic model for semiconductors. *Math. Methods Appl. Sci.*, **24**: 81–92, 2001.

[18] T. GOUDON, A. JÜNGEL, AND Y.-J. PENG. Zero-electron-mass limits in hydrodynamic models for plasmas. *Appl. Math. Lett.*, **12**: 75–79, 1999.

[19] L. HSIAO AND K.J. ZHANG. The relaxation of the hydrodynamic model for semiconductors to the drift-diffusion equations. *J. Diff. Eqs.*, **165**: 315–354, 2000.

[20] S. JUNCA AND M. RASCLE. Strong relaxation of the isothermal Euler-Poisson system to the heat equation. Preprint, Université de Nice, France, 1999.

[21] S. JUNCA AND M. RASCLE. Relaxation of the isothermal Euler-Poisson system to the drift-diffusion equations. To appear in *Quart. Appl. Math.*, 2002.

[22] A. JÜNGEL. On the existence and uniqueness of transient solutions of a degenerate nonlinear drift-diffusion model for semiconductors. *Math. Models Meth. Appl. Sci.*, **4**: 677–703, 1994.

[23] A. JÜNGEL. A nonlinear drift-diffusion system with electric convection arising in semiconductor and electrophoretic modeling. *Math. Nachr.*, **185**: 85–110, 1997.

[24] A. JÜNGEL. *Quasi-hydrodynamic Semiconductor Equations.* Progress in Nonlinear Differential Equations. Birkhäuser, Basel, 2001.

[25] A. JÜNGEL AND Y.-J. PENG. A hierarchy of hydrodynamic models for plasmas. Zero-electron-mass limits in the drift-diffusion equations. *Ann. Inst. H. Poincaré, Anal. non linéaire*, **17**: 83–118, 2000.

[26] A. JÜNGEL AND Y.-J. PENG. Zero-relaxation-time limits in hydrodynamic models for plasmas revisited. *Z. Angew. Math. Phys.*, **51**: 385–396, 2000.

[27] A. JÜNGEL AND Y.J. PENG. A hierarchy of hydrodynamic models for plasmas: quasi-neutral limits in the drift-diffusion equations. *Asympt. Anal.*, **28**: 49–73, 2001.

[28] A. JÜNGEL AND Y.J. PENG. A hierarchy of hydrodynamic models for plasmas: zero-relaxation-time limits. *Comm. P.D.E.*, **24**: 1007–1033, 1999.

[29] P.L. LIONS, B. PERTHAME, AND E. SOUGANIDIS. Existence of entropy solutions for the hyperbolic system of isentropic gas dynamics in Eulerian and Lagrangian coordinates. *Comm. Pure Appl. Math.*, **44**: 599–638, 1996.

[30] P.L. LIONS, B. PERTHAME, AND E. TADMOR. Kinetic formulation for the isentropic gas dynamics and p-system. *Comm. Math. Phys.*, **163**: 415–431, 1994.

[31] P. MARCATI AND R. NATALINI. Weak solutions to a hydrodynamic model for semiconductors and relaxation to the drift diffusion equations. *Arch. Rat. Mech. Anal.*, **129**: 129–145, 1995.

[32] P. MARCATI AND R. NATALINI. Weak solutions to a hydrodynamic model for semiconductors: The Cauchy problem. *Proc. Roy. Soc. Edinb., Sect. A*, **125**: 115–131, 1995.

[33] P.A. MARKOWICH. On steady state Euler-Poisson models for semiconductors. *Z. Angew. Math. Phys.*, **42**: 385–407, 1991.

[34] P.A. MARKOWICH, C. RINGHOFER, AND C. SCHMEISER. An asymptotic analysis of one-dimensional models of semiconductur devices. *IMA J. Appl. Math.*, **37**: 1–24, 1986.

[35] P.A. MARKOWICH AND C.A. RINGHOFER. A singularly perturbed boundary value problem modelling a semiconductor device. *SIAM J. Appl. Math.*, **44**: 231–256, 1984.

[36] R. NATALINI. The bipolar hydrodynamic model for semiconductors and the drift-diffusion equations. *J. Math. Anal. Appl.*, **198**: 262–281, 1996.

[37] Y.-J. PENG. Boundary layer analysis of the quasi-neutral limits in the drift-diffusion equations. Submitted for publication, 2000.

[38] Y.-J. PENG. Convergence of the fractional step Lax-Friedrichs scheme and Godunov scheme for a nonlinear Euler-Poisson system. *Nonlin. Anal.*, **42**: 1033–1054, 2000.

[39] F. POUPAUD, M. RASCLE, AND J. VILA. Global solutions to the isothermal Euler-Poisson system with arbitrarily large data. *J. Diff. Eqs.*, **123**: 93–121, 1995.

[40] J. SIMON. Compact sets in the space $L^p(0, T; B)$. *Ann. Math. Pura Appl.*, **146**: 65–96, 1987.

[41] M. SLEMROD AND N. STERNBERG. Quasi-neutral limit for Euler-Poisson system. *J. Nonlin. Sci.*, **11**: 193–209, 2001.

[42] L. TARTAR. Compensated compactness and applications to partial differential equations. In *Nonlinear analysis and mechanics: Heriot-Watt Symp.*, Vol. 4, Volume 39 of *Res. Notes Math.*, pp. 136–212, 1979.

[43] G.M. TROIANIELLO. *Elliptic Differential Equations and Obstacle Problems.* Plenum Press, New York, 1987.

A LANDAU–ZENER FORMULA
FOR TWO–SCALED WIGNER MEASURES

CLOTILDE FERMANIAN KAMMERER* AND PATRICK GERARD†

Abstract. The semiclassical study of multidimensional crystals leads naturally to the following question: How do Wigner measures propagate through energy level crossings?

In this contribution, we discuss a simple 2×2 system which displays such a crossing. For that purpose, we introduce two–scaled Wigner measures, which describe how the usual Wigner transforms are concentrating on trajectories passing through the crossing points. Then we derive explicit formulae for the branching of such measures. These formulae are generalizations of the so–called Landau–Zener formulae.

This contribution only contains main statements and some sketch of proofs. Details of proofs will appear in reference [6].

1. Introduction. An important mathematical problem arising in solid state physics is to understand the behaviour, as ϵ goes to 0, of solutions ψ^ϵ of the following Schrödinger equation,

$$(1) \qquad i\epsilon \frac{\partial \psi^\epsilon}{\partial t} = -\epsilon^2 \Delta_x \psi^\epsilon + \left(U\left(\frac{x}{\epsilon}\right) + V(x) \right) \psi^\epsilon$$

where $\psi^\epsilon = \psi^\epsilon(t, x)$, $t \in \mathbf{R}$, $x \in \mathbf{R}^d$, where $d = 1, 2, 3$. Equation (1) is supposed to describe the evolution of the wave function of electrons in a crystal. Here $U = U(y)$ is a periodic potential with respect to some lattice $L \subset \mathbf{R}^d$, which represents the interaction with the ions in the crystal, and $V = V(x)$ is an applied potential, which we shall assume to be smooth enough. We refer for example to the introduction of [19] for a brief discussion of this model. Of course function ψ^ϵ is assumed to satisfy some initial condition on $t = 0$, say

$$\psi^\epsilon(0, x) = \psi_0^\epsilon(x)$$

and we shall assume the following a priori estimates

$$(2) \qquad \| \psi_0^\epsilon \|_{L^2} \le C , \quad \lim_{R \to \infty} \limsup_{\epsilon \to 0} \int_{|\xi| \ge R/\epsilon} | \hat{\psi}_0^\epsilon(\xi) |^2 \, d\xi = 0,$$

where $\hat{\psi}$ denotes the Fourier transform with respect to variable x. The first above condition means the finiteness of charge, while the second one expresses that no momentum much larger than $1/\epsilon$ occurs in the spectrum of ψ^ϵ. It can be easily shown that these estimates are propagated to any finite time t.

*Mathématiques, Université de Cergy-Pontoise, 2 avenue Adolphe Chauvin, BP 222, Pontoise, 95 302 Cergy-Pontoise cedex, France. E-mail: `Clotilde.Fermanian@math.u-cergy.fr`.

†Mathématiques, Université Paris XI, Bâtiment 425, 91405 Orsay, France. E-mail: `Patrick.Gerard@math.u-psud.fr`.

Now let us try to say more about what we mean by "studying the behaviour of ψ^ϵ as ϵ goes to 0." A typical quantity of interest is of course the weak limit ρ of $|\psi^\epsilon|^2$, which describes the density of particles. Let us assume for a moment $U = 0$ in equation (1), so that we are in the usual semiclassical regime for the Schrödinger equation. In this case, it is now classical that a relevant quantity is the (spacetime) Wigner transform of ψ^ϵ,

$$W^\epsilon \psi^\epsilon(t, x, \tau, \xi) =$$

$$\int_{\mathbf{R}} \int_{\mathbf{R}^d} e^{is\tau + y \cdot \xi} \psi^\epsilon \left(t - \frac{\epsilon s}{2}, x - \frac{\epsilon y}{2} \right) \bar{\psi}^\epsilon \left(t + \frac{\epsilon s}{2}, x + \frac{\epsilon y}{2} \right) \frac{ds\, dy}{(2\pi)^{d+1}}$$

which converges weakly (up to a subsequence), as ϵ goes to 0, to a nonnegative measure μ, called a Wigner (semiclassical) measure of the family ψ^ϵ and which allows to recover the density ρ as

$$\rho(t, x) = \int_{\mathbf{R}} \int_{\mathbf{R}^d} \mu(t, x, d\tau, d\xi) .$$

We refer to [16] or [8] for a survey of main properties of these Wigner measures. Recall that, for solutions ψ^ϵ of equation (1) with $U = 0$, measure μ satisfies the following properties. First, a localization law on the energy surface,

$$(3) \qquad\qquad (\tau + |\xi|^2 + V(x))\mu = 0$$

and second, a propagation law along the classical trajectories,

$$(4) \qquad\qquad \partial_t \mu + 2\xi \cdot \nabla_x \mu - \nabla_x V(x) \cdot \nabla_\xi \mu = 0$$

with the following initial condition,

$$\mu_{|t=0} = \mu_0(x, \xi)\, \delta(\tau + |\xi|^2 + V(x)) ,$$

where μ_0 is the Wigner measure of the Cauchy data ψ_0^ϵ. As an example, if the Cauchy data has the following WKB form

$$\psi_0^\epsilon(x) = a_0(x)e^{iS_0(x)/\epsilon} ,$$

then the classical WKB theory asserts that, for small t,

$$\psi^\epsilon(t, x) = a(t, x)e^{iS(t,x)/\epsilon} + o(1) .$$

Then

$$\mu(t, x, \tau, \xi) = |a(t, x)|^2\, \delta(\xi - \nabla_x S(t, x))\, \delta(\tau - \partial_t S(t, x))$$

and condition (3) above reflects the eikonal equation

$$\partial_t S + |\nabla_x S|^2 + V(x) = 0$$

while equation (4) expresses the transport law of $\rho = \mid a \mid^2$,

$$\partial_t \rho + \nabla_x.(2\rho\nabla_x S) = 0 .$$

Let us come back to the general case where U is not identically 0. Then the above semiclassical propagation will compete with the homogenization effects induced by potential U at the same scale ϵ. In this context, it is natural to set

$$\psi^\epsilon(t, x) = u^\epsilon\left(t, x, \frac{x}{\epsilon}\right)$$

where u^ϵ is a solution to the following equation,

(5) $\qquad i\epsilon\partial_t u^\epsilon(t, x, y) = -(\epsilon\nabla_x + \nabla_y)^2 u^\epsilon + (U(y) + V(x))u^\epsilon$

Equation (5) can be viewed as a semiclassical system of PDEs in variables (t, x), with the infinite dimensional target space $L_y^2(Q)$, where Q is a cell for lattice L. Diagonalizing this system leads to the following Bloch eigenvalue problem

(6) $\quad ((\xi - i\nabla_y)^2 + U(y))\phi(y, \xi) = E(\xi)\phi(y, \xi), \quad \phi(y+l, \xi) = \phi(y, \xi), \ l \in L$

The classical theory of elliptic operators on the torus implies the existence of a sequence of eigenvalues

(7) $\qquad\qquad E_1(\xi) \leq E_2(\xi) \leq \ldots \leq E_j(\xi) \xrightarrow[j\to\infty]{} +\infty$

associated to an orthonormal base (ϕ_j) of eigenfunctions. Moreover, it is easy to check that functions $E_j(\xi)$ are L^*–periodic, where L^* denotes the dual lattice associated to L, defined by $k \in L^*$ iff $\langle k, l \rangle \in 2\pi\mathbf{Z}$ for all $l \in L$.

Now assume for a short while that the eigenvalues E_j's have constant multiplicity as ξ varies. Then it is possible to show that the Wigner measure μ of ψ^ϵ satisfies the following property. Define

$$\mu^\sharp(t, x, \tau, \xi) = \sum_{k \in L^*} \mu(t, x, \tau, \xi + k)$$

so that $\mu^\sharp$ is L^*–periodic in ξ and

$$\rho(t, x) = \int_{\tau \in \mathbf{R}} \int_{\xi \in \mathbf{R}^d/L^*} \mu^\sharp(t, x, d\tau, d\xi) .$$

Then there exists a sequence (μ_j) of positive measures on $\mathbf{R}_t \times \mathbf{R}_x^d \times \mathbf{R}_\tau \times (\mathbf{R}^d/L^*)_\xi$ such that

$$\mu = \sum_j \mu_j$$

with, for any j, the following transport equation,

$$(8) \qquad \partial_t \mu_j + \nabla_\xi E_j(\xi).\nabla_x \mu_j - \nabla_x V(x).\nabla_\xi \mu_j = 0$$

In other words, the single transport equation (4) is replaced by the sequence of transport equations (8), labelled by the energy bands, with new group velocities given as the gradients of Bloch energies E_j. For a proof of this fact for $d = 1$ — using a slightly different approach — see [2].

However, if $d > 1$, it is now customarily accepted that the multiplicities of $E_j's$ may jump, creating singularities on functions $E_j's$ (see [4] for results in that direction, in the cases of small generic two–dimensional even potentials U and of three–dimensional potentials U). In this case, the global dynamics provided by the sequence of equations (8) become doubtful. Indeed, one expects that band crossings allow electrons to jump from one band to another one. A crucial problem is to understand quantitatively such a phenomenon, in particular the role played by the applied potential V.

A notable exception is the case $V = 0$. Indeed, in this case, the transport equations (8) do not involve any propagation effect in the variable ξ, and it is possible to prove that the above dynamics are still correct, assuming the Wigner measure of the Cauchy data does not see the singular set of the E_j's (see [7, 18, 8]).

On the contrary, if ∇V is not 0, there is no reason which would prevent a jump from one band to another one. Of course, at this time the geometric spectral theory of Bloch waves (see for instance [21] and more recently [13]) does not provide us a precise description of the singularities of the E_j's, so we are forced to deal with simple models of systems displaying eigenvalue crossings, with the hope that such models reflect locally some genericity of this complicated problem. The aim of this contribution is to understand completely the simplest of these models in two space dimensions.

2. The model system. Our model system is

$$(9) \qquad i\epsilon \partial_t \psi^\epsilon = (A(-i\epsilon \nabla_x) + V(x))\psi^\epsilon,$$

where $t \in \mathbf{R}, x \in \mathbf{R}^2$, ψ^ϵ is valued in $\mathbf{C}^2$ and the matrix

$$(10) \qquad A(\xi) = \begin{pmatrix} \xi_1 & \xi_2 \\ \xi_2 & -\xi_1 \end{pmatrix}$$

has eigenvalues $\pm \mid \xi \mid$ displaying a crossing at $\xi = 0$. Observe that this system can be seen as a simple version of the Dirac system, and that special solutions of type coherent states have already been studied by Hagedorn [9] and Hagedorn–Joye [10]. For a complete study of systems in one variable, we refer to [5]. Here we are interested in general solutions of this system in two space dimensions, assuming the Cauchy data ψ_0^ϵ only satisfies conditions (2).

Outside $\xi = 0$, results of [8] provide the dynamics of the Wigner measure of ψ^ϵ, namely

$$\mu(t, x, \tau, \xi) = \sum_\pm \mu^\pm(t, x, \xi)\, \delta(\tau \pm |\xi| + V(x))$$

with the transport equations

(11) $\qquad \partial_t \mu^\pm \pm \frac{\xi}{|\xi|} \cdot \nabla_x \mu^\pm - \nabla V(x).\nabla_\xi \mu^\pm = 0 \ , \ \xi \neq 0 \ .$

Therefore the main problem is to describe μ near the singular set $\xi = 0$. Let assume that $\nabla V(x) \neq 0$. The following lemma asserts that no part of μ sticks to this singular set.

LEMMA 1. *With the above notation,*

(12) $\qquad \mu(\{\xi = 0, \ \nabla V(x) \neq 0\}) = 0.$

In other words, the mass of $\mu^\pm$ always goes through the singular set as t varies. As a consequence, the only unknown of the dynamics of μ is the mass transfer from $\mu^\pm$ to $\mu^\mp$ as ξ goes through 0. We shall focus on this task in the next two sections.

3. The particular case of a linear potential. In this section, we study system (9) in the particular case $V(x) = x_1$. Introducing the rescaled Fourier transform

$$\tilde{\psi}^\epsilon(t, \xi) = \int_{\mathbf{R}^2} e^{-ix.\xi/\epsilon} \psi^\epsilon(t, x)\, \frac{dx}{(2\pi\epsilon)}$$

this system becomes

$$i\epsilon \partial_t \tilde{\psi}^\epsilon = A(\xi)\tilde{\psi}^\epsilon + i\epsilon \partial_{\xi_1} \tilde{\psi}^h.$$

In order to eliminate parameter ϵ, introduce the following rescaled variables

$$s = \frac{\xi_1}{\sqrt{\epsilon}} , \quad \eta = \frac{\xi_2}{\sqrt{\epsilon}}$$

so that we are led to the following system of ordinary differential equations,

(13) $\qquad \frac{1}{i}\partial_s u = \begin{pmatrix} s & \eta \\ \eta & -s \end{pmatrix} u,$

by setting

(14) $\qquad u(s, t, \eta) = \tilde{\psi}^\epsilon(t - \xi_1, \xi_1, \xi_2).$

Observe that the classical trajectories passing through the singular set $\xi = 0$ at time $t = t_0$ are just

$$x_t^\pm = x_0 + (\mp |t_0 - t|, 0) \ ; \ \xi_t^\pm = (t_0 - t, 0) \ .$$

In particular, variable ξ_1 is a good time variable on those trajectories, and passing through the singular set corresponds in variable $s = \xi_1/\sqrt{\epsilon}$ to a scattering problem for system (13).

It turns out that this scattering problem had been already studied by Landau [15] and Zener [22] (see also Joye[12] for a general result in the framework of abstract adiabatic theory). The result is that solutions to (13) can be written as follows, assuming $\mid \eta \mid \ll \mid s \mid$,

$$u(s,\eta) = e^{is^2/2} \mid s \mid^{i\eta^2/2} \begin{pmatrix} \alpha^{\pm} \\ 0 \end{pmatrix} + e^{-is^2/2} \mid s \mid^{-i\eta^2/2} \begin{pmatrix} 0 \\ \beta^{\pm} \end{pmatrix} + o(1)$$

as $\quad s \to \pm\infty$

where the coefficients $\alpha^{\pm}(\eta), \beta^{\pm}(\eta)$ are related by

$$\begin{pmatrix} \alpha^+ \\ \beta^+ \end{pmatrix} = \begin{pmatrix} a(\eta) & -\bar{b}(\eta) \\ b(\eta) & a(\eta) \end{pmatrix} \begin{pmatrix} \alpha^- \\ \beta^- \end{pmatrix}$$

with the following explicit formulae

$$(15) \quad a(\eta) = e^{-\frac{\pi\eta^2}{2}}, \quad b(\eta) = \frac{2i}{\sqrt{\pi\eta}} 2^{-i\eta^2/2} e^{-\pi\eta^2/4} \Gamma(1 + i\frac{\eta^2}{2}) \operatorname{sh}(\frac{\pi\eta^2}{2}).$$

It is therefore natural to introduce a generalization of Wigner measures which takes into account the new rescaled variable $\eta = \xi_2/\sqrt{\epsilon}$. This leads to the following proposition.

PROPOSITION 1. *Denote by $\mathcal{A}$ the space of smooth functions $a = a(t, x, \tau, \xi, \eta)$, compactly supported in (t, x, τ, ξ), satisfying moreover, for some $R = R(a) > 0$,*

$$a(t, x, \tau, \xi, \eta) = a(t, x, \tau, \xi, \pm\infty) \quad if \quad \pm\eta \geq R.$$

Let μ be the Wigner measure of ψ^{ϵ}. Then there exists a positive measure $\nu = \nu(t, x, \tau, \xi_1, \eta)$ on $\{\xi_2 = 0\} \times \overline{\mathbf{R}}_{\eta}$ and a subsequence (ϵ_k) such that

$$\int_{(\mathbf{R}\times\mathbf{R}^2)^2} W^{\epsilon}\psi^{\epsilon}(t, x, \tau, \xi) \, a\left(t, x, \tau, \xi, \frac{\xi_2}{\sqrt{\epsilon}}\right) dt \, dx \, d\tau \, d\xi$$

$$\xrightarrow[\epsilon \to 0, \epsilon = \epsilon_k]{} \int_{\{\xi_2=0\}\times\overline{\mathbf{R}}_{\eta}} a(t, x, \tau, \xi_1, 0, \eta) \, d\nu(t, x, \tau, \xi_1, \eta)$$

$$+ \int_{\{\xi_2\neq 0\}} a(t, x, \tau, \xi, \operatorname{sgn}(\xi_2)\infty) \, d\mu(t, x, \tau, \xi).$$

Measure ν defined by Proposition (1) is called a two–scaled Wigner measure of the family (ψ^{ϵ}) associated to the hypersurface $I = \{\xi_2 = 0\}$ and to the scale $\sqrt{\epsilon}$. Such objects were introduced in a different context by L. Miller in [17]. Of course, knowing ν, it is possible to recover μ on I as

$$\mathbf{1}_I \mu = \int_{\overline{\mathbf{R}}_{\eta}} \nu(., d\eta) .$$

Moreover, the complement of I is preserved by the classical dynamics, so the evolution of measure μ outside I is completely characterized by the transport equations (11). Therefore we are led to describe the evolution of ν. This is the purpose of the following theorem.

THEOREM 1. *The measure ν can be decomposed as*

$$
(16) \quad
\begin{aligned}
\nu(t,x,\tau,\xi_1,\eta) &= \nu^+(t,x,\xi_1,\eta)\delta(\tau + |\xi_1| + x_1) \\
&\quad + \nu^-(t,x,\xi_1,\eta)\delta(\tau - |\xi_1| + x_1) ,
\end{aligned}
$$

where $\nu^\pm$ are positive measures satisfying the following transport equations in $\xi_1 \neq 0$,

$$
(17) \qquad \partial_t \nu^\pm \pm \operatorname{sign}(\xi_1)\partial_{x_1}\nu^\pm - \partial_{\xi_1}\nu^\pm = 0 .
$$

Moreover, if $\nu^- = 0$ in $\{\xi_1 < 0\}$, then the jump relations through $\xi_1 = 0$ are

$$
(18) \quad \nu^+ \big|_{\xi_1=0^+} = \left(1 - e^{-\pi\eta^2}\right)\nu^+ \big|_{\xi_1=0^-} \; ; \; \nu^- \big|_{\xi_1=0^+} = e^{-\pi\eta^2}\nu^+ \big|_{\xi_1=0^-} .
$$

Remarks. 1. Formulae (18) describe explicitly which part of the energy is transmitted from the mode $+$ to the mode $-$ when passing through the crossing set, in terms of the rescaled momentum η. The assumption that $\nu^- = 0$ in the past can be relaxed (see [6]). However, it is important to make some assumption which implies that the incoming particles on both modes do not interact too much.

2. In this specific example, it may seem strange to try to derive the evolution of Wigner measures, while a very precise description of the solution itself is available through the above Landau–Zener scattering result. However, in the case of a general potential which we shall study in the next section, this explicit description is nomore available, while our analysis of Wigner measures remains valid, as we shall see.

4. The general case. Let us come back to a general smooth potential function $V = V(x)$ and assume $\nabla V(x) \neq 0$ at least at points x near which we study measure μ. In order to generalize results of the previous section to this case, we first need to find a variable which will play the role of ξ_2, in order to define a rescaled momentum which fits the geometry induced by V. This can be done by solving some eikonal equation.

LEMMA 2.   *Let $x_0 \in \mathbf{R}^2$ such that $\nabla V(x_0) \neq 0$. Near $(x,\tau) = (x_0, \tau_0 = -V(x_0))$, there exists a unique solution $\phi = \phi(x,\tau)$ of the following Hamilton–Jacobi equation*

$$
(19) \qquad |\nabla_x \phi(x,\tau)|^2 = (\tau + V(x))^2
$$

with the additional prescription

$$(20) \qquad \phi(x, \tau) = (\tau + V(x))^2 \, g(x, \tau) \, ,$$

where g is a positive smooth function.

The proof of Lemma 2 relies on the possibility to solve the Cauchy problem for the classical motion on modes $\pm$,

$$\dot{x}^{\pm}(t) = \pm \frac{\xi^{\pm}}{|\xi^{\pm}|} \, , \quad \dot{\xi}^{\pm}(t) = -\nabla V(x^{\pm})$$

with singular data $x \pm (0) = x$, $\xi^{\pm}(0) = 0$, provided $\nabla V(x) \neq 0$. Observe that these curves are not smooth through $t = 0$, since

$$\dot{x}^{\pm}(0^+) = -\dot{x}^{\pm}(0^-) = \mp \nabla V(x) \, .$$

However, it turns out that the $t > 0$ part of $x^{\pm}$ matches smoothly with the $t < 0$ part of $x^{\mp}$, so that, as x varies in the curve $\{\tau + V(x) = 0\}$, we get, for every τ, two surfaces $\Lambda_{\tau}, \Lambda'_{\tau}$ in $\mathbf{R}^4_{x, \xi}$. On can show that these surfaces are the graphs of $\nabla_x \phi$ and $-\nabla_x \phi$ for some function ϕ which satisfies equation (19). Moreover, it is possible to prove that all the other smooth solutions of (19) are $\pm \phi +$ constant, so that formula (20) with $g > 0$ characterizes ϕ.

Let ϕ be the function given by Lemma 2. There exists a smooth function $\omega = \omega(x, \tau)$ such that $|\omega| = 1$ and

$$\nabla_x \phi(x, \tau) = (\tau + V(x)) \omega(x, \tau) \, .$$

Observe that

$$\omega(x, -V(x)) = \frac{\nabla V(x)}{|\nabla V(x)|} \, .$$

Then we define

$$I = \{\xi \wedge \omega = 0\}$$

This hypersurface will play the role played by $\{\xi_2 = 0\}$ in the previous section. Notice that I meets the energy surface $\{(\tau + V(x))^2 = |\xi|^2\}$ exactly on the union $\Lambda \cup \Lambda'$ of the classical trajectories issued from the singular set $\{\xi = 0\}$. We can now define the two–scaled Wigner measure associated to I.

PROPOSITION 2. *There exists a positive measure ν on $I \times \overline{\mathbf{R}}$ and a subsequence (ϵ_k) such that*

$$\int_{(\mathbf{R} \times \mathbf{R}^2)^2} W^{\epsilon} \psi^{\epsilon}(t, x, \tau, \xi) \, a\left(t, x, \tau, \xi, \frac{\xi \wedge \omega(x, \tau)}{\sqrt{\epsilon}}\right) dt \, dx \, d\tau \, d\xi$$

$$\xrightarrow[\epsilon \to 0, \epsilon = \epsilon_k]{} \int_{I \times \overline{\mathbf{R}}_{\eta}} a(t, x, \tau, \xi, \eta) \, d\nu(t, x, \tau, \xi, \eta)$$

$$+ \int_{\xi \wedge \omega \neq 0} a(t, x, \tau, \xi, \mathrm{sgn}(\xi \wedge \omega(x, \tau))\infty) \, d\mu(t, x, \tau, \xi).$$

The proof of Proposition 2 is perceptibly more involved than the one of Proposition 1, due to the fact that the new momentum variable $\xi \wedge \omega$ depends on both variables x, ξ. Observe that ν is defined locally near $x = x_0, \xi = 0, \tau = -V(x_0)$ and describes the concentration, at the scale $\sqrt{\epsilon}$, of the Wigner transform of ψ^ϵ on the classical trajectories issued from the crossing set $\{\xi = 0\}$.

We can now state our main result, which generalizes Theorem 1.

THEOREM 2. *The measure ν can be decomposed as*

(21)
$$\begin{aligned}
\nu(t, x, \tau, \xi, \eta) &= \nu^+(t, x, \xi, \eta)\delta(\tau+ \mid \xi \mid +V(x)) \\
&+ \nu^-(t, x, \xi, \eta)\delta(\tau- \mid \xi \mid +V(x)) ,
\end{aligned}$$

where $\nu^\pm$ are positive measures satisfying the following transport equations in $\xi \neq 0$,

$$\partial_t \nu^\pm \pm \frac{\xi}{\mid \xi \mid} \cdot \nabla_x \nu^\pm - \nabla V(x) \cdot \nabla_\xi \nu^\pm \pm (\nabla \cdot \omega) \partial_\eta(\eta \nu^\pm) = 0$$
$$on \quad \{\xi \cdot \nabla V(x) > 0\}$$

$$\partial_t \nu^\pm \pm \frac{\xi}{\mid \xi \mid} \cdot \nabla_x \nu^\pm - \nabla V(x) \cdot \nabla_\xi \nu^\pm \mp (\nabla \cdot \omega) \partial_\eta(\eta \nu^\pm) = 0$$
$$on \quad \{\xi \cdot \nabla V(x) < 0\}$$

Moreover, if $\nu^- = 0$ in $\{\xi \cdot \nabla V(x) < 0\}$, then the jump relations through $\xi \cdot \nabla V(x) = 0$ are

(22)
$$\begin{aligned}
\nu^+ \mid_{\xi \cdot \nabla V(x)=0+} &= (1 - T)\nu^+ \mid_{\xi \cdot \nabla V(x)=0-} ; \\
\nu^- \mid_{\xi \cdot \nabla V(x)=0+} &= T\nu^+ \mid_{\xi \cdot \nabla V(x)=0-} ,
\end{aligned}$$

with

$$T = T(x, \eta) = e^{-\pi \eta^2 / |\nabla V(x)|} .$$

Let us briefly review the main steps of the proof of Theorem 2. Propagation equations outside $\{\xi = 0\}$ can be proved essentially in the same way as equations (11) for the usual Wigner measure, so the main part is the proof of Landau–Zener type formulae (22).

The first step consists in observing that there exists a local symplectic change of coordinates from $\mathbf{R}^3_{t,x} \times \mathbf{R}^3_{\tau,\xi}$ to $\mathbf{R}^3_{s,z} \times \mathbf{R}^3_{\sigma,\zeta}$ such that

$$\begin{aligned}
s &= \lambda(x, \tau)\omega(x, \tau) \cdot \xi, & z_1 &= \tau, \\
\sigma &= -\lambda(x, \tau)(\tau + V(x)), & \zeta_2 &= \lambda(x, \tau)\mu(x, \tau)\omega(x, \tau) \wedge \xi ,
\end{aligned}$$

where $\lambda = \sqrt{2g}$, g as in Lemma 2, and μ is any positive smooth function of (x, τ) such that

$$\nabla \cdot \left(\frac{\omega}{\lambda \mu}\right) = 0 .$$

Applying to the unknown ψ^ϵ a rotation in the direction of vector ω and then a Fourier integral operator associated to the above canonical transformation, we are led to the following system near $s = \sigma = \zeta = 0$,

$$(23) \qquad \frac{\epsilon}{i}\partial_s v = \begin{pmatrix} s & \mathrm{op}_\epsilon(\alpha(z,\sigma)\zeta_2) \\ \mathrm{op}_\epsilon(\alpha(z,\sigma)\zeta_2) & -s \end{pmatrix} v,$$

where $\alpha(z,\sigma) = 1/\mu$ in the new variables, and

$$\mathrm{op}_\epsilon(\alpha(z,\sigma)\zeta_2)f = -\frac{i\epsilon}{2}\left(\alpha(z,-i\epsilon\partial_s)\partial_{z_2}f + \partial_{z_2}(\alpha(z,-i\epsilon\partial_s)f)\right).$$

Observe that the hypersurface I is now $\{\zeta_2 = 0\}$ in these new coordinates, so that the rescaled momentum is $\eta = \zeta_2/\sqrt{\epsilon}$. Unfortunately, system (23) is not yet system (13), because α is actually depending on variable σ (in particular). As a consequence, we are forced to follow two different strategies according to the values of the rescaled momentum η:

a) In the zone $\eta = \pm\infty$, we expect, in view of formulae (22), that no transmission from one mode to another one occurs. We can prove this fact in the widest generality, namely without assuming any fact about the incoming particles. The proof is a refinement of geometric optics transport equations of [8] using sharp estimates on pseudodifferential operators (see [11]).

b) In the zone $|\eta| < +\infty$, it is possible to take advantage of the smallness of variable ζ_2 by proving that system (23) can be conjugated to system (13). Similar techniques in different contexts are used in [1, 14, 20]. Using Theorem 1, this gives formulae (22) in this zone.

Finally, results of part a) and b) yields Theorem 2. Observe that, keeping track of the solution through the various changes of variables and of unknowns, it is possible to have a rather precise —but complicated— description of the solution ψ^ϵ itself in the zone $|\eta| < +\infty$. However, it is not clear how to match it with the refined geometric optics in part a) in order to get a complete description of ψ^ϵ. Hence the viewpoint of Wigner measures appears to be much more flexible.

REFERENCES

[1] S. Alinhac: Branching of singularities for a class of hyperbolic operators. *Indiana Univ. Math. J.*, **27**, N° 6 (1978), pp. 1027–1037.

[2] P. Bechouche and F. Poupaud: Semiclassical limit in a stratified medium, to appear in *Monatshefte für Mathematik* (2000).

[3] A. P. Calderón, R. Vaillancourt: On the boundedness of pseudo-differential operators. *J. Math. Soc. Japan*, **23**, N° 2 (1971), pp. 374–378.

[4] Y. Colin de Verdière: Sur les singularités de Van Hove génériques, *Bull. Soc. Math. de France*, **119**, Mémoire 46 (1991), pp. 99–109.

[5] Y. Colin de Verdière, M. Lombardi, and J. Pollet: The microlocal Landau-Zener formula. *Ann. Inst. Henri Poincaré*, **71**, N° 1 (1999), pp. 95–127.

[6] C. Fermanian Kammerer and P. Gérard: Mesures semi–classiques et croisements de modes, preprint, to appear in *Bull. S.M.F.*

[7]   P. Gérard: Mesures semi-classiques et ondes de Bloch, Exposé de l'Ecole Polytechnique, E.D.P., Exposé N°XVI (1991).

[8]   P. Gérard, P.A. Markowich, N.J. Mauser, and F. Poupaud: Homogenization Limits and Wigner Transforms. *Comm. Pure Appl. Math.*, **50**(4) (1997), pp. 323–379.

[9]   G.A. Hagedorn: Proof of the Landau-Zener formula in an adiabatic limit with small eigenvalue gaps. *Commun. Math. Phys.*, **136** (1991), pp. 433–449.

[10] G.A. Hagedorn and A. Joye: Landau-Zener transitions through small electronic eigenvalue gaps in the Born-Oppenheimer approximation. *Ann. Inst. Henri Poincaré*, **68**, N° 1 (1998), pp. 85–134.

[11] L. Hörmander: The analysis of linear Partial Differential Operators III. Springer-Verlag (1985).

[12] A. Joye: Proof of the Landau-Zener formula. *Asymptotic Analysis*, **9** (1994), pp. 209–258.

[13] Y. Karpeshina: Perturbation theory for the Schrödinger operator with a periodic potential, *Lecture Notes in Mathematics*, **1663**, Springer, 1997.

[14] N. Kaidi and M. Rouleux: Forme normale d'un hamiltonien à deux niveaux près d'un point de branchement (limite semi-classique), *C.R. Acad. Sci. Paris* Série I Math, **317** (1993), N° 4, pp. 359–364.

[15] L. Landau: *Collected papers of L. Landau*, Pergamon Press (1965).

[16] P-L. Lions and T. Paul: Sur les mesures de Wigner. *Revista Matemática Iberoamericana*, **9** (1993), pp. 553–618.

[17] L. Miller: Propagation d'onde semi-classiques à travers une interface et mesures 2-microlocales. *Thèse de l'Ecole Polytechnique*, 1996.

[18] P.A. Markowich, N.J. Mauser, and F. Poupaud: A Wigner function approach to semi-classical limits: electrons in a periodic potential, *J. Math. Phys.*, **35** (1994), pp. 1066–1094.

[19] F. Poupaud and C. Ringhofer: Semi-classical limits in a crystal with exterior potentials and effective mass theorems, *Comm. Part. Diff. Eq.*, **21** (1996), N° 11–12, pp. 1897–1918.

[20] M. Rouleux: Tunelling effects for h pseudodifferential operators, Feshbach resonances, and the Born-Oppenheimer approximation, *Evolution equations, Feshbach resonances, Hodge theory*, pp. 131–242, Math. Top., **16**, Wiley-VCH, Berlin (1999).

[21] C. Wilcox: Theory of Bloch waves, *Journal d'Analyse Mathématique*, **33** (1978), pp. 146–167.

[22] C. Zener: Non-adiabatic crossing of energy levels, *Proc. Roy. Soc. Lond.*, **137** (1932), pp. 696–702.

MESOSCOPIC MODELING OF SURFACE PROCESSES[*]

MARKOS A. KATSOULAKIS[†] AND DIONISIOS G. VLACHOS[‡]

Abstract. In this paper we discuss mesoscopic models describing a broad class of pattern formation mechanisms, focusing on a prototypical system of surface processes. These models are in principle stochastic integrodifferential equations and are derived through an exact coarse-graining, directly from microscopic lattice models, and include detailed microscopic information on particle-particle interactions and particle dynamics.

1. Introduction. Surface processes, such as catalysis, chemical vapor deposition and epitaxial growth, typically involve transport and chemistry of precursors in a gas phase; unconsumed reactants and radicals adsorb onto the surface of a substrate where numerous processes may take place simultaneously, for instance surface diffusion, reaction(s) and desorption back to the gas phase (see Fig. 1(a)). Surface processes have traditionally been modeled using continuum-type reaction diffusion models [1, 2], where the adsorptive layer has been assumed to be spatially uniform. This approach either neglects detailed interactions between particles or treats them phenomenologically, while on the other hand non-equilibrium statistical mechanics theories provide an exact microscocpic description [3]. The mathematical tools employed in the statistical mechanics models are Interacting Particle Systems (IPS), which are Markov processes set on a lattice corresponding to a solid surface; typical examples are the Ising-type systems [4], describing the evolution of an order parameter at each lattice site. Microscopic models set in the continuum space or on a lattice are solved numerically using molecular dynamics (MD) or Monte Carlo (MC) algorithms [5, 4]. Despite their widespread use, these methods are currently limited to short length and time scales, whereas morphological features seen in experiments or device sizes often invoke much larger space and/or time length scales [6]. This disparity underscores the need to develop theories for larger scales, which take in consideration microscopic details rather than relying only on phenomenology. Recently, coarse-grained, *mesoscopic*, models of Ising systems were developed, when the Hamiltonian describes spin exchange, spin flip, or combination of the two mechanisms [7–13]. Through homogenization and asymptotics, the underlying macroscopic laws of interface velocity, surface tension, mobility, and critical nucleus size have also been derived [9, 14–16] and shown to have a direct dependence on the interaction potential. Thus a direct analytical link of molecular interactions with macroscopic laws of transport and thermodynamics can

---

[*]The research of MAK is partially supported by NSF DMS-9626804, NSF DMS-9801769 and NSF DMS-0079536, and the research of DGV is partially supported by the NSF CTS-9702615, NSF CTS-9904242 and NETI.

[†]Department of Mathematics and Statistics, University of Massachusetts, Amherst, MA 01003.

[‡]Department of Chemical Engineering, University of Delaware, Newark, DE 19716.

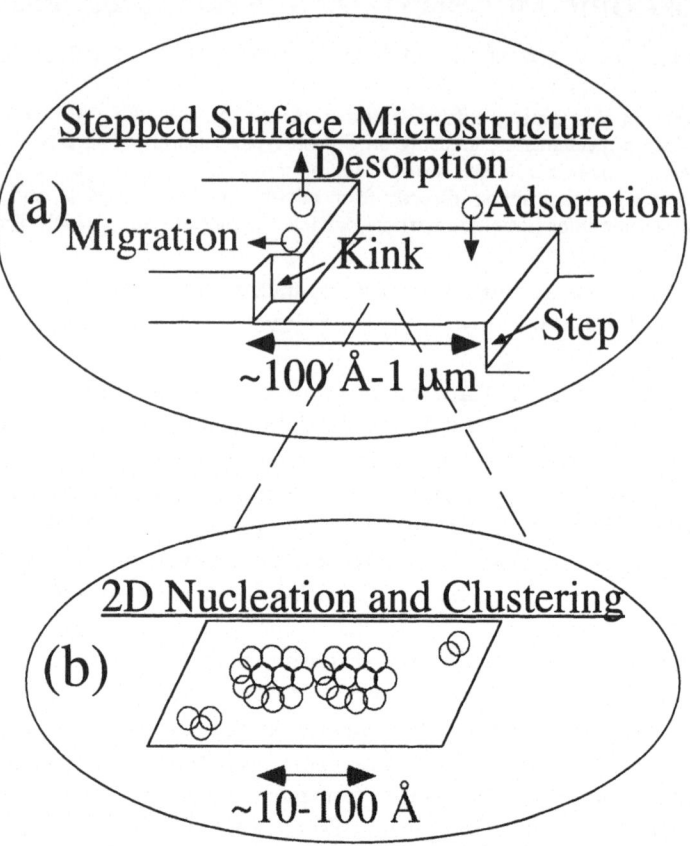

FIG. 1. *(a) Microscopic Mechanisms on a stepped surface: Adsorption of an atom (dotted sphere) from the fluid phase, desorption of the adsorbate back to the fluid, and migration of adsorbate along the surface. (b) Nucleation and clustering of particles due to the particle/particle attractive lateral interactions.*

be established, enabling comparison of experimentally measurable quantities to intermolecular forces on one hand, and very efficient evaluation of different interaction potentials on the other. In contrast to well-known phenomenological theories, e.g. [17–20] (where an ad-hoc truncation is involved), the mesoscopic models discussed here are derived *exactly* from the microscopic processes in the local mean field limit, i.e. when the interaction potential range becomes infinite, while the coarse grained coverage variable still captures local space/time variations. The issue of an infinite range potential may appear to be too restrictive, however a first comparison of results from the mesoscopic theory with gradient Monte Carlo simulations under far from equilibrium conditions indicates that only a relatively short range potential is adequate for quantitative agreement [12]. Furthermore, asymptotics indicate that the deviation of the solution of the mesoscopic theory from the infinite range potential decays exponentially, due to an underlying Large Deviation Principle, and decreases with the square root of the problem dimensionality; as a result, mesoscopic theories for 2D and 3D lattices require even shorter-range interaction potentials [21].

Since mesoscopic models are directly and explicitly related to IPS and the corresponding Monte-Carlo algorithms, they also share with them the same versatility. For instance a variety of interaction potentials including anisotropic ones (see [9] for a related study of Ising spin flip dynamics) can be employed in mesoscopic models, which is very important in numerous examples of adsorbate-adsorbate interactions on crystal surfaces and in epitaxial growth of materials. Furthermore, different microscopic dynamics of diffusion can be studied (e.g., Metropolis, Kawasaki, Arrhenius, etc.), and simultaneous microscopic processes (e.g., diffusion coupled with adsorption, desorption, and reaction) can be systematically modeled. A first testimony of the versatility of mesoscopic theories is their application to different systems ranging from nanoscale patterns in surface reaction adlayers on catalyst surfaces, to sintering of alloys, to diffusion on crystal surfaces and through membranes [12, 22, 23]. Another compelling feature of mesoscopic theories is the inclusion of random fluctuations directly derived from the underlying master equation. In contrast to the ad hoc addition of a white noise term in the conservation equation (e.g., Langevin equation), multiple noise terms are incorporated, each one associated with a single microscopic process [10, 12]. These noise terms, which are compatible with the fluctuation-dissipation principle, are crucial at the mesoscale and play an important role in phase transitions and nucleation of islands of adsorbates on surfaces, as well as pattern formation and selection, to mention a few.

Numerical methods for mesoscopic equations were developed in [10] and [24] based on finite-difference schemes that involve the direct numerical simulation of possibly long range interaction terms. To overcome such difficulties we developed in [25] spectral algorithms for the solution of the mesoscopic equations presented here, exhibiting enhanced computational efficiency for a wide range of intermolecular potentials, as well as fast relaxation to steady state.

The organization of this paper is as follows. In Section 2 we present the main microscopic mechanisms for surface processes and derive corresponding mesoscopic theories. We also discuss their validity and relations to phenomenological approaches. Scaling laws, cluster evolution and pattern formation mechanisms are discussed in Section 3 from a microscopic/mesoscopic perspective.

2. Microscopic derivations of mesoscopic models.

2.1. Microscopic mechanisms.
Surface processes are modeled at the microscopic level by employing dynamic Ising type systems. These models are IPS defined on a d-dimensional lattice $\mathbb{Z}^d$. At each lattice

site $x \in \mathbb{Z}^d$ an order parameter is allowed to take the values 0 and 1 describing vacant and occupied sites respectively. In accordance to the classical Ising model, we refer to the order parameter as spin. A spin configuration σ is an element of the configuration space $\Sigma = \{0, 1\}^{\mathbb{Z}^d}$; we write $\sigma = \{\sigma(x) : x \in \mathbb{Z}^d\}$ and call $\sigma(x)$ the spin at x. The energy H of the system, evaluated at σ, is given by a Hamiltonian

$$H(\sigma) = \sum_{x \neq y} J(x, y)\sigma(x)\sigma(y) + h \sum \sigma(x),$$

where h is attributed to an external field and $J = J^\gamma$ is the intermolecular potential defined by

(2.1) $$J(x, y) = J^\gamma(x, y) = \gamma^d J(\gamma(x - y)) \qquad x, y \in \mathbb{Z}^d,$$

$\gamma^{-1} > 0$ being the interaction range. Here J is assumed to be nonnegative, even, i.e. $J(r) = J(-r) \geq 0$, decays rapidly at infinity and is nonnegative, i.e. the interactions are *attractive*. The scaling in (2.1) guarantees the summability of the Hamiltonian H, provided $J \in L^1(\mathbf{R}^d)$. The assumption that J is nonnegative is an important one from the physical point of view, since it implies that clusters of particles are energetically preferred to totally disordered structures (see Fig. 1(b)).

Equilibrium states of the Ising model are described by the Gibbs states at the prescribed temperature T,

(2.2) $$\mu_\Lambda(d\sigma) = \frac{1}{Z_{\beta, \Lambda}} \exp(-\beta H(\sigma))d\sigma$$

where $\beta = \frac{1}{kT}$, k being the Boltzmann constant and $Z_{\beta, \Lambda}$ is a normalizing constant (partition function) so that μ_Λ is a probability measure defined on the configuration space $\Sigma = \{0, 1\}^\Lambda$, where Λ is an expanding (as $|\Lambda| \to \infty$) finite box on the infinite lattice, with specified boundary conditions. It is well known that phase transitions, i.e. non-uniqueness of the Gibbs measures, may occur at low temperatures, in the infinite volume limit [26].

The dynamics of the model consists of a sequence of flips and spin exchanges that correspond to different physical processes, as shown schematically in Figure 1.

2.1.1. Adsorption/desorption-spin flip mechanism. A spin flip at the site x is a spontaneous change in the order parameter, 1 is converted to 0 and vice versa. Physically this mechanism describes the desorption of a particle from the surface to the gas phase and conversely the adsorption of a particle from the gas phase to the surface (see Figure 1).

If σ denotes the configuration prior to a flip at x, then after the flip the configuration is denoted by σ^x where

$$\sigma^x(y) = \begin{cases} 1 - \sigma(x), & \text{when} \quad y = x, \\ \sigma(y), & \text{when} \quad y \neq x. \end{cases}$$

We assume that a flip occurs at x, when the configuration is σ, with a rate $c(x,\sigma)$ i.e. a spin flip occurs at x, during $[t, t + \Delta t]$ with probability $c(x,\sigma)\Delta t + O(\Delta t^2)$. Rigorously, the underlying stochastic process $\{\sigma_t\}_{t \geq 0}$ is a jump Markov process on $L^\infty(\Sigma; \mathbf{R})$ with generator given by

$$(2.3) \qquad L_\gamma^{ad} f(\sigma) = \sum_{x \in \mathbf{Z}^N} c(x,\sigma)[f(\sigma^x) - f(\sigma)], \qquad f \in L^\infty(\Sigma; \mathbf{R}).$$

An obvious requirement on the resulting dynamics is that, when restricted on a finite dimensional box Λ, they should leave the Gibbs measure (2.2) invariant. This condition is called a *detailed balance* law, and is equivalent to [27],

$$(2.4) \qquad c(x,\sigma) = c(x,\sigma^x) \exp(-\beta \Delta_x H(\sigma)).$$

Here $\Delta_x H(\sigma) = H(\sigma^x) - H(\sigma)$ is the energy difference after performing a spin flip at the site x. The simplest type of dynamics satisfying (2.4) is

$$(2.5) \qquad c(x,\sigma) = \Psi(-\beta \Delta_x H(\sigma)),$$

yielding the relation on Ψ, $\Psi(r) = \Psi(-r)e^{-r}$, $r \in \mathbf{R}$. Typical choices of Ψ's are $\Psi(r) = (1 + e^r)^{-1}$ (Glauber dynamics), $\Psi(r) = e^{-r/2}$ or $\Psi(r) = e^{-r^+}$ (Metropolis dynamics).

2.1.2. Surface diffusion–spin exchange dynamics. A spin exchange between the neighboring sites x and y is a spontaneous exchange of the values of the order parameter at x and y. Physically this mechanism describes the diffusion of a particle on a flat surface (see Figure 1). Note that sites cannot be occupied by more than one particle (exclusion principle). As in the spin flip dynamics, a spin exchange occurs with rate $c(x,y,\sigma)$ satisfying the detailed balance law

$$(2.6) \qquad c(x,y,\sigma) = c(x,y,\sigma^{(x,y)}) \exp(-\beta \Delta_{(x,y)} H(\sigma)),$$

where $\sigma^{(x,y)}$ is the new configuration after a spin exchange between sites x and y

$$\sigma^{(x,y)}(z) = \begin{cases} \sigma(y), & \text{when} \quad z = x, \\ \sigma(x), & \text{when} \quad z = y, \\ \sigma(z), & \text{otherwise.} \end{cases}$$

Furthermore $\Delta_{x,y} H(\sigma) = H(\sigma^{(x,y)}) - H(\sigma)$, is the energy difference after performing a spin exchange between the neighboring sites x and y. The Hamiltonian H associated with diffusion may have a different intermolecular potential J than adsorption. The resulting stochastic process $\{\sigma_t\}_{t \geq 0}$ is a jump Markov process on $L^\infty(\Sigma; \mathbf{R})$ with generator given by

$$(2.7) \qquad L_\gamma^d f(\sigma) = \sum_{x \in \mathbf{Z}^d} c(x,y,\sigma)[f(\sigma^{(x,y)}) - f(\sigma)].$$

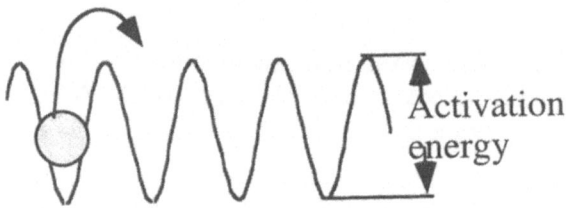

FIG. 2. *One-dimensional cut of the potential energy surface describing species-crystal interactions in the limit of zero concentration ($J_m = J_d = 0$). Species diffuse from minima of the potential energy to adjacent sites by overcoming an energy barrier known as activation energy.*

The simplest type of dynamics satisfying (2.6) is

$$(2.8) \qquad \begin{aligned} &c(x, y, \sigma) = \\ &\begin{cases} \Psi(-\beta \Delta_{x,y} H(\sigma)), & \text{when } x \text{ and } y \text{ are nearest neighbors,} \\ 0, & \text{otherwise,} \end{cases} \end{aligned}$$

where $\Psi(r) = \Psi(-r)e^{-r}, r \in \mathbf{R}$. Typical choices of Ψ's are $\Psi(r) = 2(1 + e^r)^{-1}$ (Kawasaki dynamics) and $\Psi(r) = e^{-r^+}$ (Metropolis dynamics).

2.1.3. Arrhenius dynamics. In most Monte Carlo simulations, motion of species is performed according to Metropolis or Kawasaki dynamics [4]. For such dynamics, the energy barrier for diffusion depends only on the energy difference between the initial and final states, often known as the heat of the process. Since the equilibrium-state of a system is independent of dynamics, different choices of microscopic dynamics result in the same long time solution due to the detailed balance condition. However, time dependent solutions and the time needed to approach equilibrium depend on the details of microscopic dynamics. It is then more natural to describe the activation energy of surface diffusion as the energy barrier a species has to overcome in jumping from one site to another [28, 29]. This activation energy corresponds (omitting the zero point energy difference for clarity) to the difference between the minimum and maximum energies shown in Fig. 2. Similarly are handled the adsorption/desorption mechanisms. This class of dynamics is defined as Arrhenius-type.

The Arrhenius adsorption/desorption (spin flip) rate is given by

$$(2.9) \qquad c(x, \sigma) = \begin{cases} c_0 \exp\left[-\beta(U_0 + U(x))\right], & \text{when} \quad \sigma(x) = 1. \\ c_0, & \text{when} \quad \sigma(x) = 0. \end{cases}$$

The Arrhenius surface diffusion (spin exchange) rate is given for nearest neighbors x and y by

$$(2.10) \qquad \begin{aligned} &c(x, y, \sigma) = \\ &\begin{cases} c_0 \exp\left[-\beta(U_0 + U(x))\right], & \text{when} \quad \sigma(x) = 1, \quad \sigma(y) = 0, \\ c_0 \exp\left[-\beta(U_0 + U(y))\right], & \text{when} \quad \sigma(x) = 0, \quad \sigma(y) = 1, \\ 0, & \text{otherwise}, \end{cases} \end{aligned}$$

where in both formulae

$$U(x) = \sum_{z \neq x} J(x - z)\sigma(z),$$

is the total energy contribution from the particle interactions with the particle located at the site x, while U_0 is the energy associated with the surface binding of the particle at x (c_0 is a rate constant that can be chosen arbitrarily). Both spin flip and spin exchange dynamics satisfy the detailed balance law. A more complex dependence of the activation energy on the energetics of adjacent sites is also possible, e.g. Arrhenius parabolic jump models [28, 29].

2.2. Mesoscopic models. Here we present mesoscopic theories for each of the microscopic models we introduced earlier, as well as combinations of such mechanisms. At large space/time scales and for long range potentials, it turns out that the small scale fluctuations of the Ising systems are suppressed and an almost deterministic pattern emerges described by suitable, possibly stochastic, integrodifferential equations. The passage in the limit $\gamma \to 0$, (the interaction range is γ^{-1}, see (2.1)), which in the physics literature is identified with coarse graining, of quantities like the thermodynamic pressure, total coverage, etc. is known as the Lebowitz-Penrose limit [30]. Along these lines we study the asymptotics, as $\gamma \to 0$, of the averaged coverage

$$u_\gamma(x, t) = E_{\mu^\gamma} \sigma_t(x), \qquad (x, t) \in \mathbb{Z}^d \times [0, \infty)$$

of the system, where E_{μ^γ} denotes the expectation of the IPS starting from a measure μ^γ. Similarly one may study the related asymptotic limit of a suitable averaged in space occupation

$$v_\gamma(x, t) = \frac{1}{|B_x|} \sum_{y \in B_x} \sigma_t(y)$$

where B_x is ball centered at x with radius $R = \gamma^{-a} \ll \gamma^{-1}$ where γ^{-1} is the interaction range and $0 < a < 1$. Through the scaling $R = \gamma^{-a}$, the ball B_x where the averaging is carried out contains enough points so that the random fluctuations will be suppressed due the Law of Large Numbers, while at the same time spatial variations in the coverage are captured since the averaging is performed over regions relatively smaller than the interaction range. Thus, as $\gamma \to 0$, $u_\gamma(x,t) - v_\gamma(x,t)$ converges to zero; in addition there is a normally distributed correction to $v_\gamma(x,t)$ of order $O(\gamma^{d/2})$ as in the Central Limit Theorem. We refer to the review article [31] and references therein on the rigorous asymptotic limits of Kawasaki/Metropolis-type dynamics; here we present only the formal derivation of the mesoscopic equations for each of the micromechanisms discussed earlier.

2.2.1. Adsorption/desorption-spin flip mechanism. The generator (2.3) yields that the averaged coverage $u_\gamma(x,t) = E_{\mu^\gamma}\sigma_t(x)$ solves

$$(2.11) \qquad \frac{d}{dt}E_{\mu^\gamma}\sigma_t(x) = E_{\mu^\gamma}(1 - 2\sigma_t(x))c(x,\sigma_t).$$

When the interparticle potential is weak and long range, the fluctuations of $\sigma_t(z)$ around their averages are approximately independent, the Law of Large Numbers formally applies and, as $\gamma \to 0$,

$$(2.12) \qquad \sum_{z\neq x} J(x-z)\sigma_t(z) \approx \sum_{z\neq x} J(x-z)E_{\mu^\gamma}\sigma_t(z).$$

In addition there is a normally distributed correction of order $O(\gamma^{d/2})$ in a d-dimensional lattice due to the Central Limit Theorem. Here we ignore such random corrections but in principle they would give rise to a stochastic PDE instead of the deterministic equations below (see Section 2.2.4). Back in (2.9), we substitute the spin flip rate (2.5), and using (2.10) we obtain, as $\gamma \to 0$, $u_\gamma(x,t) = E_{\mu^\gamma}\sigma_t(x) \approx u(\gamma x, t)$, and u solves [9],

$$(2.13) \qquad u_t = \Psi(-\beta(J*u+h))[1 - u - \exp(-\beta h)u\exp(-\beta J*u)].$$

Similarly, for the Arrhenius adsorption/desorption dynamics we have the mesoscopic equation [25],

$$(2.14) \qquad u_t = c_0[1 - u - \exp(-\beta h)u\exp(-\beta J*u)].$$

It is easy to see that u can be also viewed as the probability density of the coverage.

Next we review some basic properties of (2.11) and (2.12). First, both equations are equipped with a comparison principle, at least when $J \geq 0$. Steady state solutions of either equations satisfy the algebraic equation

$$(2.15) \qquad f(x) := \alpha(1-x) - xe^{-\lambda x} = 0,$$

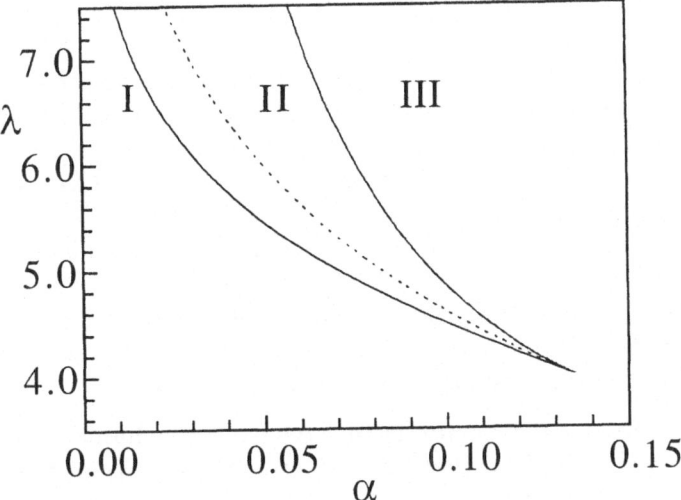

FIG. 3. *Phase diagram for the system in (2.11), (2.12), and (2.21), (2.22) when $k_r = 0$. Within the cusp (Region II), the system is bistable and thus may have regions that are dilute while other regions are dense; outside this cusp, the system tends to be either dilute or dense. The line of stationary coexistence is the dotted line within Region II.*

where $\alpha = \exp(\beta h)$ and $\lambda = J_0\beta$, $J_0 = \int J(r)dr$. Fig. 3 shows schematically the phase diagram as a function of the external field α, and the interaction parameter λ. Outside the cuspy envelope, intermolecular forces are either weak or very strong resulting in a single-valued isotherm corresponding to the single root of (2.13). Region I corresponds to a dilute phase and Region III to a dense phase. Within the cuspy envelope (Region II), both phases may exist. In this case (2.13) has three solutions $m_- = m_-(\alpha, \lambda) < m_0 = m_0(\alpha, \lambda) < m_+ = m_+(\alpha, \lambda)$, where m_+ and m_- correspond to the dense and the dilute phases of the system respectively. The dynamics of cluster growth with Region II depends on the relative parameter location with respect to the *stationary coexistence* curve (the dotted line in Fig. 3), given by $\alpha = e^{-\lambda/2}$. In this last case the roots of (2.13) become $m_0 = \frac{1}{2}$ and $m_\pm = \frac{1}{2} \pm \nu$, for some $0 < \nu < \frac{1}{2}$.

Standing and traveling waves for (2.11) and (2.12) play an important role in the long space/time asymptotics of mesoscopic equations [9, 16], since they connect high and low density phases, across a cluster boundary. They are special solutions of the type $u(r, t) = q(r \cdot e - ct)$, $r \in \mathbf{R}^n$, where e is any unit vector, $q = q(\xi)$, $\xi \in \mathbf{R}$ and $c = c(\alpha, \lambda)$ denotes the speed of the wave. Equations (2.11) and (2.12) have identical standing waves $(c = c(\alpha, \lambda) = 0)$ when the parameters (α, λ) lie on the line of stationary coexistence $\alpha = e^{-\lambda/2}$ and the corresponding wave q satisfies

$$(2.16) \qquad -\tilde{J} * q + \frac{1}{\beta} \ln\left(\frac{q}{1-q}\right) = -\frac{1}{2}J_0\,, \qquad q(\pm\infty) = m_\pm\,.$$

In addition, $\dot{q} > 0$, where $(\cdot)$ denotes the derivative with respect to the ξ variable. Here we have assumed that the interaction potential J is radially symmetric, i.e. $J(r) = J(|r|) \geq 0$, $r \in \mathbf{R}^n$ and set $\tilde{J}(\zeta) = \int_{\mathbf{R}^{n-1}} J((\zeta^2 + |r'|^2)^{1/2})dr'\,, \zeta \in \mathbf{R}$. It also turns out that (2.11) and (2.12) admit dynamics-dependent monotone traveling wave solutions. The rigorous existence, uniqueness and stability of such solutions follows from the analysis in [32], which covers a broad class of integrodifferential equations that admit a comparison principle. Other related works include [33, 34]. Note that for even potentials which are not necessarily radial we obtain direction-dependent standing and traveling waves [9].

2.2.2. Surface diffusion-spin exchange mechanism. As before, the generator (2.7) yields that the averaged coverage $E_{\mu\gamma}\sigma_t(x)$ solves

$$(2.17) \qquad \frac{d}{dt}E_{\mu\gamma}\sigma_t(x) = \sum_{y \in N(x)} E_{\mu\gamma}(\sigma_t(y) - \sigma_t(x))c(x,y,\sigma)\,,$$

where $N(x)$ denotes the nearest neighbors of x. Reasoning as in the adsorption/ desorption case we also have that $\sum_{z \neq x} J(x-z)\sigma_t(z) \approx \sum_{z \neq x} J(x - z)E_{\mu\gamma}\sigma_t(z)$. Rescaling time as $t \mapsto t\gamma^{-2}$ in (2.15) and using the approximate independence of different lattice sites for $\gamma \ll 1$, we obtain for the dynamics (2.8) that, as γ vanishes, $u_\gamma(x,t) = E_{\mu\gamma}\sigma_{t\gamma^{-2}}(x) \approx u(\gamma x,t)$, and u solves, [13],

$$(2.18) \qquad u_t - D\nabla \cdot \left[\nabla u - \beta u(1-u)\nabla J * u\right] = 0\,.$$

where $D = \Psi(0)$. For the Arrhenius diffusion mechanism we obtain the mesoscopic equation [12],

$$(2.19) \qquad u_t - \nabla \cdot \left\{ D\exp(-\beta J * u)\left[\nabla u - \beta u(1-u)\nabla J * u\right]\right\} = 0\,,$$

where $D = \exp(-\beta U_0)$. Here we also assumed for simplicity that the external field h is set to 0. By introducing the free energy,

$$(2.20) \qquad \begin{aligned} E[u] = \;&-\frac{1}{2}\int\int J(r-r')u(r)u(r')drdr' \\ &+ \int \frac{1}{\beta}[u\ln u + (1-u)\ln(1-u)]dr\,, \end{aligned}$$

(2.16) and (2.17) can be both written as the constrained gradient flow

$$(2.21) \qquad u_t - \nabla \cdot \left\{\mu[u]\nabla\left(\frac{\delta E[u]}{\delta u}\right)\right\} = 0\,.$$

In the case of (2.16) the mobility term $\mu[u]$ is $\mu[u] = D\beta u(1 - u)$, while in the Arrhenius case (2.17) it is a nonlocal function given by $\mu[u] = D\beta u(1 - u)\exp(-\beta J * u)$. Note that in both equations the coverage u satisfies $0 \leq u \leq 1$ due to the presence of the term $u(1 - u)$ in the mobilities, which enforces the exclusion principle (i.e. at most one particle per lattice site) and consequently the monolayer structure at the mesoscopic level.

Equations (2.16) and (2.17) include two competing forces: a Fickian diffusion term which competes with an uphill diffusion due to the attractive potential $J \geq 0$. We expect that when the parameter β is large the particles will tend to organize in clusters, overcoming the Fickian diffusive effects. These heuristics become more clear using a linearization argument around a constant coverage u_0, which yields a regime of *spinodal decomposition*. Indeed, we consider a solution $u = u_0 + \epsilon \exp(\omega t + i\xi \cdot x)$ of (2.16) for instance, where u_0 is a constant state and $\epsilon << 1$. The linearization of the equation around u_0 yields the dispersion relation

$$(2.22) \qquad \omega = -D|\xi|^2 \left[1 - \beta u_0 (1 - u_0)\hat{J}(\xi) \right],$$

where $J_0 = \int J(r)dr$ and $\hat{J}$ denotes the Fourier transform of J. For example, the Fourier transform of the attractive potential $J(r) = (1/\sqrt{2\pi r_0^2})\exp(-|r|^2/2r_0^2)$, is $\hat{J}(\xi) = \exp(-r_0^2|\xi|^2/2)$. Thus, suitable β, u_0 and ξ's give rise to a positive eigenvalue ω and subsequent exponential growth of the coverage u (eventually controlled by the exclusion principle), which leads to the formation of clusters. By comparison, the dispersion relation for the Arrhenius dynamics (2.17) is

$$\omega = -D|\xi|^2 \exp(-\beta J_0 u_0)\left[1 - \beta u_0(1 - u_0)\hat{J}(\xi) \right].$$

Thus, exactly the same modes ξ for both Arrhenius and Metropolis/Kawasaki dynamics, give rise to positive eigenvalues ω and subsequent exponential growth. However in the case of the Arrhenius dynamics, at low temperatures T, the eigenvalues ω are exponentially small, of order $O\big((1/T)\exp(-1/T)\big)$.

2.2.3. Mesoscopic theories for multiple micromechanisms. Typically multiple surface processes take place simultaneously (Fig. 1) and one of the practical advantages of the mesoscopic theories is that they can be easily modified, combining the various spin flip/exchange mechanisms described earlier. Here we present a straightforward generalization of the mesoscopic theory developed in [10] (see also [16]). We consider Arrhenius adsorption/desorption dynamics, Metropolis surface diffusion and a simple unimolecular reaction; the corresponding mesoscopic local mean field equation is:

$$(2.23) \qquad \begin{aligned} &u_t - D\nabla \cdot \left[\nabla u - \beta u(1 - u)\nabla J_m * u \right] \\ &- \left[k_a p(1 - u) - k_d u \exp\left(-\beta J_d * u \right) \right] + k_r u = 0. \end{aligned}$$

Here J_d and J_m are the intermolecular potentials for surface desorption and migration. Furthermore, D is the diffusion constant, k_r, k_d, and k_a denote, respectively the reaction, desorption and adsorption constants and p is the partial pressure of the gaseous species (k_d, k_a and p are algebraically related to the parameters c_0 and h in (2.12)). Finally u denotes the surface coverage of the adsorbed species.

The Fickian diffusion case is an interesting extreme ($J_m = 0$) typically adopted in diffusion-reaction models at the continuum and microscopic levels [2, 35, 36].

$$(2.24) \qquad u_t - D\Delta u - \left[k_a p (1 - u) - k_d u \exp\left(- \beta J_d * u \right) \right] + k_r u = 0 \,.$$

The steady states of equations (2.21) and (2.22) are the same as of equations (2.11) and (2.12) and thus the phase diagram in Figure 3 also holds here with $\alpha = \frac{k_a p}{k_d}$ and $\lambda = J_0 \beta$. When $J_d = J_m$ and $k_r = 0$, (2.11), (2.12) and (2.21) also share the same standing wave. However, there are no general rigorous results available on the existence of traveling waves for (2.21); some numerical simulations for identical interaction potentials $J_m = J_d = J$, were carried out in [10] indicating the existence of non-monotone traveling waves. Finally, it is easy to see that the free energy $E[u]$ is a Lyapunov functional for (2.21). Clearly we can also consider variants of (2.21) with Arrhenius diffusion dynamics.

2.2.4. Mesoscopic equations with stochastic fluctuations. As we indicated earlier in the formal derivations of the mesoscopic equations, it is possible to include random fluctuations in these models as higher order corrections. The stochastic terms are derived directly from each of the micromechanisms and satisfy fluctuation-dissipation relations [10, 12]. For instance when we include the stochastic correction term for the Arrhenius diffusion dynamics, (2.17) becomes, [12],

$$(2.25) \qquad \begin{aligned} &u_t - \nabla \cdot \left\{ D \exp(-\beta J * u) \left[\nabla u - \beta u (1 - u) \nabla J * u \right] \right\} \\ &- \gamma^{d/2} \nabla \cdot \left\{ [2D \exp(-\beta J * u) u (1 - u)]^{1/2} \dot{W} \right\} = 0 \,, \end{aligned}$$

where d is the space dimension, γ^{-1} the interaction radius of the microscopic potential and $\dot{W} = \left(\dot{W}_1(x, t), ..., \dot{W}_d(x, t) \right)$ is a d-dimensional space/time white noise, i.e.

$$E\dot{W}_i(x, t) = 0, \quad i = 1, ..., d,$$

and

$$E\left(\dot{W}_i(x, t) \dot{W}_j(y, s) \right) = \delta(i - j) \delta(x - y) \delta(t - s), \quad i, j = 1, ..., d \,.$$

Due to the multiplicative term $[2D \exp(-\beta J * u) u (1 - u)]^{1/2}$ in the noise, (2.23) has formally an invariant (Gibbs) measure

$$\mu(du) = \frac{1}{Z_\beta} \exp\left(- \beta E[u] \right) du \,,$$

in analogy to (2.2), where Z_β is the corresponding partition function and $E[u]$ is given by (2.18). Furthermore the multiplicative term guarantees that similarly to the underlying microscopic order parameter σ, the coverage u will remain bounded, $0 \le u \le 1$, in spite of the fluctuations of the white noise. Stochastic correction terms with the same characteristics can also be derived for the other micromechanisms. We present these stochastic mesoscopic models, their relations to the models in [10] as well as related numerical spectral schemes, in an upcoming publication [37]. We also refer to [31] for the rigorous derivation of linearized SPDE from the microscopics, around deterministic solutions of the mesoscopic equations (2.11) and (2.16); note that (2.23) is the nonlinear version of such linear SPDE. The role of random fluctuations is critical in the phenomenology of surface processes, for instance in nucleation and spinodal decomposition. Furthermore they provide a selection principle for instabilities arising in deterministic cluster evolution (e.g. interface "fattening") [38].

2.2.5. Relations to the Cahn-Hilliard and Allen-Cahn models. In this subsection we briefly discuss the connections of the mesoscopic equations with well known models for phase separation such as the Allen-Cahn and the Cahn-Hilliard models. If we rescale space as $x \mapsto x/\epsilon$ the potential J gives rise to the approximation of the Dirac distribution $J^\epsilon(x) = \epsilon^{-d}J(\frac{x}{\epsilon})$. Then after a simple change of variables and formally expanding in Taylor series,

$$
\begin{aligned}
J^\epsilon * u(x) &= \int J(z)u(x + \epsilon z)dz \\
&= \int J(z)\Big[u(x) + \epsilon \nabla u(x) \cdot z + \frac{\epsilon^2}{2}z^T \nabla^2 u(x)z + O(\epsilon^3)\Big]dz .
\end{aligned}
$$

Ignoring the $O(\epsilon^3)$ terms and assuming that J is radially symmetric, i.e. $J(r) = J(|r|)$, we have that

$$(2.26) \qquad J^\epsilon * u(x) \approx J_0 u(x) + \frac{\epsilon^2}{2}J_2 \Delta u(x),$$

where $J_0 = \int J(r)dr$ and $J_2 = \int |r|^2 J(r)dr$. Then, for instance (2.12), is approximated by a "porous medium" version of the Allen-Cahn equation

$$u_t = Du\exp(-\beta J_0 u)\Delta u + c_0\big[1 - u - \exp(-\beta h)u\exp(-\beta J_0 u)\big],$$

where $D = c_0\frac{\epsilon^2}{2}\beta J_0 \exp(-\beta h)$. Note that, as discussed in Section 2.2.1, the function $\lambda^{-1}f(u) = 1 - u - \exp(-\beta h)u\exp(-\beta J_0 u)$ is bistable or equivalently is the derivative of a double well potential when the parameters lie in Region III (see Fig. 2). We remind the reader that the Allen-Cahn equation has the non-dimensional form

$$u_t = \Delta u + W'(u),$$

where W is the double well potential $W(u) = (u^2 - 1)^2$, [39].

In the case of the surface diffusion we can rewrite the free energy (2.18) as

$$E[u] = \frac{1}{4} \int \int J(r - r')[u(r) - u(r')]^2 dr dr' + \int W_\beta(u) dr$$

$W_\beta(u) = \frac{1}{2} J_0 u(1 - u) + \frac{1}{\beta}[u \ln u + (1 - u) \ln(1 - u)]$. W_β is a double-well potential provided $\beta > \beta_c = 4/J_0$. Note that W_β is also known in the polymer science literature as the Flory-Huggins free energy, see for instance [19]. Then, rescaling and expanding the convolution as before we have that

$$E[u] \approx \tilde{E}[u] := \int \frac{\epsilon^2 J_2}{8} |\nabla u|^2 + W_\beta(u) dr,$$

after omitting the higher order terms. This is the standard Ginsburg-Landau functional, in which case (2.16) becomes a Cahn-Hilliard-type equation

(2.27) $$u_t - \nabla \cdot \left\{ \mu[u] \nabla \left(\frac{\delta \tilde{E}[u]}{\delta u} \right) \right\} = 0,$$

with nontrivial mobility $\mu[u] = Du(1-u)$; recall that in the standard Cahn-Hilliard model $\mu(u) = 1$, [17]. Similarly we may simplify the Arrhenius dynamics equation (2.17) which will have effective mobility $\mu(u) = Du(1 - u) \exp(-\beta J_0 u)$. Notice that the truncations in the gradient expansions used here disregard higher order effects as well as possible anisotropies in the potential J. However, in the vicinity of the critical temperature the Allen-Cahn and Cahn-Hilliard equations become exact rescaled limits of the mesoscopic models and the underlying particle systems. Similar models to (2.24), with or without chemical reaction, have been used in the modeling of phase separation in polymer blends, see [18, 19, 40] and references therein.

2.2.6. Comparisons of Monte Carlo simulations and mesoscopic models. The mesoscopic equations studied here are derived by a rigorous coarse graining in the limit where the interaction potential range γ^{-1} becomes infinite. Although there are certainly many physical potentials with such long range interactions, it may appear that this condition is too restrictive. An initial comparison of simulations of the deterministic mesoscopic equations (2.16) and (2.17) for surface diffusion Metropolis and Arrhenius dynamics, with Gradient Continuous Time Monte Carlo (G-CTMC) simulations indicates that only a relatively short range potential (approx. $10 - -20$ neighbors in one dimension) is sufficient for *quantitative* agreement [12]. See Fig. 4 for one dimensional comparisons of MC simulations and solutions of the mesoscopic equations. One notices from the scaling of the noise term in (2.23) that our 1D numerical comparisons represent the worst case scenario for the validity of the mesoscopic theory

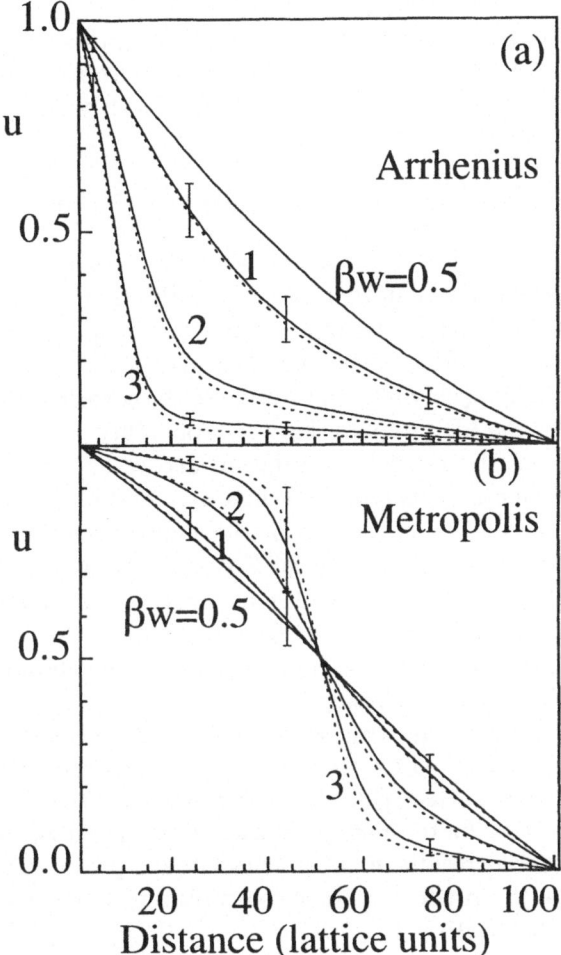

FIG. 4. *Concentration, u, vs. distance (in lattice units) from gradient Monte Carlo simulations (solid lines) and the mesoscopic theories (dashed lines) at various values of the intermolecular potential strength βw indicated; $w = J_0 = \int J(r)dr$. (a) corresponds to G-CTMC with Arrhenius dynamics and the mesoscopic equation (2.17), and (b) to Metropolis dynamics and (2.16). The potential used is a piecewise constant one that becomes zero after 10 lattice units and the interactions are attractive. The concentration at the boundaries is fixed to 0 and 1. The error bars indicate the standard deviation in MC results at selected spatial locations.*

tested against Monte Carlo simulations.   Similar numerical experiments conducted in two dimensions [21], exhibit the same qualitative behavior, but in this case a much smaller interaction radius is required, as is roughly predicted by (2.23). The need of such relatively small potential range for quantitative agreement between G-CTMC and mesoscopic solutions implies that mesoscopic theories can be an invaluable tool in the modeling of large length and time scale problems.

From a mathematical perspective the aforementioned matching of the microscopic with the mesoscopic models for a finite number of neighbors, can be rigorously justified by a conjectured underlying Large Deviation Principle (LDP), i.e.

$$\text{Prob}\big(v_\gamma \text{ is close to } u\big) \approx \exp\big(-\gamma^{-d} I_T[u]\big),$$

where v_γ is the average occupation number defined earlier, and $u = u(r, t)$ is any given function defined in $\mathbf{R}^d \times [0, T]$; note also that γ^{-d} is the number of neighbors in a potential of interaction radius γ^{-1}, in d dimensions. The rate functional $I_T[u]$–that needs to be identified–is expected to satisfy $I_T[u] \geq 0$, with equality holding only if u is a solution to the corresponding mesoscopic equation. Thus the LDP implies that the probability of a microscopic state being away from the mesoscopic solution decays exponentially with the number of neighbors. This type of results were first obtained for IPS in an infinite-dimensional context in [41] for the simple exclusion process, while in [42] the authors proved an LDP for Kawasaki dynamics with long range interactions. However the picture for non-conservative (and non-reversible) systems such as (2.21) is much less clear since in this case the H^{-1} Sobolev norm used in the construction of $I_T[u]$ in [41, 42] is not suitable, and further work in this direction is necessary.

Recently, a potential, with 36 in total significant neighbors has been experimentally calculated for the diffusion and reaction of O atoms on a Ru(0001) catalytic surface [43]. In view of the quantitative agreement of MC simulations for finite range potentials with the mesoscopic models, it is plausible that the results in [43] may eventually provide an experimental benchmark for the predictions of mesoscopic theories equipped with the O/Ru(0001) potential.

3. Macroscopic limits and cluster dynamics. Some aspects of the complex relations between micro-, meso- and macroscopic models were explored in [9, 14, 44] (see also references therein), where the authors have derived macroscopic PDE describing evolving clusters formed under the influence of attractive microscopic interactions. In [9] the authors showed that stochastic Ising models with spin flip dynamics yield evolving clusters moving with normal velocity which is a (possibly anisotropic) function of the principal curvatures of their boundaries. This function is given by a Kubo-Green-type formula which also specifies the relationship between the mobility and the surface energy of the propagating cluster boundaries on one hand, and the microscopic interaction potential and microscopic dynamics of the Ising model on the other. All these results are valid globally in time, the motion of the boundary being interpreted in the viscosity sense after the onset of the geometric singularities. In [14] the authors formally show that the large space/time asymptotics of (2.16) with isotropic potentials is described by a Mullins-Sekerka free boundary problem with surface tension identified through the microscopic Hamiltonian. This last result was rigorously established for smooth Mullins-Sekerka flows in [44]. Furthermore, at earlier times Ostwald ripening is observed; we numerically validated using spectral density functions that in this regime, clusters grow with a Lifsitz-Slyosov $t^{1/3}$ law [25].

In [16] we derived macroscopic governing laws of growth velocity, surface tension, mobility, critical nucleus size, and morphological evolution of clusters for (2.21), when multiple surface mechanisms are present and interact, and within the bistable regime of the equation. In particular, when $J_m = 0$ we proved that due to the attractive interactions J_d, cluster of particles form, and their boundaries evolve with normal velocity

$$(3.1) \qquad V = -\mu\sigma\kappa + \Lambda \,,$$

where μ and σ are the mobility and surface tension of the clusters, κ denotes the mean curvature of their boundaries, and Λ is the speed of the underlying traveling wave. The cluster motion is interpreted past singularities in the viscosity sense, via the level set formulation of the geometric law (3.1) [45, 46, 47]. In addition, μ and σ are given by Kubo-Green formulae. The added difficulty here–as well as in the case $J_m \geq 0$–is that due to the presence of multiple surface mechanisms, the linearized operator for (2.21) around a standing wave is not self-adjoint, so it is nontrivial to construct its kernel, which in turn yields the corrector necessary to derive the Kubo-Green formulae. This is accomplished by a combination of the implicit function theorem applied around $D = 0$, and the comparison principle for (2.21), which holds when $J_m = 0$.

In [16] we also derived, formally this time, the same general picture for curvature-driven cluster evolution when $J_m \geq 0$, although at earlier times our simulations indicate the existence of a Lifsitz-Slyosov-type law. The main obstacle here towards a rigorous proof using viscosity solution methods [48, 15], is that (2.21) does not have a comparison principle unless $J_m = 0$, or $D = 0$. Thus, as the formal asymptotics indicate, (2.21) is a *non-monotone* approximation of the normal velocity law (3.1), which is a monotone equation when interpreted in the viscosity sense [46, 47].

To better illustrate the effects of the multiple mechanisms we consider a simplification of (2.21) which retains its fundamental structure and can be exactly obtained from rescalings of (2.21) close to the critical temperature, when $k_r = 0$ and $J_m = J_d = J$:

$$(3.2) \qquad u_t = D\Delta\big(-\Delta u + W'(u)\big) + \Delta u - W'(u) \,,$$

W is a double-well potential, the Cahn-Hilliard term corresponds to surface diffusion, while the Allen-Cahn to adsorption/desorption, by analogy to (2.16) and (2.11). As in (2.21) the Ginsburg-Landau free energy $E[u] := \int |\nabla u|^2 + W(u)dr$, is a Lyapunov functional for (3.2). One would expect that the formal long time asymptotics of (3.2) would give rise to $V = -\kappa$ as in the Allen-Cahn case, for equal depth wells. Nevertheless due to the presence of the surface diffusion effects we obtain using a formal WKB expansion (see [16] for the more complex case (2.21)), the velocity law $V = -\mu\sigma\kappa$, where $\mu\sigma$ is given by the Kubo-Green formula

$$\mu\sigma = \int_{-\infty}^{\infty} \dot{q}^2 / \int_{-\infty}^{\infty} \dot{q}\chi,$$

q is the standing wave $\ddot{q} = W'(q)$ and $\chi : \mathbf{R} \to \mathbf{R}$ solves

$$-D\ddot{\chi} + \chi = \dot{q}, \qquad \chi(\pm\infty) = 0.$$

Clearly when $D = 0$, $\mu\sigma = 1$, i.e. the Allen-Cahn case. Thus an added difficulty in the rigorous asymptotics of (2.21) or (3.2), is the identification of the transport coefficient $\mu\sigma$. Further work on these issues is currently under way. We also refer to [25] for a numerical validation of cluster growth and scaling laws such as (3.1), in the various long time asymptotic regimes discussed in this section.

The long time behavior of (2.21) outside the bistable regime and in the presence of reaction yields a surprisingly rich morphology. In [49] the authors numerically observed that when the reaction rate k_r in (2.21) is strong enough, then it destabilizes the uniform steady states described earlier, leading to nonequilibrium structures with shorter typical wavelength than the diffusion length and on the order of micro- or nanometers. These labyrinthine pattern formations evolve slowly and are reminiscent of Ostwald ripening; however they do not obey a Lifsitz-Slyosov growth law, but instead they have a fixed characteristic length. Thus it appears that the chemical reaction terminates the growth of the spatial domains and induces self-organized nanostructures; this patterning is reminiscent of a Turing-type instability [50]. Pattern formation is commonly seen on catalytic surfaces during reaction or upon adsorption and surface diffusion processes at both micron and nanometer scales [2, 43]. Reaction-induced nanopatterns have been also predicted numerically for binary mixtures modeled by a Cahn-Hilliard equation with reaction [19, 20] and experimentally observed in polymer blends [18].

REFERENCES

[1] R. IMBIHL AND G. ERTL, Oscillatory kinetics in heterogeneous catalysis, Chem. Rev. **95**, 697 (1995).

[2] G. ERTL, Oscillatory kinetics and spatio-temporal self-organization in reactions at solid surfaces, Science **254**, 1750 (1991).

[3] G.H. GILMER AND P. BENNEMA, Simulation of crystal growth with surface diffusion, J. Appl. Phys. **43**, 1347 (1972).

[4] K. BINDER (ed.), Monte Carlo Methods in Statistical Physics. Springer-Verlag, Berlin (1986).

[5] M.P. ALLEN AND D.J. TILDESLEY, Computer Simulation of Liquids. Oxford Science Publications, Oxford (1989).

[6] S. JAKUBITH, H.H. ROTERMUND, W. ENGEL, A. VON OERTZEN, AND G. ERTL, Spatiotemporal concentration patterns in a surface reaction: Propagating and standing waves, rotating spirals, and turbulence, Phys. Rev. Letters 65, 3013 (1990).

[7] J.L. LEBOWITZ, E. ORLANDI, AND E. PRESUTTI, A particle model for spinodal decomposition, J. Stat. Phys. 63, 933 (1991).

[8] A. DE MASI, E. ORLANDI, E. PRESUTTI, AND L. TRIOLO, Glauber evolution with Kac potentials 1: mesoscopic and macroscopic limits, interface dynamics, Nonlinearity 7, 633 (1994).

[9] M.A. KATSOULAKIS AND P.E. SOUGANIDIS, Stochastic Ising models and anisotropic front propagation, J. Stat. Phys. 87, 63 (1997).

[10] M. HILDEBRAND AND A.S. MIKHAILOV, Mesoscopic modeling in the kinetic theory of adsorbates, J. Phys. Chem. 100, 19089 (1996).

[11] G. GIACOMIN AND J.L. LEBOWITZ, Exact macroscopic description of phase segregation in model alloys with long range interactions, Phys. Rev. Letters 76, 1094 (1996).

[12] D.G. VLACHOS AND M.A. KATSOULAKIS, Derivation and validation of mesoscopic theories for diffusion-reaction of interacting molecules, Phys. Rev. Lett. 85, 3898 (2000).

[13] G. GIACOMIN AND J.L. LEBOWITZ, Phase segregation dynamics in particle systems with long range interactions. I. Macroscopic limits, J. Stat. Phys. 87, 37 (1997).

[14] G. GIACOMIN AND J.L. LEBOWITZ, Phase segregation dynamics in particle systems with long range interactions. II. Interface motion, SIAM J. Appl. Math. 58, 1707 (1998).

[15] M.A. KATSOULAKIS AND P.E. SOUGANIDIS, Generalized motion by mean curvature as a macroscopic limit of stochastic Ising models with long range interactions and Glauber dynamics, Comm. Math. Phys. 169, 61 (1995).

[16] M.A. KATSOULAKIS AND D.G. VLACHOS, From microscopic interactions to macroscopic laws of cluster evolution, Phys. Rev. Letters 84, 1511 (2000).

[17] J.W. CAHN AND J.E. HILLIARD, Free energy of a nonuniform system I: Interfacial free energy, J. Chem. Phys. 28, 258 (1958).

[18] Q. TRAN-CONG AND A. HARADA, Reaction-induced ordering phenomena in binary polymer mixtures, Phys. Rev. Letters 76, 1162 (1996).

[19] S.C. GLOTZER, E.A. DI MARZIO, AND M. MUTHUKUMAR, Reaction-controlled morphology of phase-separating mixtures, Phys. Rev. Letters 74, 2034 (1995).

[20] M. MOTOYAMA AND T. OHTA, Morphology of phase-separating binary mixtures with chemical reaction, J. Phys. Soc. Jpn. 66, 2715 (1997).

[21] T. BASAK, D.G. VLACHOS, AND M.A. KATSOULAKIS, Mesoscopic theories for diffusion-reaction of interacting molecules: An alternative to Monte Carlo simulations, in preparation.

[22] M.N. KUPERMAN AND H.E. TROIANI, Pore formation during dezincification of Zn-based alloys, Applied Surf. Science 148, 56 (1999).

[23] M. HILDEBRAND, A.S. MIKHAILOV, AND G. ERTL, Nonequilibrium stationary microstructures in surface chemical reactions, Phys. Rev. E 58, 5483 (1998).

[24] N. MAURITS, P. ALTEVOGT, O. EVERS, AND J. FRAAIJE, Simple numerical quadrature rules for Gaussian chain polymer density functional calculations in 3D and implementation on parallel platforms, Comp. Polymer Sci. 6, 1 (1996).

[25] D.J. HORNTROP, M.A. KATSOULAKIS, AND D.G. VLACHOS, Spectral Methods for Mesoscopic Models In Pattern Formation, submitted.

[26] D. RUELLE, Statistical Mechanics: Rigorous Results. W.A. Benjamin, Inc., New York-Amsterdam (1969).

[27] H. SPOHN, Large Scale Dynamics of Interacting Particles. Springer-Verlag, New York (1991).

[28] H.C. KANG AND W.H. WEINBERG, Dynamic Monte Carlo with a proper energy barrier: Surface diffusion and two-dimensional domain orderings, J. Chem. Phys. **90**, 2824 (1988).

[29] H.C. KANG AND W.H. WEINBERG, Modeling the kinetics of heterogeneous catalysis, Chem. Rev. **95**, 667 (1995).

[30] A. DE MASI AND E. PRESUTTI, Mathematical Methods for Hydrodynamic Limits. Lecture Notes in Mathematics, **1501**. Springer-Verlag, Berlin (1991).

[31] G. GIACOMIN, J. LEBOWITZ, AND E. PRESUTTI, Deterministic and stochastic hydrodynamic equations arising from simple microscopic model systems, in Stochastic Partial Differential Equations: Six Perspectives. Edited by R. Carmona and B. Rozovskii, Math. Surveys Monogr., Vol. **64**, p. 107, Amer. Math. Soc., Providence, RI (1999).

[32] X. CHEN, EXISTENCE, uniqueness, and asymptotic stability of traveling waves in nonlocal evolution equations, Adv. Differential Equations **2**, 125 (1997).

[33] A. DE MASI, T. GOBRON, AND E. PRESUTTI, Travelling fronts in non-local evolution equations. Arch. Rational Mech. Anal. **132**, 143 (1995).

[34] P. BATES, P. FIFE, X. REN, AND X. WANG, Traveling waves in a convolution model for phase transitions. Arch. Rational Mech. Anal. **138**, 105 (1997).

[35] D.G. VLACHOS, L.D. SCHMIDT, AND R. ARIS, Effect of phase transitions, surface diffusion, and defects on heterogeneous reactions: multiplicities and fluctuations, Surf. Science **249**, 248 (1991).

[36] J.W. EVANS, Kinetic phase transition in catalytic reaction models, Langmuir **7**, 2514–2519 (1991).

[37] D. HORNTROP, M. KATSOULAKIS, D. VLACHOS, in preparation.

[38] M.A. KATSOULAKIS AND A.T. KHO, Stochastic Curvature Flows: Asymptotic Derivation, Level Set Formulation and Numerical Experiments, to appear in J. Interfaces and Free Boundaries.

[39] S.M. ALLEN AND J. CAHN, A microscopic theory for antiphase boundary motion and its application to antiphase domain coarsening, Act. Metall. **27**, 1089 (1979).

[40] J. FRAAIJE, B. VAN VLIMMEREN, N. MAURITS, M. POSTMA, O. EVERS, C. HOFFMANN, P. ALTEVOGT, G. GOLDBECK-WOOD, The dynamic mean-field density functional method and its application to the mesoscopic dynamics of quenched block copolymer melts, J. Chem. Phys. **106**, 4260 (1997).

[41] C. KIPNIS, S. OLLA, AND S.R.S. VARADHAN, Hydrodynamics and large deviation for simple exclusion processes, Comm. Pure Appl. Math. **42**, 115 (1989).

[42] A. ASSELAH AND G. GIACOMIN, Metastability for the exclusion process with mean-field interaction, J. Statist. Phys. **93**, 1051 (1998).

[43] S. RENISCH, R. SCHUSTER, J. WINTTERLIN, AND G. ERTL, Dynamics of adatom motion under the influence of mutual interactions: O/Ru(0001), Phys. Rev. Lett. **82**, 3839 (1999).

[44] E. CARLEN, M. CARVALHO, AND E. ORLANDI, in preparation.

[45] S. OSHER AND J. A. SETHIAN, Fronts propagating with curvature dependent speed: algorithms based on Hamilton-Jacobi formulations, J. Comp. Phys. **78**, 12 (1988).

[46] L.C. EVANS AND J. SPRUCK J., Motion of level sets by mean curvature I, J. Diff. Geom. **33**, 635 (1991).

[47] Y.-G. CHEN, Y. GIGA Y., AND S. GOTO, Uniqueness and existence of viscosity solutions of generalized mean curvature flow equations, J. Diff. Geom. **33**, 749 (1991).

[48] G. BARLES AND P. E. SOUGANIDIS, A new approach to front propagation problems: theory and applications, Arch. Rational Mech. Anal. **141**, 237 (1998).

[49] M. HILDEBRAND, A.S. MIKHAILOV, AND G. ERTL, Nonequilibrium stationary microstructures in surface chemical reactions, Phys. Rev. E **58**, 5483 (1998).

[50] A.M. TURING, The chemical basis of morphogenesis, Phil. Trans. R. Soc. (London) Ser. B **237**, 37 (1952).

HOMOGENOUS AND HETEROGENEOUS MODELS FOR SILICON OXIDATION

J.R. KING*

Abstract. Asymptotic methods can be used to derive surface reaction models as limiting cases of bulk reaction formulations, a key issue in such reductions being how the surface reaction kinetics depend on the bulk ones. An analysis is given of certain aspects of such limiting processes, as well as of the resulting heterogeneous formulations (which take the form of moving boundary problems), the focus being on anisotropic reaction effects.

1. Introduction. The relationship between heterogeneous and homogeneous models for chemical processes is of interest both because homogeneous formulations are often used, for example for numerical convenience, to smooth out the moving boundaries present in heterogeneous models (as in phase field approaches, for instance) and because the reverse process of applying asymptotic methods to a homogeneous model to derive a heterogeneous version as a limit case can be extremely valuable, particularly as a means of deriving analytical understanding of the former (the latter class very often being more amenable to explicit solution methods, scaling arguments, similarity reductions, and so forth). The primary motivation for the models we describe here comes from consideration of the kinetics of silicon oxidation, but it is hoped that the approaches described are equally applicable in many other contexts involving reaction and diffusion effects.

Background to the analysis we describe below and numerous relevant references (which we shall not repeat here) are given in the papers [1] (which concerns the derivation of heterogeneous kinetics from a homogeneous model) and [2] (which describes properties of a special case of the resulting heterogeneous (moving boundary problem) formulation). We shall seek so far as is possible to avoid repetition of results from [1–2], instead describing in more detail aspects which were only touched upon there. We focus on deriving (but not for the most part solving) reduced problems in which orientation effects play a central role, seeking to summarise the corresponding generalisations of the results of [1–2].

The remainder of the paper is organised as follows. In Section 2 we describe aspects of the derivation of surface reaction kinetics from a bulk reaction model, concentrating on the case in which the discreteness of the crystal features on the 'inner' (reaction zone) lengthscale; this represents a situation in which multiscale effects are very pronounced, the behaviour at a microscopic level within the reaction zone having a significant effect on the macroscopic ('outer') behaviour, in particular through the orientation dependence of the surface reaction coefficient. In Section 3 we focus on

---

*Division of Theoretical Mechanics, School of Mathematical Sciences, University of Nottingham, Nottingham NG7 2RD, England.

certain aspects of surface evolution in the context of the resulting heterogeneous reaction formulation. We conclude in Section 4 with discussion.

2. Discrete homogeneous model. Partly because our focus here is
on anisotropic formulations, we concentrate on a generalisation of the approach of [1] to a discrete reaction-diffusion model which explicitly accounts
for the crystal structure. For simplicity we consider a two-dimensional cubic lattice, though the approach readily generalises (to three dimensions
and to other types of crystal); we also make the unrealistic assumption
that reaction product retains the crystal structure of the unreacted material, with no volume change taking place (and with no need therefore
to take account of material deformation). Generalising the formulation to
describe cases in which the product is amorphous would be very far from
trivial; nevertheless, we believe the current highly simplified approach to
be instructive in indicating how an orientation dependent heterogeneous
reaction rate can arise out of a homogeneous model.

We consider the following two-dimensional system, in dimensionless
form:

$$\varepsilon^2 \frac{d}{dt} \alpha_{i,j} = R((\alpha), c_{i,j}),$$

(2.1)

$$\frac{\varepsilon^2}{\lambda} \frac{d}{dt} c_{i,j} = \varepsilon^2 \left(\nu_{i+1,j;i,j} - \nu_{i,j;i+1,j} + \nu_{i-1,j;i,j} \right.$$
$$- \nu_{i,j;i-1,j} + \nu_{i,j+1;i,j} - \nu_{i,j;i,j+1}$$
$$\left. + \nu_{i,j-1;i,j} - \nu_{i,j;i,j-1} \right) - R((\alpha), c_{i,j}),$$

$$\alpha_{i,j} + \beta_{i,j} = 1;$$

here $\alpha_{i,j}, \beta_{i,j}$ and $c_{i,j}$ represent, respectively, the concentrations of unreacted material, reacted material and oxidant at lattice site (i,j). The reac-
tion rate expression $R((\alpha), c_{i,j})$ is shorthand for $R(\alpha_{i,j}, c_{i,j}; \alpha_{i+1,j}, \alpha_{i-1,j},$
$\alpha_{i,j+1}, \alpha_{i,j-1})$, it being assumed that when the particle at (i,j) reacts,
the nature of the chemical bonds with its four nearest neighbours (at
$(i+1,j), (i-1,j), (i,j+1)$ and $(i,j-1)$), and hence which species oc-
cupies each of those sites, influences the activation energy of reaction; we
have $R((\alpha), c_{i,j}) = 0$ for $\alpha_{i,j} = 0$ and for $c_{i,j} = 0$. Regarding $\alpha_{i,j}$ as a probability of occupancy by the unreacted species, the $\alpha_{i,j}$ for different lattice
sites are treated as independent; this will not in practice be the case but
is the simplest closure assumption and is in keeping with the usual derivations of the corresponding continuous formulations (cf. (2.3) below). The
quantity $\nu_{i,j:k,l}$ is the rate at which oxidant jumps from site (i,j) to the
neighbouring site (k,l); for reasons outlined in [1] it is important for our
purposes that the diffusion rate of the oxidant depend on the composition,
so we take

(2.2) $\nu_{i,j;k,l} = N(\alpha_{i,j}, \alpha_{k,l}; \varepsilon) c_{i,j},$

where the function N determines the jump frequency from (i, j) to (k, l) (in practice the oxidant is expected to reside at interstitial rather than lattice sites; we adopt the current formulation for brevity, the additional complexity of incorporating interstitial sites being largely a matter of notation).

Writing $x = i\Delta x$, $y = j\Delta x$, where $\Delta x \ll 1$ is the dimensionless lattice spacing, the leading order continuum approximation to (2.1)–(2.2) (i.e. the $\Delta x \to 0$ limit) reads (writing $R(\alpha, c; \alpha, \alpha, \alpha, \alpha) = r(\alpha, c)$)

$$\varepsilon^2 \frac{\partial \alpha}{\partial t} = -r(\alpha, c),$$

$$(2.3) \quad \frac{\varepsilon^2}{\lambda} \frac{\partial c}{\partial t} = \varepsilon^2 (\Delta x)^2 \nabla \cdot (N\nabla c + (N_1 - N_2) c \nabla \alpha) - r(\alpha, c),$$

$$\alpha + \beta = 1,$$

where

$$N = N(\alpha, \alpha; \varepsilon), \quad N_1 = \frac{\partial}{\partial \alpha} \left(N(\alpha, \alpha'; \varepsilon) \right) \Big|_{\alpha' = \alpha},$$
$$N_2 = \frac{\partial}{\partial \alpha} \left(N(\alpha', \alpha; \varepsilon) \right) \Big|_{\alpha' = \alpha};$$

the model of [1], with $\gamma = 1$ and $v = 0$ (no volume expansion), corresponds to the one-dimensional version of (2.3) when $N_1 = N_2$ with $D(\alpha; \varepsilon) = (\Delta x)^2 N(\alpha, \alpha; \varepsilon)$ (for the continuum limit to be appropriate it is necessary that $(\Delta x)^2 N$ be $O(1)$ as $\Delta x \to 0$). The precise forms of the diffusion and convection terms in the second of (2.3) depend on what assumptions are made regarding the nature of the jumps (cf. the discussion above regarding interstitial sites), but such issues do not materially affect the analysis which follows. Note that the constants ε and λ are described in [1] and that the isotropy of diffusion in (2.3) follows from the assumption of a cubic lattice; in the limit $\varepsilon \to 0^+$ the reaction is confined to a narrow interior layer.

The regime we consider relates to that discussed in Section 4 of [1]. The leading order outer solutions satisfy limits of the continuous version of the equations (i.e. (2.3)), so are essentially as described in [1]. The appropriate scalings for the reaction layer separating the two outer regions then enable the distinguished limit (in which Δx and ε both tend to zero) to be identified. Given the above relation between D and N, the regime identified in [1] corresponds to $(\Delta x)^2 N = O(1)$ for α small but $(\Delta x)^2 N = O(\varepsilon^2)$ for $\alpha = O(1)$; moreover, the reaction layer thickness in Section 4 of [1] is $O(\varepsilon^2)$, so we require $\varepsilon = \mu(\Delta x)^{\frac{1}{2}}$ with $\mu = O(1)$ in order for discreteness to appear at leading order. Writing

$$N(\alpha_{i,j}, \alpha_{k,l}; \varepsilon) \sim \frac{\varepsilon^2}{(\Delta x)^2} D_1(\alpha_{i,j}, \alpha_{k,l}),$$

then on the scale of the reaction zone (which is comparable to the lattice spacing) the leading order problem takes the form

$$-V_n \frac{\partial}{\partial z}\alpha_{i,j} = -R((\alpha), c_{i,j}),$$

(2.4)
$$-\frac{V_n}{\lambda}\frac{\partial}{\partial z}c_{i,j} = \mu^4\Big(\omega_{i+1,j;i,j} - \omega_{i,j;i+1,j} + \omega_{i-1,j;i,j}$$
$$-\omega_{i,j;i-1,j} + \omega_{i,j+1;i,j} - \omega_{i,j;i,j+1}$$
$$+\omega_{i,j-1;i,j} - \omega_{i,j;i,j-1}\Big) - R((\alpha), c_{i,j}),$$

where

$$\omega_{i,j;k,l} = D_1(\alpha_{i,j}, \alpha_{k,l})c_{i,j}.$$

In (2.4), $V_n > 0$ is the leading order outward normal velocity of the interface within the outer problem, about which the reaction zone is located (this interface is denoted by $y = -f$ in [1]), the continuous variable z is in the direction of the outward normal to the interface and the origin of (i,j) is now located within the reaction layer, with

(2.5) $$z = i \sin \psi - j \cos \psi,$$

where ψ is the angle the interface (on the outer scale) makes with the horizontal. Hence $V_n \equiv V_n(s,t)$, where s is the arc length along the interface; apart from this parametric dependence on s and t, (2.4) corresponds to a travelling wave reduction to (2.1) with the wave propagating with speed V_n in the direction determined by ψ in (2.5). Despite appearances, the formulation (2.4) is actually one-dimensional, this reduction in dimensionality being noteworthy; suppressing the parametric dependence on s and t, one can write $\alpha_{i,j} = \hat{\alpha}(z), c_{i,j} = \hat{c}(z), \alpha_{i+1,j} = \hat{\alpha}(z + \sin\psi), \alpha_{i,j+1} = \hat{\alpha}(z - \cos\psi)$, and so on. As in the continuous case (see [1]), the crucial issue for (2.4) is the relationship it determines between the wave speed V_n and the concentration c in the limit $z \to -\infty$, well behind the reaction front; unlike the continuous case (cf. the rotation invariance of (2.3)) this relation will depend on the direction of propagation and hence will introduce an orientation dependence into the heterogeneous reaction rate that occurs in the relevant boundary condition for the outer problem. Such issues of the orientation dependence of the wave speed in difference and differential-difference systems also arises in many other contexts; here it represents a key aspect of multiscale phenomena, whereby the experimentally observed macroscopic orientation dependence of the heterogeneous reaction rate arises from the reaction layer having a thickness of the order of the atomic spacing (the latter being amenable to experimental confirmation through measurements of the scale over which α varies). This orientation dependence is suppressed in the limit $\mu \to \infty$ in which the leading order reaction layer formulation reverts to a continuous form.

For $0 < \psi < \pi/2$, the matching conditions on (2.4) are that

$$(2.6) \quad \begin{array}{lll} \alpha_{i,j} \to 0, & \dfrac{\partial c_{i,j}}{\partial z} \to 0 & \text{as } i \to -\infty \text{ or } j \to +\infty \quad (z \to -\infty), \\[2mm] \alpha_{i,j} \to 1, & c_{i,j} \to 0 & \text{as } i \to +\infty \text{ or } j \to -\infty \quad (z \to +\infty). \end{array}$$

For a solution to (2.4) which satisfies these conditions to exist, we require, among other constraints, that D_1 blow-up sufficiently rapidly as $\alpha_{i,j} \to 0$ (cf. [1]). The crucial quantity determined by (2.4), (2.6) is

$$(2.7) \qquad \Phi(V_n; \psi) \equiv \lim_{z \to -\infty} \hat{c}(z)$$

which feeds back into the leading order outer (moving boundary) problem via the condition

$$(2.8) \qquad c = \Phi(V_n; \psi)$$

on the moving boundary; the function Φ thus expresses the macroscopic influence of the microscopic problem. In (2.7), the function Φ depends on μ, λ, D_1 and R but not on z. In the continuum limit Φ does not depend on ψ, some of its properties being addressed in [1]. In the discrete case less can be said without solving (2.4), (2.6) numerically, though the dependence of Φ on ψ will obviously share the discrete rotation symmetries of the lattice. In the limit $\lambda \to \infty$ with $R((\alpha), c_{i,j}) = R_1((\alpha))c_{i,j}$ it may be deduced by a scaling argument that $c_{i,j}$ is proportional to V_n, so that

$$(2.9) \qquad \Phi(V_n; \psi) = \Phi_1(\psi)V_n.$$

If $R((\alpha), c_{i,j}) = R_1((\alpha))c_{i,j}^q$ with $\lambda = \infty$, then scalings arguments show that i and j scale as the large quantity $V_n^{(1-q)/(1+q)}$ for $V_n \to +\infty$, $q < 1$ and for $V_n \to 0^+$, $q > 1$, whereby we recover the continuum limit in (2.4) with c proportional to $V_n^{2/(q+1)}$ and

$$(2.10) \quad \Phi(V_n; \psi) \sim (V_n/\hat{k})^{\frac{2}{q+1}} \quad \text{for } V_n \to +\infty, \; q<1 \text{ and } V_n \to 0^+, \; q>1,$$

where the constant $\hat{k}$ is determined from the continuum formulation and therefore does not depend on ψ. In the opposite limit ($V_n \to 0^+$ when $q < 1$ and $V_n \to +\infty$ when $q > 1$), the behaviour of (2.4) with $\lambda = \infty$ is expected to be dominated by a regime in which $\alpha_{i,j}$ is small; taking R_1 and D_1 to be, respectively, homogeneous of degree p and $-m$ in the relevant α in the limit in which those α tend to zero (with $p+q+qm-1 > 0$), then for small $\alpha_{i,j}$ we anticipate that (provided a solution exists) $\alpha_{i,j}$ and $c_{i,j}$ are respectively proportional to $V_n^{(1-q)/(p+q+qm-1)}$ and $V_n^{(m+p)/(p+q+qm-1)}$, so that

$$(2.11) \qquad \Phi(V_n; \psi) \sim \Phi_1(\psi) \, V_n^{\frac{m+p}{p+q+qm-1}}$$

$$\text{for } V_n \to 0^+, \; q < 1 \text{ and } V_n \to +\infty, \; q > 1$$

for some function Φ_1, with a discrete leading order balance holding in (2.4) (with $\lambda = \infty$).

3. Moving boundary problems.

3.1. Formulation. Motivated by the results of the previous section, we study here the Stefan problem

(3.1)
$$\frac{\partial c}{\partial t} = \nabla^2 c \qquad \text{for } \boldsymbol{x} \in \Omega(t),$$

$$\lambda V_n = -\left(\frac{\partial c}{\partial n} + V_n c\right) = k(c, \hat{\boldsymbol{n}}) \quad \text{for } \boldsymbol{x} \in \Gamma(t),$$

where $\Gamma(t) \subset \partial\Omega(t)$ is the moving boundary, with appropriate conditions being imposed on the remaining (fixed) parts of $\partial\Omega$. Note that we have neglected volume expansion effects in (3.1) in order to avoid discussion of the constitutive modelling of oxide deformation; nevertheless, many of the results we outline remain pertinent when account is taken of changes of volume on reaction. The relationship $\lambda V_n = k(c, \hat{\boldsymbol{n}})$ corresponds in the two-dimensional case to (2.8). Writing $\Gamma(t)$ as $F(\boldsymbol{x}, t) = 0$, with ∇F being in the direction of the outward normal, then

$$V_n = -\frac{\partial F}{\partial t} \Big/ |\nabla F|$$

is the outward normal velocity of $\Gamma(t)$,

$$\frac{\partial c}{\partial n} = \frac{1}{|\nabla F|} \nabla c \cdot \nabla F$$

is the outward normal derivative of c and

$$\hat{\boldsymbol{n}} = \nabla F \Big/ |\nabla F|$$

is the outward normal. The formulation (3.1) thus accounts for general orientation and surface concentration dependence of the heterogeneous reaction rate; we take $k(0, \hat{\boldsymbol{n}}) = 0$. The Baiocchi transformed version of (3.1) with $k(c, \hat{\boldsymbol{n}}) = k_1(\hat{\boldsymbol{n}})c$ is worth recording. Writing $F = \omega(\boldsymbol{x}) - t$ and

$$w = \int_\omega^t c(\boldsymbol{x}, t') \, dt'$$

yields

$$\frac{\partial w}{\partial t} = \nabla^2 w - \lambda \left(1 - \nabla \cdot \left(\Omega\left(\frac{\nabla \omega}{|\nabla \omega|}\right)\right)\right)$$

$$\text{on } t = \omega \qquad w = 0, \qquad \frac{\partial w}{\partial n} = -\lambda/k_1 \left(\frac{\nabla \omega}{|\nabla \omega|}\right)$$

where $\boldsymbol{\Omega}(\hat{\boldsymbol{n}}) = \hat{\boldsymbol{n}}/k_1(\hat{\boldsymbol{n}})$.   In general (3.1) requires numerical solution, but significant analytical progress is possible in limiting cases. For example, as the Stefan number, λ, tends to infinity the rescaling

$$t \to \lambda t, \quad V_n \to V_n/\lambda$$

is appropriate, reducing (3.1) at leading order to a Hele-Shaw type problem with anisotropic kinetic undercooling, namely

$$\nabla^2 c \;=\; 0 \qquad\qquad \text{for } \boldsymbol{x} \in \Omega(t),$$

(3.2)

$$V_n \;=\; -\frac{\partial c}{\partial n} \;=\; k(c, \hat{\boldsymbol{n}}) \quad \text{for } \boldsymbol{x} \in \Gamma(t),$$

while, depending on the conditions on the fixed parts of $\partial\Omega$, the behaviour as $t \to \infty$ is in practice often diffusion controlled, reducing (3.1) (with $V_n \to 0$ as $t \to \infty$) to

$$\frac{\partial c}{\partial t} \;=\; \nabla^2 c \qquad\qquad \text{for } \boldsymbol{x} \in \Omega(t),$$

(3.3)

$$\lambda V_n \;=\; -\frac{\partial c}{\partial n}, \quad c = 0 \quad \text{for } \boldsymbol{x} \in \Gamma(t)$$

which admits the similarity reduction

$$c = c\left(\boldsymbol{x}\big/t^{\frac{1}{2}}\right), \qquad F = F\left(\boldsymbol{x}\big/t^{\frac{1}{2}}\right),$$

this being a self-consistent $t \to \infty$ limit solution to (3.1) which is often relevant in applications. Here, however, we focus on the reaction-controlled limit $k \to 0$ and the small-Stefan-number limit $\lambda \to 0$, in both of which (3.1) can be reduced at leading order to an interfacial dynamics problem in which the interface location is essentially the only unknown.

3.2. Reaction control. This limit can be relevant to high pressure oxidation, for example, and provides a convenient means of assessing reaction kinetics from experimentally measured multidimensional interface profiles. We write

$$k(c, \hat{\boldsymbol{n}}) = \varepsilon^2 K(c, \hat{\boldsymbol{n}})$$

with $0 < \varepsilon \ll 1$ and $K = O(1)$. For definiteness we impose $c = 1$ on at least some of the fixed part of $\partial\Omega$ and $\partial c/\partial n = 0$ on the remainder, though the results readily generalise. Introducing the rescalings

(3.4) $$t \to \varepsilon^{-2}t, \quad V_n \to \varepsilon^2 V_n$$

to leading order we have as $\varepsilon \to 0$ that

$$\nabla^2 c \;=\; 0 \qquad\qquad \text{for } \boldsymbol{x} \in \Omega(t),$$

(3.5)

$$\frac{\partial c}{\partial n} \;=\; 0, \quad \lambda V_n = K(c, \hat{\boldsymbol{n}}) \quad \text{for } \boldsymbol{x} \in \Gamma(t),$$

so that

$$c = 1$$

and the interface evolves according to the interfacial dynamics prescription (in which the only unknown is the moving boundary location)

$$(3.6) \qquad \lambda V_n = K(1, \hat{n})$$

or, equivalently, as the contour $F = 0$ of

$$(3.7) \qquad \lambda \frac{\partial F}{\partial t} = -K\left(1, \frac{\nabla F}{|\nabla F|}\right) |\nabla F|,$$

which is readily amenable to level set techniques (see Sethian [5], for instance) or characteristic methods (certain anisotropies K being studied in [4] in the two-dimensional case, with phenomena such as corner formation occurring in convex curves with sufficiently strong orientation dependence), for example; the similarity reduction

$$(3.8) \qquad F \sim G(\boldsymbol{x}/t) \qquad \text{as } t \to \infty$$

typically describes the large-time behaviour.

If the domain has an aspect ratio of $O(\varepsilon)$ then a different leading order balance arises; such scenarios can arise in practice when the oxidation proceeds from grain boundaries, for example. Unlike (3.6)–(3.7), the resulting formulations do not take a purely interfacial dynamics form. We first consider the case when the scalings

$$(3.9) \qquad x \to \varepsilon^{-1}x, \quad y \to y, \quad z \to z,$$

together with (3.4), pertain and denote the cross-section of the surface $\Gamma(t)$ at each (relevant) x by the closed curve $\gamma(x,t)$. We then have

$$(3.10) \qquad \begin{aligned} c &\sim c_0(x,t) + \varepsilon^2 c_1(x,y,z,t), \qquad V_n \sim V_{0_n}, \quad \gamma \sim \gamma_0(x,t), \\ F &\sim F_0(x,y,z,t) \quad \text{as } \varepsilon \to 0 \end{aligned}$$

with

$$(3.11) \qquad \frac{\partial^2 c_1}{\partial y^2} + \frac{\partial^2 c_1}{\partial z^2} = \frac{\partial c_0}{\partial t} - \frac{\partial^2 c_0}{\partial x^2},$$

so that

$$(3.12) \qquad \oint_{\gamma_0} \hat{n}_0 \cdot \nabla c_1 \, ds = A_0\left(\frac{\partial c_0}{\partial t} - \frac{\partial^2 c_0}{\partial x^2}\right),$$

where s is the arc length around γ_0, $A_0(x,t)$ is the area enclosed by $\gamma_0(x,t)$ and

$$(3.13) \qquad \hat{n}_0 = \hat{\nabla} F_0 \Big/ |\hat{\nabla} F_0|, \qquad V_{0_n} = -\frac{\partial F_0}{\partial t} \Big/ |\hat{\nabla} F_0|,$$

where $\hat{\nabla} = (0, \partial/\partial y, \partial/\partial z)$. The moving boundary conditions yield

$$\lambda V_{0_n} = -\hat{n}_0 \cdot \nabla c_1 - \frac{1}{|\hat{\nabla} F_0|} \frac{\partial F_0}{\partial x} \frac{\partial c_0}{\partial x} - V_{0_n} c_0 = K(c_0, \hat{n}_0)$$

so that F_0 satisfies

$$(3.14) \qquad \lambda \frac{\partial F_0}{\partial t} = -K\left(c_0, \frac{\hat{\nabla} F_0}{|\hat{\nabla} F_0|}\right) |\hat{\nabla} F_0|,$$

We note that, regarding $c_0(x,t)$ as known, F_0 is constant along the characteristics of (3.14), so that different level sets of F_0 decouple, just as they do for (3.7); this is to be expected. Using (3.12) and

$$(3.15) \qquad \frac{\partial A_0}{\partial t} = \oint_{\gamma_0} V_{0_n}\, ds, \qquad \frac{\partial A_0}{\partial x} = -\oint_{\gamma_0} \frac{1}{|\hat{\nabla} F_0|} \frac{\partial F_0}{\partial x}\, ds,$$

we then obtain

$$(3.16) \qquad \frac{\partial}{\partial t}\left(A_0(\lambda + c_0)\right) = \frac{\partial}{\partial x}\left(A_0 \frac{\partial c_0}{\partial x}\right)$$

as the equation relating $A_0(x,t)$ and $c_0(x,t)$, at least if γ_0 is suitably smooth. This coupled formulation is quite an interesting one — the evolution of the cross-section parallel to the (y, z) plane is for each x determined by (3.14), which is similar in form to (3.7), and this feeds into the slow variation in the x direction via the appearance of $A_0(x,t)$ (the area enclosed by curve $\gamma_0(x,t)$) in (3.16); previously only the two-dimensional version of this problem (no z-dependence) has been formulated (see [2]). In the isotropic case, in which K does not depend on $\hat{n}$, (3.14) implies that

$$\lambda \frac{\partial A_0}{\partial t} = K(c_0) L_0,$$

where $L_0(x,t)$ is the length of γ_0, so with radial symmetry the problem reduces to (3.15) and

$$(3.17) \qquad \lambda \frac{\partial A_0}{\partial t} = 2\sqrt{\pi A_0} K(c_0).$$

In typical applications the large-time behaviour of (3.14), (3.16) takes the self-similar form

$$(3.18) \qquad c_0 \sim C(x/t^{\frac{1}{2}}), \qquad A_0 \sim t^2 a(x/t^{\frac{1}{2}}), \qquad F_0 \sim g(x/t^{\frac{1}{2}}, y/t, z/t).$$

In the case of linear kinetics, i.e.

$$(3.19) \qquad K(c, \hat{n}) = K_1(\hat{n})c,$$

we note that the $\lambda \to \infty$ limit problem (3.2) is, for any ε, invariant under the infinite-dimensional group

(3.20) $$\hat{t} = T(t), \quad c = \dot{T}(t)\hat{c}, \quad V_n = \dot{T}(t)\hat{V}_n$$

in which T is an arbitrary function with $\dot{T} > 0$ (the choice $T = -t$ corresponds to the time-reversibility of the Hele-Shaw problem with kinetic undercooling); the corresponding (rescaled, i.e. $t \to \lambda t$) version of (3.14), (3.16) reads

(3.21)
$$\frac{\partial F_0}{\partial t} = -K_1\left(\frac{\hat{\nabla}F_0}{|\hat{\nabla}F_0|}\right)|\hat{\nabla}F_0|c_0,$$
$$\frac{\partial A_0}{\partial t} = \frac{\partial}{\partial x}\left(A_0\frac{\partial c_0}{\partial x}\right),$$

and inherits this property, as does the evolution equation

(3.22) $$\frac{\partial A_0}{\partial t} = \frac{1}{\sqrt{\pi K_1}}\frac{\partial}{\partial x}\left(A_0\frac{\partial^2}{\partial x\partial t}\left(\sqrt{A_0}\right)\right), \quad c_0 = \frac{1}{\sqrt{\pi K_1}}\frac{\partial}{\partial t}\left(\sqrt{A_0}\right),$$

which results when radial symmetry holds with K_1 constant (cf. (3.17)). A particular consequence is that (3.21) and (3.22) admit solutions of the form

(3.23) $$A_0 = A_0(x - s(t)), \quad c_0 = \dot{s}(t)C_0(x - s(t)), \quad F_0 = F_0(x - s(t), y, z),$$

in which $s(t)$ is arbitrary. The reduction (3.23) can be useful in applications; moreover a remarkable amount of analytical progress is possible, as we now illustrate. Writing

(3.24) $$\zeta = x - s(t)$$

(note that s does not denote arc length here) and imposing

(3.25)
$$\text{at } t = 0 \quad F_0 = f_\infty(y, z), \quad A_0 = a_\infty,$$
$$\text{at } x = 0 \quad c_0 = 1,$$
$$\text{as } x \to +\infty \quad c_0 \to 0,$$

where the constant a_∞ is the area enclosed by the level set $f_\infty = 0$ (note that we assume here that the cross-section of the oxide domain does not initially depend on x), we have from (3.21) that $A_0(\zeta)$, $C_0(\zeta)$ and $F_0(\zeta, y, z)$ satisfy

(3.26)
$$\frac{\partial F_0}{\partial \zeta} = K_1\left(\frac{\hat{\nabla}F_0}{|\hat{\nabla}F_0|}\right)|\hat{\nabla}F_0|C_0,$$
$$-(A_0 - a_\infty) = A_0\frac{dC_0}{d\zeta},$$

with C_0 decaying exponentially as $\zeta \to +\infty$. Introducing

$$\xi = \int_\zeta^\infty C_0(\zeta') \, d\zeta'$$

decouples the first of (3.26), giving

(3.27)
$$\frac{\partial F_0}{\partial \xi} = -K_1 \left(\frac{\hat{\nabla} F_0}{|\hat{\nabla} F_0|} \right) |\hat{\nabla} F_0|,$$

$$\text{at } \xi = 0 \qquad F_0 = f_\infty(y, z),$$

which amounts to a two-dimensional version of (3.7). Solving (3.27), for example by charac-teristics, gives $F_0(\xi, y, z)$ and (hence $A_0(\xi)$) and the second of (3.26) yields an ordinary differential equation for $\xi(\zeta)$, namely

$$\frac{d^2 \xi}{d\zeta^2} = 1 - \frac{a_\infty}{A_0(\xi)}$$

$$\text{as } \zeta \to +\infty \qquad \xi \to 0,$$

so that

(3.28)
$$\frac{d\xi}{d\zeta} = - \left(2 \left(\xi - a_\infty \int_0^\xi \frac{d\xi'}{A_0(\xi')} \right) \right)^{\frac{1}{2}},$$

which determines $\xi(\zeta)$, up to translations of ζ and hence of s in (3.24); using this freedom in the choice of the origin of s we can without loss of generality set

$$\text{at } \zeta = 0 \qquad \xi = 1,$$

for example. Finally, $s(t)$ is determined via the second condition of (3.25) by integrating

(3.29)
$$C_0(-s)\dot{s} = 1,$$

$$\text{as } t \to 0^+ \qquad s \to -\infty,$$

where

$$C_0(\zeta) = -\frac{d\xi}{d\zeta}.$$

The above analysis assumes that $a_\infty > 0$; when $a_\infty = 0$ the required solution is compactly supported in ζ. The special case (3.22) can be used to make these considerations more transparent, with $A_0(\xi) = \pi(K_1\xi + r_0)^2$ where $f_\infty = 0$ corresponds to $y^2 + z^2 = r_0^2$.

Some further progress is more generally possible in the separable case $K = K_1(\hat{n})K_2(c)$. We write

$$K_1(\hat{n}_0) = \hat{K}_1(\psi),$$

where $\psi(s; x, t)$ is the angle γ_0 makes with the horizontal and the arc length s is measured anticlockwise, so the curvature of γ_0 is given by $\kappa = \partial\psi/\partial s$, which we assume to be positive. It follows from results in [3] that

(3.30)
$$\frac{\partial L_0}{\partial t} = \oint_{\gamma_0} \kappa V_{0_n} \, ds = \frac{K_2(c_0)}{\lambda} \int_0^{2\pi} \hat{K}_1(\psi) \, d\psi,$$

$$\frac{\partial A_0}{\partial t} = \oint_{\gamma_0} V_{0_n} \, ds, \qquad \frac{\partial^2 A_0}{\partial t^2} = \frac{1}{K_2} \frac{\partial K_2}{\partial t} \frac{\partial A_0}{\partial t} + \frac{K_2^2}{\lambda^2} \Lambda,$$

where the constant Λ is defined by

$$\Lambda = \int_0^{2\pi} \left(\hat{K}_1^2 - \left(\frac{d\hat{K}_1}{d\psi} \right)^2 \right) d\psi.$$

The last relation in (3.30) is noteworthy and is satisfied by

(3.31)
$$K_2(c_0) = \frac{\lambda}{\sqrt{2\Lambda A_0}} \frac{\partial A_0}{\partial t}$$

(cf. (3.17)), but this requires that the initial data have

$$\sqrt{A_0} = \frac{1}{\sqrt{2\Lambda}} \oint_{\gamma_0} \hat{K}_1 \, ds$$

for all x which need not be true except in the isotropic radially symmetric case in which it is satisfied identically; if it does hold then (3.31) enables the coupled formulation (3.14), (3.16) to be significantly simplified.

The other relevant formulation of the above type applies when the problem is slowly varying in two directions, so that (3.8) is replaced by

(3.32)
$$x \to \varepsilon^{-1} x, \quad y \to \varepsilon^{-1} y, \quad z \to z$$

together with (3.4). Now we have

$$c \sim c_0(x, y, t) + \varepsilon^2 c_1(x, y, z, t)$$

with

(3.33)
$$\frac{\partial^2 c_1}{\partial z^2} = \frac{\partial c_0}{\partial t} - \left(\frac{\partial^2 c_0}{\partial x^2} + \frac{\partial^2 c_0}{\partial y^2} \right).$$

For definiteness we take the problem to be symmetric about $z = 0$ and write $F = z - f(x,y,t)$ in rescaled co-ordinates; since

$$\hat{n} = \left(-\varepsilon\frac{\partial f}{\partial x}, -\varepsilon\frac{\partial f}{\partial y}, 1\right) \bigg/ \left(1 + \varepsilon^2\left(\frac{\partial f}{\partial x}\right)^2 + \varepsilon^2\left(\frac{\partial f}{\partial y}\right)^2\right)^{\frac{1}{2}},$$

anisotropy appears at leading order only if k is strongly dependent on $\hat{n}$, in the sense that

$$(3.34) \qquad k(c, \hat{n}) \sim \varepsilon^2 K\left(c_0, \frac{\partial f_0}{\partial x}, \frac{\partial f_0}{\partial y}\right),$$

and for this distinguished limit we obtain from (3.1) and (3.33) the system

$$(3.35) \qquad \begin{aligned} \frac{\partial}{\partial t}\left(f_0(\lambda + c_0)\right) &= \frac{\partial}{\partial x}\left(f_0\frac{\partial c_0}{\partial x}\right) + \frac{\partial}{\partial y}\left(f_0\frac{\partial c_0}{\partial y}\right), \\ \lambda\frac{\partial f_0}{\partial t} &= K\left(c_0, \frac{\partial f_0}{\partial x}, \frac{\partial f_0}{\partial y}\right). \end{aligned}$$

Considering the case $K = K(c_0)$ (the inclusion of orientation dependence leads to a number of extra complications), boundary value problems relevant in applications include, writing $x = r\cos\theta$, $y = r\sin\theta$,

$$c_0 = 1 \text{ on } \theta = 0 \text{ and } \theta = \pi/2, \qquad c_0 \to 0 \text{ as } r \to \infty, \qquad 0 < \theta < \pi/2,$$

and

$$c_0 = 1 \text{ on } \theta = \pi \text{ and } \theta = -\pi/2, \qquad c_0 \to 0 \text{ as } r \to \infty, \qquad -\pi/2 < \theta < \pi;$$

in either case we have as $t \to \infty$ the self-similar forms

$$c_0 \sim C\left(x/t^{\frac{1}{2}}, y/t^{\frac{1}{2}}\right), \qquad f_0 \sim t\phi\left(x/t^{\frac{1}{2}}, y/t^{\frac{1}{2}}\right),$$

cf. (3.18). The rescaled evolution equation for $\lambda \to \infty$ with $K(c_0) = K_1 c_0$ reads

$$\frac{\partial f_0}{\partial t} = \frac{1}{K_1}\left(\frac{\partial}{\partial x}\left(f_0\frac{\partial^2 f_0}{\partial x \partial t}\right) + \frac{\partial}{\partial y}\left(f_0\frac{\partial^2 f_0}{\partial y \partial t}\right)\right),$$

which, for the reason outlined above, admits similarity reductions such as

$$f_0 = f_0(x - T(t), y), \quad f_0 = f_0(r, \theta - T(t)), \quad f_0 = T^2(t)\phi\left(x/T(t), y/T(t)\right)$$

in which $T(t)$ is an arbitrary function.

3.3. Small Stefan number. We now consider the limit $\lambda \to 0$. The appropriate rescalings is

$$(3.36) \qquad\qquad t = \ln(1/\lambda)\bar{t};$$

as outlined in [2], this is not the only timescale needing to be discussed to give the full picture, but it suffices for our purposes here. We also restrict attention to the case in which $k(c, \hat{n}) \sim k_1(\hat{n})c$ as $c \to 0$. There is an interior layer about the moving boundary with scalings

$$(3.37) \quad \boldsymbol{x} = \boldsymbol{X}(\bar{t}) + \hat{n}z, \quad c \sim \lambda c_0, \quad V_n \sim V_{0_n}, \quad \hat{n} \sim \hat{n}_0, \quad \boldsymbol{X} \sim \boldsymbol{X}_0,$$

where $F = 0$ for $\boldsymbol{x} = \boldsymbol{X}$, with

$$\frac{d\boldsymbol{X}}{d\bar{t}} = \ln(1/\lambda)V_n(\hat{\boldsymbol{X}}, \bar{t})\hat{n}$$

so that $|\boldsymbol{X}| = O(\ln(1/\lambda))$; hence

$$(3.38) \qquad\qquad -V_{0_n}\frac{\partial c_0}{\partial z} = \frac{\partial^2 c_0}{\partial z^2}$$

$$\text{at } \bar{z} = 0 \qquad V_{0_n} = -\left(\frac{\partial c_0}{\partial z} + V_{0_n}c_0\right) = k_1(\hat{n}_0)c_0,$$

where $V_{0_n} = V_{0_n}(\boldsymbol{X}_0, \bar{t})$, so that

$$(3.39) \qquad\qquad c_0 = \left(1 + V_{0_n}\big/k_1(\hat{n}_0)\right)e^{-V_{0_n}z} - 1.$$

The outer region has $\boldsymbol{x} = \ln(1/\lambda)\bar{\boldsymbol{x}}$ and the WKBJ ansatz

$$\ln c \sim -\ln(1/\lambda)f(\bar{\boldsymbol{x}}, \bar{t})$$

applies, leading to

$$(3.40) \qquad\qquad \frac{\partial f}{\partial \bar{t}} = -|\bar{\nabla}f|^2.$$

The most important matching information comes from the scaling of c in (3.37), which implies that the moving boundary $F = 0$ corresponds to leading order to the level set $f = 1$ of (3.40); matching with the exponential in (3.39) implies that we also require

$$(3.41) \qquad\qquad \frac{\partial f}{\partial \bar{n}} = V_{0_n}$$

for $\bar{\boldsymbol{x}} = \boldsymbol{X}_0$, but this is automatic for a level set of (3.40). To complete the specification of (3.40) we need to impose conditions on the fixed part of $\partial\Omega$ and we shall not elaborate on this. We note, however, that f is not constant on characteristics of (3.40), so that level sets do not decouple; nevertheless the required solution to (3.40) in practice very often (for example, for the exterior problem, whereby $f = 0$ is prescribed on the fixed boundary with $f \to +\infty$ as $\bar{t} \to 0^+$ and as $|\bar{x}| \to +\infty$) takes the separable form

$$(3.42) \qquad f(\bar{x}, \bar{t}) = \phi^2(\bar{x})/4\bar{t},$$

where ϕ satisfies the eikonal equation

$$(3.43) \qquad |\bar{\nabla}\phi|^2 = 1.$$

We now show that (3.43) implies the interfacial dynamics law

$$(3.44) \qquad V_{0_n} = 1/\bar{t}^{\frac{1}{2}},$$

whereby we may write

$$(3.45) \qquad \frac{\partial F_0}{\partial \bar{t}} = -\frac{1}{\bar{t}^{\frac{1}{2}}}|\bar{\nabla}F_0|,$$

for which level sets do decouple; writing $F_0 = \phi(\bar{x}) - 2\bar{t}^{\frac{1}{2}}$, so the level set $F_0 = 0$ of (3.45) corresponds via (3.42) to $f = 1$ in (3.40), we recover (3.43), as required. Under such circumstances, at leading order the interface evolves isotropically, according to (3.45). However if k_1 is strongly anisotropic, in the sense that

$$(3.46) \qquad k_1(\hat{n}; \lambda) = K_1(\hat{n})e^{-\ln(1/\lambda)K_0(\hat{n})},$$

say, with $K_{0_{\min}} = 0$ and (for reasons which will shortly become apparent) $K_{0_{\max}} < 1$, then the inner scaling

$$(3.47) \qquad c \sim \lambda e^{\ln(1/\lambda)K_0(\hat{n})}c_0(\boldsymbol{X}, z, t)$$

(note that $\hat{n}$ does not depend on z) leads to a rather more complicated balance than (3.38) and the moving boundary is now determined at leading order via

$$(3.48) \qquad f = 1 - K_0(\hat{n}),$$

where f is the solution to (3.40); $K_{0_{\max}} < 1$ is thus required to obviate the need for a different formulation. Here one has a competition between the highly anisotropic form of the reaction rate and the strong isotropic effect of diffusion. Since $f(\bar{x}, \bar{t})$, as determined by (3.40), does not depend on the location of the moving boundary (it in fact gives the far-field behaviour of the solution to the corresponding purely diffusive problem in which no moving boundary is present), $F_0 = \omega_0(\bar{x}) - \bar{t}$ (and hence the moving boundary) can be constructed by first solving (3.40) and then determining ω_0 from

$$(3.49) \qquad K_0\left(\frac{\nabla\omega_0}{|\nabla\omega_0|}\right) = 1 - f(\bar{x}, \omega_0).$$

It does not in general seem possible to decouple level sets in such formulations, even when (3.42) applies, so the reduction to (3.44)–(3.45) given above is rather special; this 'non-local' aspect of moving boundary evolution is another consequence of the controlling role played by diffusion in the small-Stefan-number limit. While there are evident similarities between the reaction-controlled and small-Stefan-number limits (compare the isotropic version of (3.6) with (3.44)), in the former case even the orientation dependent reduced problem, (3.7), requires the tracking of only one level set, in contrast to (3.40).

4. Discussion. We conclude by noting some of the more interesting features of the homogeneous and heterogeneous models discussed in, respectively, Sections 2 and 3.

We start with a simpler (bistable) differential-difference system, on a square lattice, namely

$$(4.1) \qquad \frac{d}{dt}\phi_{i,j} = \frac{1}{\varepsilon^2}\left(\phi_{i+1,j} + \phi_{i-1,j} + \phi_{i,j+1} + \phi_{i,j-1} - 4\phi_{i,j}\right) + f(\phi_{i,j})$$

with

$$f(\phi_{\pm}) = 0, \qquad f'(\phi_{\pm}) < 0,$$

ϕ_+ and ϕ_- being separated by a single unstable equilibrium ϕ_u (i.e. $f(\phi_u) = 0$, $f'(\phi_u) > 0$). Writing $x = \varepsilon i$, $y = \varepsilon j$, $\phi_{i,j}(t) = \phi(x, y, t)$, (4.1) is formally equivalent to

$$(4.2) \qquad \frac{\partial \phi}{\partial t} = 2\sum_{m=1}^{\infty} \frac{\varepsilon^{2(m-1)}}{(2m)!}\left(\frac{\partial^{2m}\phi}{\partial x^{2m}} + \frac{\partial^{2m}\phi}{\partial y^{2m}}\right) + f(\phi).$$

Considering now travelling wave solutions

$$\phi = \phi(z), \quad z = x\sin\psi - y\cos\psi - qt$$

with ψ and q constant, we have

$$(4.3) \qquad -q\frac{d\phi}{dz}(z) = \frac{1}{\varepsilon^2}\Big(\phi(z + \varepsilon\sin\psi) + \phi(z - \varepsilon\sin\psi) + \phi(z + \varepsilon\cos\psi) + \phi(z - \varepsilon\cos\psi) - 4\phi(z)\Big) + f(\phi(z))$$

or

$$(4.4) \qquad -q\frac{d\phi}{dz} = 2\sum_{m=1}^{\infty} \frac{\varepsilon^{2(m-1)}}{(2m)!}\left(\sin^{2m}\psi + \cos^{2m}\psi\right)\frac{d^{2m}\phi}{dz^{2m}} + f(\phi).$$

The identity

$$\sin^{2m}\psi + \cos^{2m}\psi = \frac{(2m)!}{2^{2m}}\left(\frac{1}{(m!)^2} + 2\sum_{l=1}^{[m/2]} \frac{1}{(m-2l)!\,(m+2l)!}\cos 4l\psi\right)$$

holds, with (4.4) exhibiting the invariance of (4.3) under translations of ψ by $\pi/2$. Requiring

$$\text{as } z \to \pm\infty \qquad \phi \to \phi_\pm,$$

the wavespeed q in (4.3) or (4.4) is determined as part of the solution and we now elaborate on the continuum limit $\varepsilon \to 0^+$ in order to clarify the orientation dependence of q; similar procedures can also be applied in the more complicated scenario described in Section 2. Writing

$$\phi \sim \phi_0(z) + \varepsilon^2 \phi_1(z), \quad q \sim q_0 + \varepsilon^2 q_1 \qquad \text{as } \varepsilon \to 0,$$

we have at leading order the usual bistable reaction-diffusion problem

(4.5)
$$\frac{d^2\phi_0}{dz^2} + q_0 \frac{d\phi_0}{dz} + f(\phi_0) = 0,$$

$$\text{as } z \to \pm\infty \qquad \phi_0 \to \phi_\pm,$$

whereby ϕ_0 is independent of ψ and

$$q_0 = - \int_{\phi_-}^{\phi_+} f(\phi) \, d\phi \Big/ \int_{-\infty}^{\infty} \left(\frac{d\phi_0}{dz} \right)^2 dz;$$

we shall treat the case $q_0 \neq 0$. The first correction terms satisfy

$$\frac{d^2\phi_1}{dz^2} + q_0 \frac{d\phi_1}{dz} + f'(\phi_0)\phi_1 = -q_1 \frac{d\phi_0}{dz} - \frac{1}{48}(3 + \cos 4\psi)\frac{d^4\phi_0}{dz^4},$$

$$\text{as } z \to \pm\infty \quad \phi_1 \to 0,$$

so that

$$\frac{d}{dz}\left(e^{q_0 z} \left(\frac{d\phi_0}{dz}\frac{d\phi_1}{dz} - \phi_1 \frac{d^2\phi_0}{dz^2} \right) \right)$$
$$= -e^{q_0 z}\frac{d\phi_0}{dz}\left(q_1 \frac{d\phi_0}{dz} + \frac{1}{48}(3 + \cos 4\psi)\frac{d^4\phi_0}{dz^4} \right),$$

implying the solvability condition

(4.6)
$$q_1 = \frac{q_0}{96}(3 + \cos 4\psi)\left(q_0^2 - 3\frac{\int_{-\infty}^{\infty} e^{q_0 z}\left(\frac{d^2\phi_0}{dz^2} \right)^2 dz}{\int_{-\infty}^{\infty} e^{q_0 z}\left(\frac{d\phi_0}{dz} \right)^2 dz} \right),$$

making explicit the orientation dependence of the wave speed. The analysis just given is intended to clarify by means of a simpler example the role of discreteness in leading to anisotropic propagation laws. In the type of context outlined in Section 2, it is only when the width of the reaction layer is not substantially greater than the lattice spacing (i.e. when

$\varepsilon = O((\Delta x)^{\frac{1}{2}}))$ that such orientation effects come into play at leading order on the macroscopic (outer) scale; reaction zones exhibiting such very rapid spatial variation are observed in practice and, as outlined above, when this occurs an anisotropic macroscopic reaction rate automatically results, even though the continuum limit of the original formulation is isotropic. Such issues of discrete interior layers (and the way in which they match into continuum outer regions) are thus of interest from both asymptotic and practical points of view. We note from the analysis just given that quasi-continuum (rather than genuinely discrete) models may have a useful role to play in giving insight into such matters; in particular, perturbation results such as (4.6) (giving small orientation dependence) can be reproduced by truncating the summation in (4.2) after a finite number of terms (an odd number of terms presumably being needed for well-posedness).

Inner analyses such as those of Section 2 lead to rather general kinetic laws (cf. (2.8)) motivating the consideration of more general kinetic under-cooling expressions than the linear one considered in [2], this being work in progress. The homogeneous models obviate the need to track an interface explicitly (a property shared with phase field approaches, though the physics involved is significantly different) and can be expected to resolve the non-uniqueness issues which arise in the heterogeneous (moving boundary) formulations (cf. [2]). Stefan problems with kinetic undercooling (as discussed in [2] and in Section 3 above) are noteworthy not only for such non-uniqueness difficulties but also because of the ability of an interface to propagate corners and discontinuities in curvature (for example), even in the isotropic version of the model (this is discussed in [4]); both the non-uniqueness and the interface singularity propagation phenomena just alluded to relate here to a hyperbolic (eikonal) equation arising in the reaction-controlled limit, so neither occurs in the classical Stefan problem. In the framework of a homogeneous model, smoothing of a corner in the outer solution will be achieved over an inner region which is genuinely multidimensional at leading order, rather than being one-dimensional as in [1] and Section 2 above. Non-uniqueness can be illustrated via similarity solutions to (3.7) of the form

$$F = y - tg(x/t)$$

(cf. (3.8)), whereby $g(\eta)$ satisfies

$$(4.7) \qquad \lambda\left(g - \eta\frac{dg}{d\eta}\right) = \hat{K}\left(\frac{dg}{d\eta}\right)\left(1 + \left(\frac{dg}{d\eta}\right)^2\right)^{\frac{1}{2}},$$

where

$$\hat{K}\left(\frac{dg}{d\eta}\right) = K\left(1, \frac{\nabla F}{|\nabla F|}\right).$$

Equation (4.7) is of Clairaut's form and the non-uniqueness here relates to the existence of a singular (envelope) solution, as well as the general one (cf. [4] and [2], details of the isotropic case being given in the latter); the singular solution plays a crucial role in describing the large-time behaviour of closed curves.

The introduction of discreteness into a reaction-diffusion model is well known to lead to additional phenomena, such as the possible failure of wave propagation. The aspect we have chosen to emphasise here is the way in which it can lead to anisotropy in the corresponding macroscopic model, this being an example of the interaction between very different spatial scales.

REFERENCES

[1] J.D. EVANS AND J.R. KING. On the derivation of heterogeneous reaction kinetics from a homogeneous reaction model. *SIAM J. Appl. Math.* **60**, 1977–1996, 2000.

[2] J.D. EVANS AND J.R. KING. Asymptotic results for the Stefan problem with kinetic undercooling. *Q. J. Mech. Appl. Math.* **53**, 449–473, 2000.

[3] M.E. GURTIN. Thermomechanics of evolving phase boundaries in the plane. Clarendon Press, Oxford, 1993.

[4] J.R. KING. DPhil Thesis, University of Oxford, 1986.

[5] J.A. SETHIAN. Level set methods. Cambridge University Press, Cambridge, 1996.

FEATURE-SCALE TO WAFER-SCALE MODELING AND SIMULATION OF PHYSICAL VAPOR DEPOSITION

PETER L. O'SULLIVAN*, FRIEDER H. BAUMANN*, GEORGE H. GILMER*, JACQUES DALLA TORRE†, CHAN-SOO SHIN‡, IVAN PETROV‡, AND TAE-YOON LEE‡

Abstract. We present results of modeling and simulation of sputter-deposited thin films used for barrier and seed layers for the metalization stage of integrated circuits. We employ a continuum model in conjunction with the level set method for simulating the motion of the interface. An important physical input in our code is the angular distribution of the arriving material flux. For this we can incorporate empirical data for target erosion as well as the effects of gas scattering during transport in the low pressure sputter chamber. The chamber dimensions are of the same order of magnitude as the mean free path of sputtered atoms and the vapor transport is simulated in a pre-processing Monte Carlo module. First, we compare results for different angular distributions with experimental data for titanium. Second, we report on variations in step coverage across the wafer. Third, we report on validation with experiment for long-throw deposition of tantalum. Finally, we report preliminary work on simulating columnar growth.

Key words. Physical vapor deposition, sputtering, numerical simulation, feature-scale, wafer-scale, angular distribution.

AMS(MOS) subject classifications. Primary 82B21, 82B80, 81T80, 82C70, 82B24, 65-06, 65M99, 76M28.

1. Introduction. Metalization is the "back-end" of the integrated circuit (IC) fabrication process where the transistor interconnections are formed. Fig. 1 shows an example of this for a static random access memory chip. Current and future generations of ICs always require the deposition of thin diffusion barrier layers (such as Ta, TaN, Ti or TiN) as well as thin seed layers (mainly Cu) before the bulk of the metal is deposited. Barrier layers prevent in-diffusion into the dielectric or the underlying conducting regions of the transistors while seed layers promote adhesion. An example is shown in Fig. 1. Both types of layer are essential to the successful performance of the circuits and are even more crucial since they are among the later steps of a long and expensive industrial process.

By and large, these layers are deposited using low pressure physical vapor deposition (PVD) by direct current (DC) magnetron sputtering of polycrystalline metal targets. This process is illustrated schematically in Fig. 2. The magnet assembly behind the target increases the sputtering efficiency and is rotated to provide axisymmetric erosion of the target (and

---

*Bell Laboratories, Lucent Technologies, Murray Hill, NJ 07974. Emails: pos/frieder/ ghg@bell-labs.com.

†LAAS-CNRS, Université Paul Sabatier, 31077 Toulouse, France. Email: dalla@ laas.fr.

‡Materials Research Laboratory, University of Illinois, Urbana, IL 61801.

hence axisymmetric deposited films on the wafer). At 1–2 mTorr (0.13–0.27 Pa) the mean free path of the sputtered atoms is around 5–10 cm which is also the typical "throw distance" from the target to the substrate. Therefore, the vapor transport through the chamber is in the transitional regime of kinetic theory.

The integrity of these thin barrier layers is crucial for the reliability of the metal interconnects and hence, predictive modeling and simulation can be a valuable tool. In this paper we employ a continuum model for the thin film which ignores microstructure (although, as device sizes shrink this assumption is breaking down). The formation of microstructural voids or pores can potentially lead to in-diffusion of the conducting metal (Al or Cu) towards the electrical contacts or the SiO_2 inter-metal dielectric layer leading to performance degradation and possible device failure. An example of this is shown in Fig. 3 where the atomistic Monte Carlo code of Huang *et al.* [15] has been used to simulate Ta deposition at oblique incidence, as happens on the sidewalls of trenches and vias used in metalization. The porous microstructure is a result of a geometric self-shadowing instability in conjunction with the low surface mobility of Ta. We will address this issue briefly at the end of the paper.

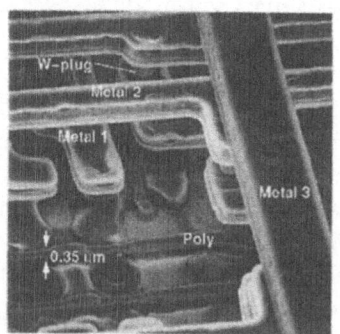
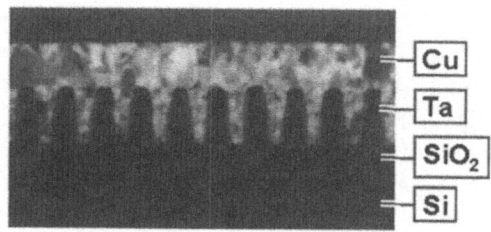

FIG. 1. *Left: Scanning electron micrograph (SEM) of an SRAM chip with dielectric etched away to expose multiple levels of metalization (courtesy F. Baumann, Lucent Technologies Inc.). Right: Cross-sectional SEM of a test structure showing the underlying Si, the insulating SiO_2, the barrier layer of Ta and finally the main Cu layer.*

A second industrial concern is the ability to predict the minimum "field" coverage (namely, film thickness on the flat parts of the wafer) in order to obtain a given minimum "step" coverage (i.e. film thickness) inside the features (trenches or vias). If such information were accurately known it would lead to increased throughput and a more efficient process. In order to achieve this goal we must know the relative sputter rate at the target since the action of the magnets is to provide non-uniform target erosion. We must also take into account collisions of the sputtered atoms with the background ambient gas in the chamber in order to obtain the

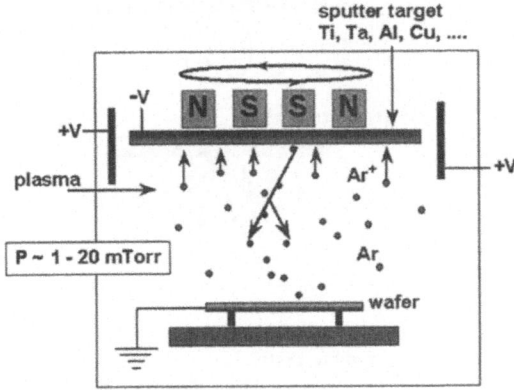

FIG. 2. *Schematic diagram of magnetron chamber. Argon ions (for example) are accelerated towards the target and kick out metal atoms which travel towards the wafer possibly suffering a few collisions with the background Ar gas. Note: pressure of 1 mTorr equals 0.133 Pa.*

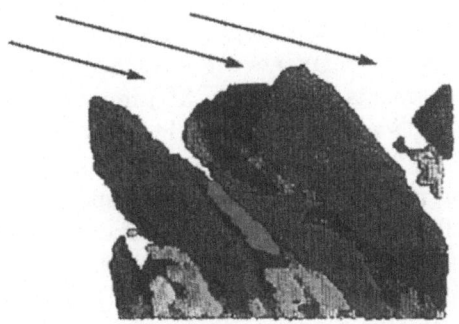

FIG. 3. *Atomistic lattice Monte Carlo simulation of Ta deposited at oblique incidence. Gray shades denote different grains (of a single [001] texture).*

local angular distribution of arriving flux at any location on the wafer. These two components of our modeling effort will be described in § 2.2.

The most rapid method of predicting thin film deposition over submicron trenches and vias is the continuum model where microstructural information is ignored. In this approximation, the macroscopic deposition rates throughout features are computed based on self-shadowing and line-of-sight ballistic deposition from target to substrate (wafer). The target can either be interpreted literally (i.e. we ignore scattering during transport) or, if we do account for scattering then it can be represented using an angular distribution of an effective "virtual" target located immediately above the wafer as shown in Fig. 4.

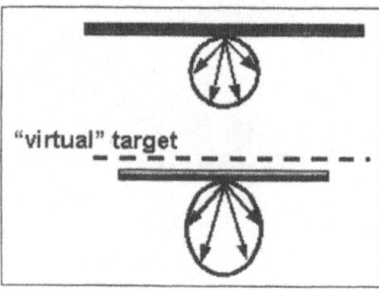

FIG. 4. *Effect of gas scattering on sputtered atoms. Angular distribution (shown as polar plot) at target relaxes to angular distribution immediately above wafer. Transport is from top to bottom.*

The present state-of-the-art of continuum models for film evolution and a more complete list of references is described in the review article by Cale *et al.* [9]. The continuum model of topographic evolution for moving interfaces has traditionally been simulated numerically using marker-particle/segment-based methods [5, 23] or front-tracking methods [14, 12, 13]. One major drawback in these approaches is the formation of "swallow-tails" when adjoining line segments are advanced and cross one another. These overlapping segments (surfaces in 3D) must then be "de-looped" in order to obtain a single-valued surface. This de-looping involves complex decision rules and, in 3D, requires considerable programming effort and excessive computing time. The level set technique was introduced by Osher and Sethian [20] as a fast alternative to front-tracking (see Ref. [26] for a comprehensive introduction) and has been used by Adalsteinsson and Sethian for microelectronics applications [1–3]. Since the level set approach is *not* a segment-based method, self-intersections of curves and surfaces never occur and therefore, there is no need for de-looping. In fact, topographic break-up or merger occurs naturally without any need for programmer-intervention for special cases.

In this paper we describe results on sputter deposition using level set simulations for the topography evolution. We have several results. Firstly, we demonstrate the need for accurate angular distributions if one wants to accurately predict thin film coverage over topography. Secondly, we show that, using a 3D level set code, the film thickness across the wafer varies significantly in an applied setting. Thirdly, we show a failing of continuum modeling in certain situations (based on validation with experiment). Finally, we demonstrate that it is possible to simulate columnar microstructure using the continuum model in conjunction with level set simulations.

2. Continuum model. In this section we describe the standard continuum model of thin film deposition and the numerical algorithms associated with advancing the topography in time. The essential details of this problem are described in Refs. [19, 5, 23] where segment-based algorithms are used to perform two dimensional simulations of sputter deposition. In these earlier works the initial feature profile was discretized into a finite number of linear segments. The flux of sputtered material was then computed on each segment based upon the parts of the target which were visible (i.e. not shadowed by other parts of the feature profile). The transport of atoms from the target to the wafer can be assumed to be collisionless which is a zeroth order approximation for low pressure PVD. Alternatively, if gas-phase scattering is taken into account then a different angular distribution (from that at the target) must be provided as input to the topography simulator. We will describe a Monte Carlo code which models this vapor transport in § 2.2.

We distinguish now between the model and the numerical algorithm. The model determines how the flux, or local growth rate, is obtained. In the references cited above the material (Al, Ti, Ta, etc.) was arbitrary and surface diffusion, resputtering and atomic length-scale effects were all neglected. Non-unity sticking probability can be incorporated [7, 8] although this is still within the ambit of a continuum model.

The segment-based methods represent a numerical method of advancing the film-vapor interface in time, subject to some given flux at each point on the feature profile. An alternative to segment-based algorithms are level set methods which were developed by Osher and Sethian [20] as a general numerical analytic method of propagating interfaces without recourse to *ad hoc* decision making. The propagation of the interface (which we will denote as Γ) is modeled as a partial differential equation (PDE)

$$(1) \qquad \frac{\partial \phi}{\partial t} + \mathrm{F} \left| \nabla \phi \right| = 0,$$

where F is the growth rate and ϕ is the level set function. The function ϕ is taken to equal the signed distance from the interface, Γ (positive in the vapor and negative inside the material). Hence, $\phi = 0$ corresponds to Γ. Numerically, a fixed rectangular Eulerian grid and finite difference schemes are employed to solve this PDE. Clearly, the growth rate, F, is only defined on Γ but can be "extended" to the other level sets (and the rest of the domain) mathematically and numerically. For two dimensional problems this PDE is solved in the plane for the motion of curves and in 3D it is solved in space for the motion of surfaces. Once the flux, or local growth rate, on Γ has been computed, it is extended to the entire domain and finally the PDE is solved. In some problems Γ need never be computed explicitly but in this problem Γ *must* be computed at every time step because the flux depends non-locally on Γ itself. This computation is achieved using a contour plotter given ϕ on the finite difference grid. For brevity we omit any more details of these methods but refer to [26].

The level set method essentially applies Huyghens' principle to the advancing front thereby producing a unique solution at corners or at locations where the flux is discontinuous. The method has an enormous advantage in three dimensions where topological merger, pinch-off etc. occurs naturally without any user intervention. Also, the computer-programming effort is minimal compared with front-tracking methods [12, 13].

2.1. Level set simulations. We have previously implemented the level set method for simulating sputter deposition in both two and three dimensions [21] including the effects of self- shadowing. We have also validated all of our codes using analytic formulae for step coverage inside trenches and vias. In the case of ballistic transport of sputtered material from the target to the wafer the flux on the interface, F, is computed using

$$(2) \qquad F(\mathbf{x}) = \int \int V(\mathbf{x}, \mathbf{x}') \frac{w(\rho) f(\theta) \cos \gamma}{|\mathbf{x} - \mathbf{x}'|^2} dA.$$

Here, $\mathbf{x}$ is on the interface, $\mathbf{x}'$ is on the target, $V(\mathbf{x}, \mathbf{x}')$ is a visibility factor, $w(\rho)$ is a weight factor for non-uniform sputtering at the target, θ is the angle of emission from the target, γ is the angle made by the arriving material with the local unit normal, $r = |\mathbf{x} - \mathbf{x}'|$, and dA is the area element on the target. ρ denotes the radial distance of $\mathbf{x}'$ from the axis of the target. The visibility factor, $V(\mathbf{x}, \mathbf{x}')$, of each target area element is computed for every point on the interface (at each time step) using ray tracing algorithms, where $V(\mathbf{x}, \mathbf{x}') = 1(0)$, if $\mathbf{x}'$ is visible (not visible) from $\mathbf{x}$. For 3D problems the ray tracing consumes most of the computation (we will not discuss these methods here though). The function $f(\theta)$ represents the angular distribution from the target and equals the amount of material per unit solid angle emitted at an angle θ to the downward normal as shown in Fig. 5.

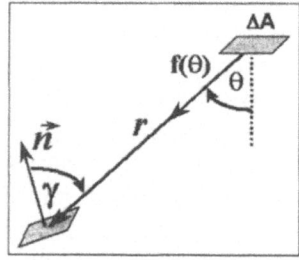

FIG. 5. *Geometric flux computation based on visibility of target area element, ΔA.*

For isotropic emission $f(\theta) = \cos \theta$ which is often assumed in practice. However, for many materials under low energy sputtering this cosine distribution is merely a first approximation. If we take into account gas-phase scattering inside the magnetron chamber then the flux is computed by setting $w(\rho) = 1$ but with $f(\theta)$ computed in a pre-processing module which *does* utilize the non-uniform target erosion weight function, as described below.

2.2. Vapor transport. The probability of travelling a distance x without suffering a collision with a background gas atom is $\exp(-x/\lambda)$, where λ is the mean free path. For typical operation of a magnetron at 2 mTorr the mean free path is around 5 cm. With a typical target-substrate distance of 6 cm it follows that approximately 40% of the target flux undergoes collisionless transport. Those that do suffer collisions are, on average, deflected by 10° and 38° in the case of Ta and Ti, respectively, based on their mass ratios with Ar. This last estimate indicates that gas-phase scattering may have some non-trivial effects in the case of Ti PVD even at this low pressure but that sputtered Ta is much less influenced.

We have developed a Monte Carlo code to simulate this vapor transport. We model the process as that of two hard spheres colliding elastically. The physical input to the simulation is the mean free path and the atomic masses of the sputtered atom and that of the background gas (typically Ar). The simulation uses pseudo-random numbers to generate a location on the target (suitably weighted by $w(\rho)$ for non-uniform target erosion) and a direction for the initial trajectory of a sputtered atom. Another pseudo-random number determines if and when a collision takes place according to

$$(3) \qquad\qquad x = -\lambda \ln r,$$

where λ is the mean free path, r is the random number ($0 \leq r < 1$) and x is the distance travelled. This results in the correct Poisson process for collisions. If the trajectory intersects the wafer then the point of intersection and angle of arrival is stored. If the trajectory takes the atom outside the chamber or back above the target then we discard the information and start again. If the trajectory remains inside the chamber and above the wafer then we perform a collision calculation where again, we select the impact parameter and the plane of the collision based on pseudo-random numbers. When the statistics are sufficiently converged we stop the calculation and compute the probability density for the arriving angle data. This yields $f(\theta)$ which will then be used in the level set code. In this manner, the real target is replaced conceptually by a virtual target immediately above the wafer with this function as the angular distribution. For more details on this approach see e. g. Refs. [11, 16] and [18].

2.3. Surface kinetics. We have used a sticking coefficient of 1.0 in all of the simulations we report here since this is appropriate for refractory materials at room temperature (see e.g. Ref. [22] for data on Ti PVD). Also, recent molecular dynamics simulations by Coronell *et al.* [10] for Cu show that low energy atoms ($E < 15$ eV) have a sticking probability very close to 1.0, regardless of their impact angle. Therefore, higher order complex physical effects, such as re-sputtering, have not been included in this work.

3. Comparison with experiment (Ti). We now compare simulation with experimental data for sputter deposited Ti/TiN in axisymmetric circular vias [4]. The experiments were conducted in an industrial magnetron sputter chamber with Ar plasma at a pressure of 2 mTorr (0.27 Pa). In the simulations we varied the feature aspect ratio (AR) from 0.5 to 2.0 keeping the feature depth, d, fixed at 1.4 μm (the same depth as in the experiments). We performed three sets of simulations. First, we employed the commonly used cosine angular distribution. Second, we employed the angular distribution obtained from our vapor transport code described earlier. Thirdly, we used a semi-empirical subcosine distribution suggested by Malaurie and Bessaudou [16]. In Fig. 6(a) we show these three angular distributions. The dashed line in the figure indicates the cut-off angle in the case of collisionless transport based on the radius of the target and its separation from the wafer. No flux arrives at angles above this value which is around 68° for a target of diameter 30 cm at a distance of 6 cm from the wafer. We see that the MC data for Ti at 2m Torr are "overcosine" meaning that there is more flux at lower angles compared with the cosine distribution. There is also some "spill" above the cut-off angle due to scattering. Conversely, the subcosine distribution yields higher flux at higher angles and less at normal incidence from the target. This is shown in the accompanying polar plot where 0° corresponds to the vertical line.

In Fig. 6(b) we have plotted the experimental data (circles plus error bars) together with the simulation results. The symbols represent the final bottom coverage (BC) after 12.5% of the feature depth has been deposited on top in each case. Also, the data are plotted as a function of the initial aspect ratio in each case. For the cosine distribution, we observe reasonable agreement but with a consistent over-prediction. Quantitatively, we find, for example, that the cosine distribution over-predicts bottom coverage by 30% at AR=1. The MC-based distribution over-predicts bottom coverage even more due to the gas-phase collimation effect. The subcosine distribution (without scattering) underestimates the bottom coverage. Therefore, we conclude that the subcosine distribution is qualitatively more accurate as the true angular distribution for low energy sputtering of Ti. In future work we will perform a scattering computation using this distribution although ideally it would be more satisfactory to use either experimental data or molecular dynamics data for the true (intrinsic) angular distribution from the target.

4. Across-wafer Non-uniformity. We now demonstrate the utility of a 3D topographic simulator in predicting step coverage as a function of radial location, x_{off}, on a wafer. In Fig. 7 we show the target-wafer geometry. Also shown are surface plots of the initial-final topography of a deposition into a contact via with aspect ratio, AR, of 2 located at the edge

(a)

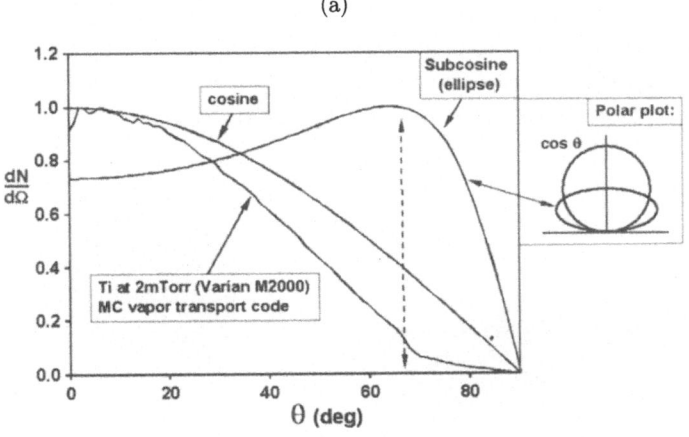

(b)

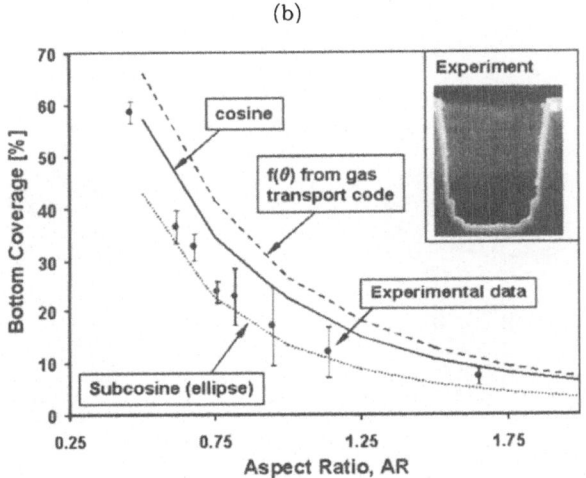

FIG. 6. (a) Angular distributions used to validate simulations with experiment. Vertical dashed line indicates maximum angle based on target diameter for ballistic transport. (b) Bottom coverage in contact vias as function of aspect ratio for different angular distributions: cosine (solid line), result from scattering code (dashed line), subcosine (dotted) and experimental data for Ti/TiN (symbols). Inset shows high resolution SEM.

of the wafer ($x_{off} = 10$ cm). In Fig. 8(a) we plot the relative sputter rate, $w(\rho)$, based on empirical data taken from a partially eroded target. Using this data we then computed the relative deposition rate on the wafer as a function of radial location, x_{off}, for a flat unpatterned wafer. This result is shown Fig. 8(b) where we also show the analagous result for the case of uniform target erosion. (We did not account for gas scattering in any of the results in this section). The non-uniform erosion leads to an improvement in uniformity across the wafer with deviations of only 6% compared with 16% for the uniform erosion case. In fact, there is an art to achieving uniform deposition rates at the wafer using the magnet configuration and this is, in general, an objective of back-end processing.

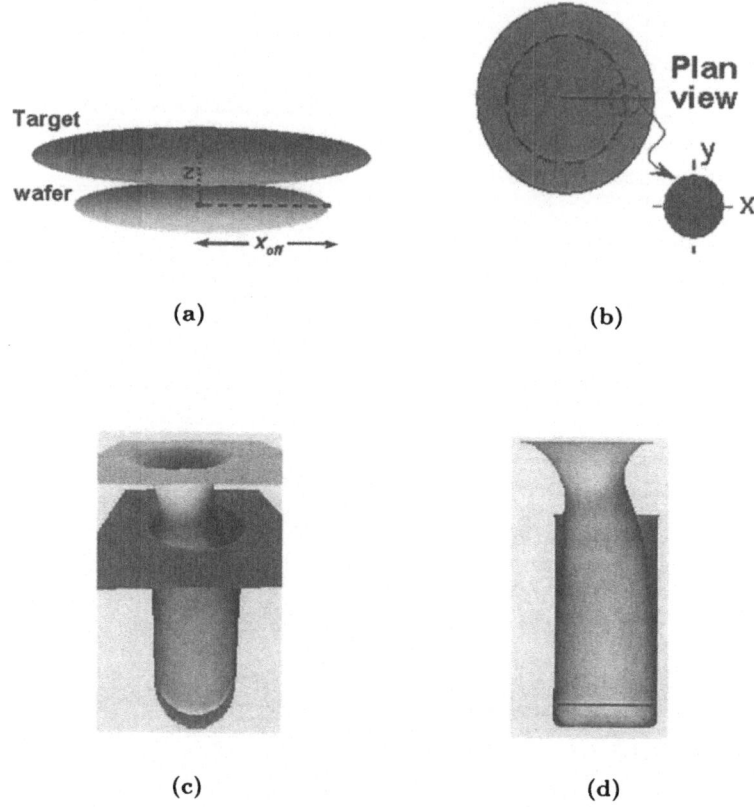

FIG. 7. *Schematic diagram for off-axis deposition (a) and in plan view (b). At (c), two cut-away views of simulated film in circular via located at $x_{off} = 10$ cm (aspect ratio 2:1). Solid line in image at (d) indicates location of minimum step coverage.*

Next, we ran our 3D simulation for several values of x_{off} with both uniform and non-uniform target erosion. In each case we deposited 100 nm of material on the top (field) and then computed the ratio of the film thickness to this field thickness. These results are shown in Fig. 9(a) and (b). The solid lines in each case denote the minimum coverage in the $x - z$, or radial, plane while the dotted lines are for the $y - z$, or tangential, plane. We see that both the minimum bottom coverage, BC, and the minimum sidewall coverage does not vary significantly as a function of x_{off} in the case of uniform target erosion. However, with non-uniform erosion there are large variations depending on location on the wafer. In Fig. 9(a) we see that BC can be reduced from 5% to less than 3% (50

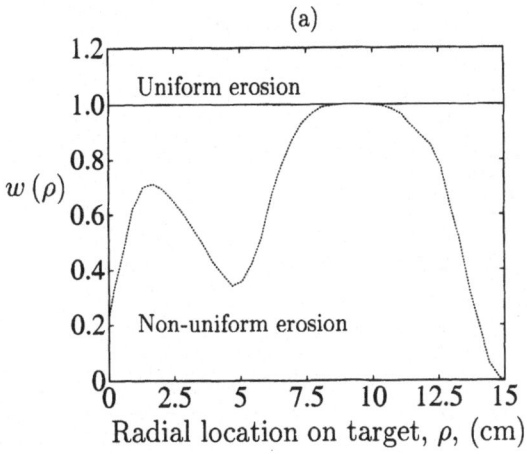

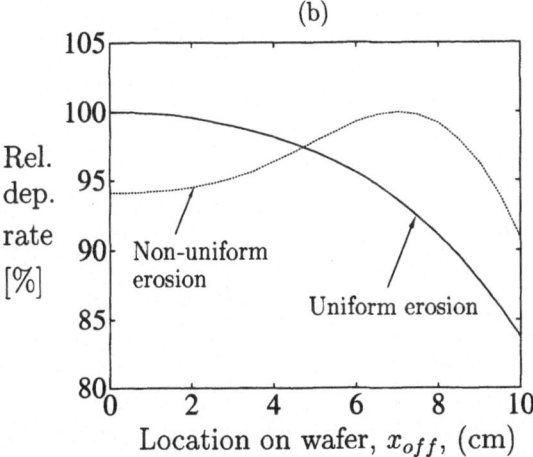

FIG. 8. *(a) Relative sputter rate (non-dimensional) typical of industrial magnetron targets (courtesy of J. E. Bower and M. Morris, Lucent Technologies, Inc.). Horizontal line denotes uniform target erosion. (b) Resulting blanket coverage for unpatterned wafer for both uniform (solid line) and non-uniform target erosion (dotted line).*

Å to around 25 Å physically, in the worst case). The sidewall coverage is not reduced so much (1.4 % to 1.1 %, or close to 10 Å in the worst case). In both cases the step coverage is improved in the region from $x_{off} = 8 - 10$ cm. Hence, in the case of barrier and seed layers deposited over topography, the uniformity deteriorates across the wafer *despite* the manufacturing objective of increasing uniformity using non-uniform target erosion.

5. Comparison with experiment (Ta). As an additional validation exercise we sputter-deposited thin films of Ta onto patterned substrates of SiO_2 in a DC magnetron apparatus at the University of Illinois. The

(a)

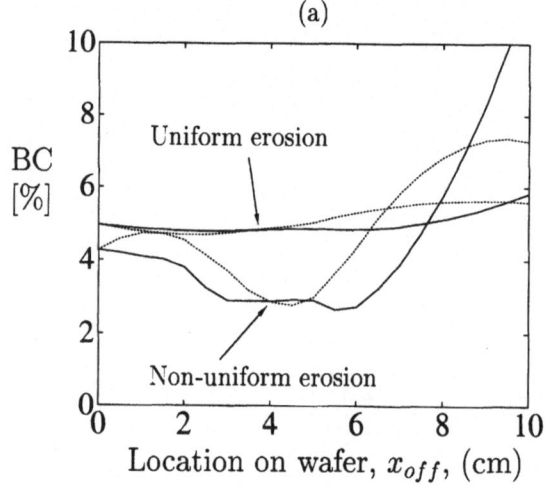

(b)

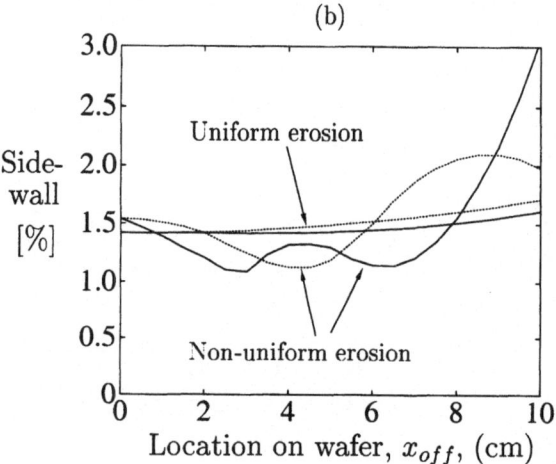

FIG. 9. *Step coverage in circular vias of aspect ratio 2:1 vs. radial location on wafer for both uniform and non-uniform target erosion. (a) Bottom coverage (BC) and (b) sidewall coverage. Solid lines denote minima in $x - z$ plane (radial direction) and dashed lines in $y - z$ plane (tangential direction).*

sputter gas was Ar at pressure of 1 mTorr. The magnetron was operated at 400 eV at room temperature and there was zero electrical bias applied to the substrate. The target-substrate geometry is depicted schematically in Fig. 10(a). To simulate this experiment we first modeled the target erosion profile based on empirical data. This provided the weight function, $w(\rho)$, which was used in our vapor transport code and is shown in Fig. 10(b). There is a peak in the sputter rate at the location of the "racetrack". Note that in this experimental apparatus the co-annular magnets did not rotate but still provided axisymmetric erosion. The resultant angular distribution after using this data in our vapor transport code is shown in Fig. 10(c), together with the cosine distribution for comparison.

An example of one such deposition into a contact via is shown in the SEM of Fig. 11(a), where the aspect ratio is 1.09 and approximately 160 nm of Ta was deposited on the field. The throw distance is 10 cm while the target diameter is 6.35 cm. Hence, at low pressure where there is negligible scattering the flux at the substrate is confined to a cone of angles less than $\theta_C = \tan^{-1}(3.175/10.0) \approx 18°$. That is to say, the flux of Ta is fairly collimated.

In Fig. 11(a) we see three important features of the film morphology. First, the sidewalls exhibit a highly columnar microstructure. This is a straightforward consequence of the high angles of incidence of the largely collimated flux together with the extremely low surface mobility of Ta at room temperature. Second, the film appears to be fully dense on those parts of the via which are approximately normal to the incident flux, namely on the bottom of the via and on the field. The film in these places is very smooth which is a consequence of the almost-normal incidence but also perhaps because of the energy of the impinging atoms. The energy of the sputtered Ta is in the range 5–20 eV on average while the energy of reflected Ar is around 100 eV on average. (The reflected Ar accounts for roughly 20% of the arriving flux). Third, the minimum step coverage occurs at the upper end of the sidewall rather than the more usual location at the bottom of the sidewall in less collimated deposition. Immediately above this "pinched" region the film bulges sideways forming an almost semi-circular overhang which is the cause of the pinch at the upper sidewall. This overhang is known to engineers as "breadloafing" and is observed in long-throw sputter deposition [25, 6, 17], collimated deposition (see e. g. Fig. 14 of Ref. [24] and Fig. 2 of Ref. [17]) and also in low energy ionized PVD [24] (at higher energies resputtering causes faceting). The lateral overgrowth should be distinguished from the more usual "keyhole" formation since here, there is the associated pinch at the corner.

In Fig. 11(b) and (c) we show two simulations in cross-section using the cosine distribution (no scattering) and the distribution from our scattering code (see Fig. 10(c)). Using the cosine distribution alone does not reproduce the experiment satisfactorily. By taking into account the scattering (Fig. 11(c)), the agreement improves markedly in terms of the bottom coverage. However, we see that the morphology at the upper corner of the via is still not reproduced in the simulations. This is an area of ongoing research which we believe is related to atomic scale effects. With the advent of low energy ionized PVD this failing of the continuum model becomes more significant if one wants to develop a robust and general-purpose topography simulation capability.

6. Columnar growth. Here, we present some preliminary numerical experiments related to columnar growth. We performed 2D simulations

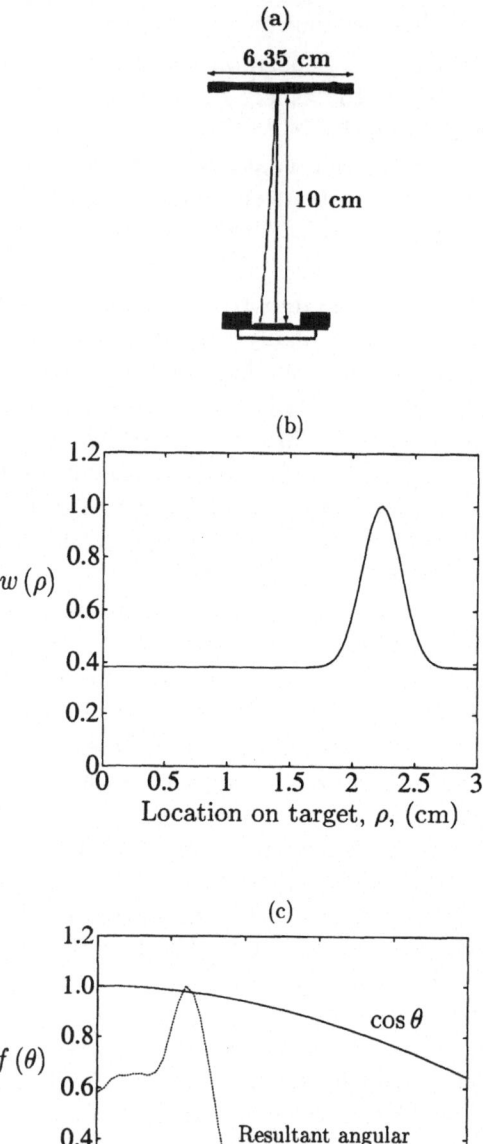

FIG. 10. *(a) Schematic drawing of Ta deposition experiment. Target erosion (as shown) is incorporated into simulations. (b) Relative sputter rate vs. radial distance on target (derived from empirical target erosion data). (c) Computed angular distribution at wafer from MC vapor transport code.*

(a)

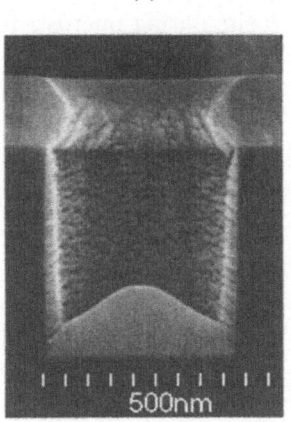

(b)   (c)

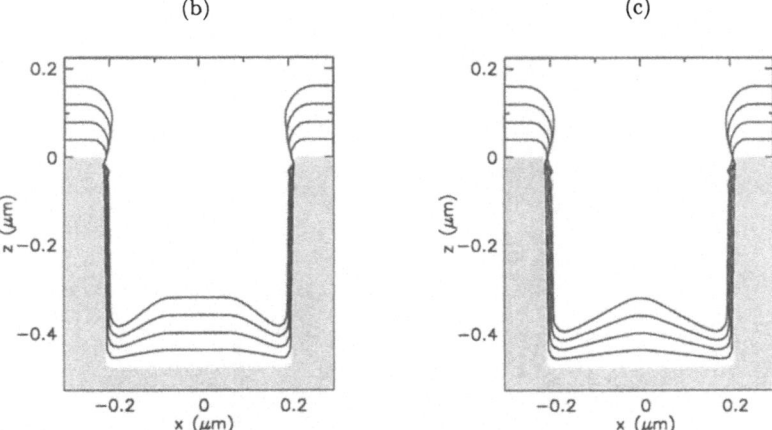

FIG. 11. *(a) Cross-sectional SEM of axisymmetric contact via with Ta barrier layer. Dark areas indicate SiO₂ substrate and light areas indicate Ta. (b) Simulation using cosine angular distribution. (c) Simulation using angular distribution taken from MC vapor transport code of Fig. 10(c).*

of deposition into trenches where the initial feature profile was perturbed using a sawtooth form. The results are shown in Fig. 12. The objective is to see if the continuum model is capable of producing columnar growth. Clearly, the model and simulation do produce a microstructure resembling the sidewalls of the SEM shown previously. The angle of growth is approximately in the direction of the incoming flux (around 20° in this case based on a (2D) target width of 3.175 cm at 10 cm from the feature. The perturbations have been imposed artificially and there is no physical mechanism to sustain the growth (such as statistical fluctuations which are present in the experiment). Nevertheless, this result points to the fact that, in principle, there is no obstacle to simulating microstructure in cases such as this using a continuum model.

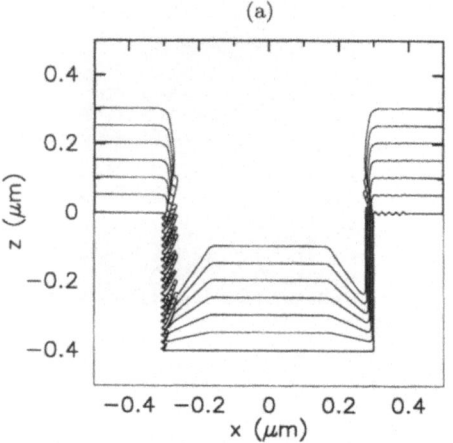

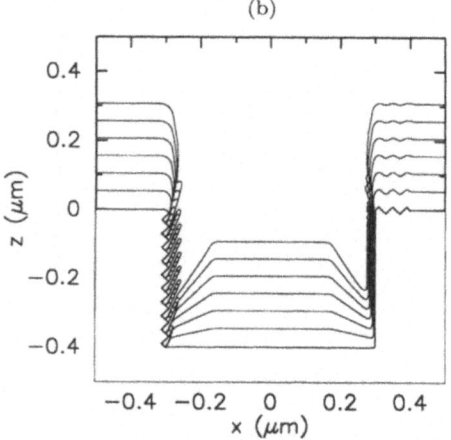

FIG. 12. *Simulated deposition into a trench with sawtooth perturbations of the initial feature profile. Left: amplitude of perturbations is* 4Δ *and right: amplitude of perturbations is* 8Δ, *where* Δ *is the grid size. Domain is* $1\mu m \times 1\mu m$ *and grid resolution is* 400×400 *giving* $\Delta = 0.0025\mu m$.

7. Summary. We have described our work on feature-scale to wafer-scale modeling and simulation for predicting thin film topography and step coverage in sub-micron features. A crucial component of this kind of work is the link between gas scattering calculations and the topography evolution. We have used a simple hard sphere binary collision model and the Monte Carlo technique to generate angular distributions which feed into the continuum simulations. Undoubtedly, there is scope for refining this module. Nevertheless, even such a simple model together with target erosion modeling leads to improved agreement with experiment. Our validation work for titanium and tantalum points to the need for more accurate data for the "intrinsic" angular distribution of these materials (at the target before scattering occurs). It also points to a deficiency in the continuum model in certain cases, namely collimated, long-throw and ionized PVD. Lastly, we performed a parametric study of variation in step coverage across the wafer using a 3D level set code. Contrary to intuition perhaps, the step coverage varies *more* rapidly under the "optimized" conditions of non-uniform target erosion which are designed to produce a more uniform blanket film.

Acknowledgement. This work has been funded by an NSF/DARPA grant in Virtual Integrated Prototyping administered through the University of Illinois. We wish to thank Eric Bower and Mark Morris (Bell Labs.) for providing the target erosion data of Fig. 8 as well as Walter Brown, Rich Liu and C.S. Pai (Bell Labs.) for many helpful discussions on tantalum deposition.

REFERENCES

[1] D. Adalsteinsson and J.A. Sethian, *A level set approach to a unified model for etching, deposition and lithography I: Algorithms and two-dimensional simulations*, Journal of Computational Physics, **120** (1995), pp. 128–124.

[2] ——, *A level set approach to a unified model for etching, deposition and lithography II: Three-dimensional simulations*, Journal of Computational Physics, **122** (1995), pp. 348–366.

[3] ——, *A level set approach to a unified model for etching, deposition and lithography III: Redeposition, reemission, surface diffusion and complex simulations*, Journal of Computational Physics, **138** (1997), pp. 193–223.

[4] F.H. Baumann, R. Liu, C.B. Case, and W.Y.-C. Lai, *3D modeling of contact material deposition and its impact on equipment design parameters*, International Electron Devices Meeting Technical Digest, (1993), pp. 861–864.

[5] I.A. Blech and H.A. Vander Plas, *Step coverage simulation and measurement in a dc planar magnetron sputtering system*, Journal of Applied Physics, **54** (1983), pp. 3489–3496.

[6] J.N. Broughton, M.J. Brett, S.K. Dew, and G. Este, *Titanium sputter deposition at low pressures and long throw distances*, IEEE Transactions on Semiconductor Manufacturing, **9** (1996), pp. 122–127.

[7] T.S. Cale and G.B. Raupp, *Free molecular transport and deposition in cylindrical features*, Journal of Vacuum Science and Technology B, **8** (1990), pp. 649–655.

[8] T.S. Cale, G.B. Raupp, and T.H. Gandy, *Free molecular transport and deposition in long rectangular trenches*, Journal of Applied Physics, **68** (1990), pp. 3645–3652.

[9] T.S. Cale, B.R. Rogers, T.P. Merchant, and L.J. Borucki, *Deposition and etch processes: continuum film evolution in microelectronics*, Computational Materials Science, **12** (1998), pp. 333–353.

[10] D.G. Coronell, D.E. Hansen, A.F. Voter, C.-L. Liu, X.-Y. Liu, and J.D. Kress, *Molecular dynamics-based ion-surface interaction models for ionized physical vapor deposition feature scale simulations*, Applied Physics Letters, **73** (1998), pp. 3860–3862.

[11] C. EISENMENGER-SITTNER, R. BEYERKNECHT, A. BERGAUER, W. BAUER, AND G. BETZ, *Angular distribution of sputtered neutrals in a post magnetron geometry: Measurement and Monte Carlo simulation*, Journal of Vacuum Science and Technology A, **13** (1995), pp. 2435–2443.

[12] J. GLIMM, J.W. GROVE, X.L. LI, K.-M. SHYUE, Y. ZENG, AND Q. ZHANG, *Three-dimensional front tracking*, SIAM Journal on Scientific Computing, **19** (1998), pp. 703–727.

[13] J. GLIMM, S.R. SIMANCA, D. TAN, F.M. TANGERMAN, AND G. VANDERWOUDE, *Front tracking simulations of ion deposition and resputtering*, SIAM Journal on Scientific Computing, **20** (1999), pp. 1905–1920.

[14] S. HAMAGUCHI, M. DALVIE, R.T. FAROUKI, AND S. SETHURAMAN, *A shock-tracking algorithm for surface evolution under reactive-ion etching*, Journal of Applied Physics, **74** (1993), pp. 5172–5184.

[15] H. HUANG, G.H. GILMER, AND T. DÍAZ DE LA RUBIA, *An atomistic simulator for thin film deposition in three dimensions*, Journal of Applied Physics, **84** (1998), p. 3636.

[16] A. MALAURIE AND A. BESSAUDOU, *Numerical simulation of the characteristics of the different metallic species falling on the growing film in d. c. magnetron sputtering*, Thin Solid Films, **286** (1996), pp. 305–316.

[17] A.A. MAYO, S. HAMAGUCHI, J.H. JOO, AND S.M. ROSSNAGEL, *Across-wafer nonuniformity of long throw sputter deposition*, Journal of Vacuum Science and Technology B, **15** (1997), pp. 1788–1793.

[18] T. OHTA AND H. YAMADA, *A sputter equipment simulation system for VLSI device*, Vacuum, **51** (1998), pp. 479–484.

[19] W.G. OLDHAM, A.R. NEUREUTHER, C. SUNG, J.L. REYNOLDS, AND S.N. NANDGAONKAR, *A general simulator for VLSI lithography and etching processes: Part II-application to deposition and etching*, IEEE Transactions on Electron Devices, **27** (1980), pp. 145–1459.

[20] S. OSHER AND J.A. SETHIAN, *Fronts propagating with curvature-dependent speed: Algorithms based on Hamilton-Jacobi formulations*, Journal of Computational Physics, **79** (1988), pp. 12–49.

[21] P.L. O'SULLIVAN, F.H. BAUMANN, AND G.H. GILMER, *Simulations of physical vapor deposition into trenches and vias: validation and comparison with experiment*, Journal of Applied Physics, **88** (2000), pp. 4061–4068.

[22] B.R. ROGERS, C.J. TRACY, AND T.S. CALE, *Compositional variation in sputtered Ti-W films due to re-emission*, Journal of Vacuum Science and Technology B, **12** (1994), pp. 2980–2984.

[23] R.C. ROSS AND J.L. VOSSEN, *Plasma-deposited thin-film step coverage calculated by computer simulation*, Applied Physics Letters, **45** (1984), pp. 239–240.

[24] S. . ROSSNAGEL, *Directional and ionized physical vapor deposition for microelectronics applications*, Journal of Vacuum Science and Technology B, **16** (1998), pp. 2585–2608.

[25] S.M. ROSSNAGEL, C. NICHOLS, S. HAMAGUCHI, D. RUZIC, AND R. TURKOT, *Thin, high atomic weight refractory film deposition for diffusion barrier, adhesion layer, and seed layer applications*, Journal of Vacuum Science and Technology B, **14** (1996), pp. 1819–1827.

[26] J.A. SETHIAN, *Level Set Methods and Fast Marching Methods*, Cambridge University Press, New York, 1999.

WKB ANALYSIS IN THE SEMICLASSICAL LIMIT OF A DISCRETE NLS SYSTEM

STEPHEN P. SHIPMAN*

Abstract. The linear spectral problem associated with the inverse solution of a finite discrete nonlinear Schrödinger chain is studied in the semiclassical limit. The discrete spectral problem is a recursion relation for a vector quantity, with boundary conditions, depending on initial data and a spectral parameter. WKB analysis is performed and then interpreted for the case that the quantities in the chain are less than one in modulus. In this case, the spectrum lies on the unit circle and an asymptotic density is obtained. The density is supported by known facts about the discrete spectra, numerical results, and rigorous results concerning the asymptotics of the solution of the spectral boundary-value problem. In addition, the norming constants in the spectral transform are positive in this special case, and a proposed asymptotic norming exponent is corroborated by numerical data.

1. Introduction. This article examines the spectral transform associated with an inverse solution of a finite defocusing discrete nonlinear Schrödinger (DNLS) system of ordinary differential equations in the semiclassical limit. The problem possesses a dichotomy of behavior depending on initial data characterized by the unitarity or non-unitarity of the linear spectral problem. Formal, rigorous, and numerical results lead to an understanding of the asymptotics of the unitary case. The non-unitary case is not addressed and is as yet not understood. In the unitary case, the spectrum of eigenvalues lies on the unit circle of the complex plane, and in the semiclassical limit, the dimension of the linear problem is unbounded and we seek an asymptotic density of eigenvalues. Naive WKB analysis leads to a candidate for this density, which is then confirmed by numerical calculations, comparison with properties of the spectrum of the discrete problem, and rigorous asymptotics of the unitary eigenvalue problem. In addition, the proposed density has been applied successfully in [S] to the study of the semiclassical limit of the solution of the DNLS system. In the WKB analysis, the discrete index in the system of ODEs approaches a continuous variable and the typical intervals of "oscillatory" and "exponential" behavior of the solution arise. The density, as usual, involves an integral over an oscillatory interval. A candidate for the asymptotics of the associated norming constant has been proposed in [S] in light of analysis there of the semiclassical limit of the inverse spectral solution. The candidate, as is typical in such asymptotic problems, involves an integral over the exponential intervals for a special class of data, and it was chosen to provide the correct results in that analysis. It is not understood how it may arise directly from asymptotic analysis. In this article, however, it is corroborated by numerical results and by comparison with properties of the norming constant for the discrete system.

---

*Department of Mathematics, Duke University, Durham, NC 27708.

Previous work on continuum limits of discrete systems solvable by inverse methods and the asymptotic (WKB) analysis of the associated linear spectral problem includes the analysis by Deift and McLaughlin [DM] of a continuum limit of the Toda lattice. Using candidates arising from formal WKB analysis, they rigorously established the asymptotics of the solutions and the spectral density and norming constants. These results were preceded by rigorous results by Geronimo and Smith [GS] on asymptotic solutions to second-order recursion relations. Costin [C] has made rigorous the WKB results for finite-order recursion relations. Akin to the WKB analysis in this article but for the non-unitary case is the non-self-adjoint Zakharov-Shabat eigenvalue problem in the semiclassical limit studied by Bronski [B].

2. The spectral problem. The defocusing discrete nonlinear Schrödinger (DNLS) system

$$i\dot{Q}_n + Q_{n-1} - 2Q_n + Q_{n+1} - |Q_n|^2(Q_{n-1} + Q_{n+1}) = 0$$

is transformed under the change of dependent variable $Q_n \mapsto Q_n e^{-2it}$ into the system

$$(1) \qquad i\dot{Q}_n + (1 - |Q_n|^2)(Q_{n-1} + Q_{n+1}) = 0.$$

If one puts

$$(2) \qquad |Q_0(0)| = |Q_N(0)| = 1$$

into (1), then Q_0 and Q_N are constant in time and a finite subchain becomes detached from the rest of the chain. One then has a finite system of ordinary differential equations for $Q_1 \ldots Q_{N-1}$. This system is solvable by an inverse spectral method [V].

In the semiclassical limit of the finite system, one considers initial data of the form

$$(3) \qquad Q_n(0) = q(n\epsilon) \exp\left(\frac{i}{\epsilon} \phi(n\epsilon)\right),$$

in which q and ϕ are fixed functions on the real unit interval such that $q(0) = q(1) = 1$ and $\epsilon = 1/N$, and considers the limiting behavior of the modulus and phase as ϵ tends to zero. As we will see, the WKB analysis lends itself to a meaningful interpretation with regard to the asymptotic distribution of eigenvalues in this special case. However, if the condition $|Q_n| < 1$ is violated, there is no satisfactory interpretation (so far). The reason for this is that the spectrum, in the case $|Q_n| < 1$, is constrained to the unit circle of the complex plane, whereas otherwise such a constraint is not known.

We now discuss the eigenvalue problem associated with the inverse spectral solution for the finite discrete system (1, 2). Let $\{Q_n\}_{n=0}^N$ be given such that $|Q_0| = |Q_N| = 1$ and normalized such that $Q_0 = 1$. Denote Q_N by ξ:

$$\xi = Q_N, \quad |\xi| = 1.$$

Let z be an arbitrary complex parameter, and define the matrices

$$U_n(z) = \begin{bmatrix} z & \bar{Q}_n \\ Q_n & z^{-1} \end{bmatrix},$$

and the resulting "transfer matrices"

$$T_n(z) = \begin{bmatrix} \xi^{\frac{1}{2}} & 0 \\ 0 & \bar{\xi}^{\frac{1}{2}} \end{bmatrix} U_N(z) \dots U_n(z).$$

The eigenvalues in the spectral transform are the roots of the trace of T_1 as a function of z. We denote

$$J(z) = \operatorname{tr} T_1(z).$$

Let $F(z)$ denote the upper left entry of $T_1(z)$. The coefficients in the partial-fraction decomposition of F/J are the norming constants in the spectral transform. One shows that

$$J(z) = \xi^{\frac{1}{2}} z^{-N} \prod_{k=1}^{N} (z^2 - z_k^2) \qquad \text{(eigenvalues } z_k),$$

$$\frac{F(z)}{J(z)} = z^2 \sum_{k=1}^{N} \frac{W_k}{(z^2 - z_k^2)} \qquad \text{(norming constants } W_k).$$

In fact, the roots of J are equal to the eigenvalues of the following boundary-value problem for the discrete evolution of a complex vector $\mathbf{u}_n$ in $\mathbb{C}^2$:

$$(4) \qquad \mathbf{u}_{n+1}(z) = U_n(z)\mathbf{u}_n(z); \quad \mathbf{u}_0(z) = \begin{bmatrix} z \\ 1 \end{bmatrix}, \quad \mathbf{u}_{N+1}(z) = \begin{bmatrix} 0 \\ 0 \end{bmatrix}.$$

The following proposition lists a number of facts about the spectral problem. We use the notation $\hat{f}(z) := \overline{f(\bar{z}^{-1})}$. Proofs are omitted.

PROPOSTION 1. **Facts on the spectral problem**

1. *On the spectrum:*
    (a) *There are $2N$ eigenvalues, counting multiplicities.*
    (b) *The eigenvalues exist in plus-minus pairs.*
    (c) *If z is an eigenvalue, then so is $\bar{z}^{-1}$.*
    (d) *If the values of Q_n are all real, then the eigenvalues exist in conjugate pairs.*

(e) If $\{Q_n\}_{n=0}^N$ has spectrum $\{\pm z_k\}_{k=1}^N$, then, for any real constant χ, $\{Q_n e^{in\chi}\}_{n=0}^N$ has spectrum $\{\pm z_k e^{-i\chi/2}\}_{k=1}^N$.

(f) If $|Q_n| < 1$ for $n = 1, \ldots, N-1$, then $|z_k| = 1$ for $k = 1, \ldots, N$ and the eigenvalues are distinct.

2. *On the norming constants:*

(a) $\sum_{k=1}^N W_k = 1$.

(b) If $\bar{z}_k^{-1} = z_l$ then $W_l = \overline{W}_k$.

(c) If $|Q_n| < 1$ for $n = 1, \ldots, N-1$, then the norming constants W_k are real and positive and

$$W_k = \frac{G_k := |F(z_k)|}{\prod_{k' \neq k} |z_{k'}^2 - z_k^2|} \qquad (|Q_n| < 1).$$

(d) If the Q_n are all real and $z_{k'} = \bar{z}_k$, then $W_{k'} = \bar{W}_k$. In particular, if $|Q_n| < 1$ for $n = 1, \ldots, N-1$, then $W_{k'} = W_k > 0$ and $G_{k'} = G_k$.

(e) Using the notation in (2e), if $\{Q_n\}_{n=0}^N$ has norming constants $\{W_k\}$ and $z_{k'} = z_k e^{-i\chi/2}$, then $W_{k'} = W_k$.

(f) If $|Q_n| < 1$ for $n = 1, \ldots, N-1$ and the G_k are all equal, then the Q_n have the property that $Q_{N-n} = \xi \bar{Q}_n$.

3. The asymptotics of the spectral transform. We consider the eigenvalue condition in the semiclassical limit. The dependence on the spectral parameter will usually be suppressed. Let continuous functions q and ϕ be given such that q has two continuous derivatives and ϕ has three continuous derivatives and

$$q : [0,1] \to [0,1], \quad 0 \leq q(x) < 1 \quad \text{for} \quad x \in (0,1);$$
$$\phi : [0,1] \to \mathbb{R}, \qquad \phi(0) = 0;$$

and put $Q_n = q(n\epsilon) \exp(\frac{i}{\epsilon}\phi(n\epsilon))$. The eigenvalue condition is (4), in which

$$U_n = \begin{bmatrix} z & q(n\epsilon) \exp\left(-\frac{i}{\epsilon}\phi(n\epsilon)\right) \\ q(n\epsilon) \exp\left(\frac{i}{\epsilon}\phi(n\epsilon)\right) & z^{-1} \end{bmatrix}.$$

To make the problem amenable to WKB analysis, we can remove the large exponent from U_n by means of the change of coordinates

$$\mathbf{u}_n = \begin{bmatrix} u_n^1 \\ u_n^2 \end{bmatrix} = \begin{bmatrix} e^{-i\frac{\phi(n\epsilon)}{2\epsilon}} \tilde{u}_n^1 \\ e^{i\frac{\phi(n\epsilon)}{2\epsilon}} \tilde{u}_n^2 \end{bmatrix};$$

then the vectors $\check{\mathbf{u}}_n = \begin{bmatrix} \check{u}_n^1 & \check{u}_n^2 \end{bmatrix}^t$ satisfy

(5) $\qquad \begin{bmatrix} \check{u}_{n+1}^1 \\ \check{u}_{n+1}^2 \end{bmatrix} = \check{U}_n \begin{bmatrix} \check{u}_n^1 \\ \check{u}_n^2 \end{bmatrix}, \quad \check{U}_n = \begin{bmatrix} ze^{i\frac{\psi_n}{2}} & q_n e^{i\frac{\psi_n}{2}} \\ q_n e^{-i\frac{\psi_n}{2}} & z^{-1}e^{-i\frac{\psi_n}{2}} \end{bmatrix},$

in which $\psi_n = \frac{\phi(n\epsilon+\epsilon)-\phi(n\epsilon)}{\epsilon}$ and $q_n = q(n\epsilon)$. Let $\lambda_n^{\pm}$ be the eigenvalues of $\check{U}_n$ and $\mathbf{p}_n^{\pm}$ corresponding eigenvectors, and set $\theta_n = \arg \frac{\lambda_n^+}{\lambda_n^-}$. Then the following expansions are valid:

(6) $\qquad \begin{aligned} \check{U}_n &= \underline{\check{U}}^{\epsilon}(n\epsilon) & \text{where} \quad \underline{\check{U}}^{\epsilon}(x) &= \underline{\check{U}}(x) + \epsilon\underline{\check{U}}_1(x) + \mathcal{O}(\epsilon^2), \\ \lambda_n^{\pm} &= \underline{\lambda}^{\pm\epsilon}(n\epsilon) & \text{where} \quad \underline{\lambda}^{\pm\epsilon}(x) &= \underline{\lambda}^{\pm}(x) + \epsilon\underline{\lambda}_1^{\pm}(x) + \mathcal{O}(\epsilon^2), \\ \theta_n &= \underline{\theta}^{\epsilon}(n\epsilon) & \text{where} \quad \underline{\theta}^{\epsilon}(x) &= \underline{\theta}(x) + \epsilon\underline{\theta}_1(x) + \mathcal{O}(\epsilon^2), \\ \mathbf{p}_n^{\pm} &= \underline{\mathbf{p}}^{\pm\epsilon}(n\epsilon) & \text{where} \quad \underline{\mathbf{p}}^{\pm\epsilon}(x) &= \underline{\mathbf{p}}^{\pm}(x) + \epsilon\underline{\mathbf{p}}_1^{\pm} + \mathcal{O}(\epsilon^2). \end{aligned}$

So the underscore signifies functions of the continuous variable x. $\underline{\lambda}^{\pm\epsilon}(x)$ and $\underline{\mathbf{p}}^{\pm\epsilon}(x)$ are the eigenvalues and eigenvectors of $\underline{\check{U}}^{\epsilon}(x)$, and $\underline{\theta}^{\epsilon}(x) = \arg \frac{\underline{\lambda}^{+\epsilon}(x)}{\underline{\lambda}^{-\epsilon}(x)}$. One sees that

(7) $\qquad \underline{\check{U}}(x) = \begin{bmatrix} ze^{i\frac{\phi'(x)}{2}} & q(x)e^{i\frac{\phi'(x)}{2}} \\ q(x)e^{-i\frac{\phi'(x)}{2}} & z^{-1}e^{-i\frac{\phi'(x)}{2}} \end{bmatrix},$

and, for unitary spectral values $z = e^{i\eta}$,

(8) $\qquad \lambda_n^{\pm} = \cos\left(\eta + \frac{\psi_n}{2}\right) \pm \sqrt{q_n^2 - \sin^2\left(\eta + \frac{\psi_n}{2}\right)},$

(9) $\quad \underline{\lambda}^{\pm}(x, e^{i\eta}) = \cos\left(\eta + \frac{\phi'(x)}{2}\right) \pm \sqrt{q(x)^2 - \sin^2\left(\eta + \frac{\phi'(x)}{2}\right)}.$

3.1. WKB anaylsis. We begin the asymptotic analysis with a naive WKB approach to determine the leading-order behavior of the vector $\begin{bmatrix} \check{u}_n^1 & \check{u}_n^2 \end{bmatrix}^t$. We consider the approximate problem for vectors $\mathbf{v}_n$ given by

$$\mathbf{v}_{n+1} = \underline{\check{U}}(n\epsilon)\mathbf{v}_n$$

and perform leading-order WKB analysis on the components of $\mathbf{v}_n$ with respect to the basis of eigenvectors $\underline{\mathbf{p}}^{\pm}(n\epsilon)$ using the ansatz

(10) $\qquad \mathbf{v}_n = \exp\left(\frac{1}{\epsilon}S_+(n\epsilon)\right)\underline{\mathbf{p}}^+(n\epsilon) + \exp\left(\frac{1}{\epsilon}S_-(n\epsilon)\right)\underline{\mathbf{p}}^-(n\epsilon)$

in which S_+ and S_- are functions of x that are to be determined. We write $\mathbf{v}_{n+1}$ in two ways: On one hand,

$$\mathbf{v}_{n+1} = \sum_{\pm} \exp\left(\frac{1}{\epsilon}\left(S_\pm(n\epsilon) + \epsilon S'_\pm(n\epsilon)\right) + \mathcal{O}(\epsilon)\right) \underline{\mathbf{p}}^\pm(n\epsilon + \epsilon)$$

$$= \sum_{\pm} \exp\left(\frac{1}{\epsilon}S_\pm(n\epsilon)\right) \exp\left(S'_\pm(n\epsilon)\right)(1 + \mathcal{O}(\epsilon)) \underline{\mathbf{p}}^\pm(n\epsilon + \epsilon).$$

On the other hand, from the evolution of $\mathbf{v}_n$,

(11)
$$\mathbf{v}_{n+1} = \sum_{\pm} \lambda^\pm(n\epsilon) \exp\left(\frac{1}{\epsilon}S_\pm(n\epsilon)\right) \underline{\mathbf{p}}^\pm(n\epsilon)$$

$$= \sum_{\pm} \exp\left(\frac{1}{\epsilon}S_\pm(n\epsilon)\right) \lambda^\pm(n\epsilon)\left(\underline{\mathbf{p}}^\pm(n\epsilon + \epsilon) + \vec{\mathcal{O}}(\epsilon)\right).$$

Comparing the two representations of $\mathbf{v}_{n+1}$, one obtains the formal result

$$S'_\pm(x) = \log(\underline{\lambda}^\pm(x)), \quad \text{or} \quad S_\pm(x) = \int^x \log(\underline{\lambda}^\pm(y))\, dy.$$

Let us consider the implications of this result in the case that $0 \leq q(x) < 1$ for $0 < x < 1$. By Statement (2c) of Proposition 1, this condition constrains the spectrum to the unit circle. Thus, let us put $z = e^{i\eta}$. The ratio of the WKB components of $\mathbf{v}_n$ with respect to an eigenvector basis, which will be relevant in proposing the spectral density, is

$$\mathcal{R}(x, \eta) := \frac{\exp\left(\frac{1}{\epsilon}S_+(x, e^{i\eta})\right)}{\exp\left(\frac{1}{\epsilon}S_-(x, e^{i\eta})\right)} = \exp\left[\frac{1}{\epsilon}\int^x \log\frac{\lambda^+(y, e^{i\eta})}{\lambda^-(y, e^{i\eta})}\, dy\right].$$

We make some observations about the values of $\underline{\lambda}^\pm$ and this ratio: $\underline{\lambda}^\pm$ are either both real with the same sign or complex conjugates of each other. For a given value of η, x-regions with these different properties are separated from each other by "turning points" x_* for which $q^2(x_*) = \sin^2\left(\eta + \frac{\phi'(x_*)}{2}\right)$. In an x-interval in which $\underline{\lambda}^\pm(x)$ are both real, we find that $\mathcal{R}(x, \eta)$ is a real-valued function of x (plus a complex constant), and in an x-interval in which $\underline{\lambda}^\pm(x)$ are complex conjugate, $\mathcal{R}(x, \eta)$ is a unitary complex function of x (plus complex a constant). Thus the interval $[0, 1]$ is divided into "exponential" and "oscillatory" intervals separated by turning points, which depend on the value of η. For generic values of η, the x-values 0 and 1 are endpoints of exponential regions.

3.2. The spectral density. We now use the formal WKB result to propose an asymptotic distribution of eigenvalues. Letting $\begin{bmatrix} c_n^1 & c_n^2 \end{bmatrix}$ represent the vector $\breve{\mathbf{u}}_n$ with respect to the basis $\{\underline{\mathbf{p}}^\pm\}$, the boundary-value problem (4) sets conditions on the quantities $\arg\left(\frac{c_n^1}{c_n^2}\right)$ at $n = 1$ and $n = N$. Since we know that the eigenvalues are unitary, the problem is to specify those values of z, as z traverses the unit circle, for which the total increment of $\arg\left(\frac{c_n^1}{c_n^2}\right)$ is equal to $\arg\left(\frac{c_{N-1}^1}{c_{N-1}^2}\right) - \arg\left(\frac{c_1^1}{c_1^2}\right) + 2\pi k$ for some integer k. We already have the leading order behavior of $\arg\left(\frac{c_n^1}{c_n^2}\right)$: it is constant in an exponential region and equal to $\frac{1}{\epsilon}\int^{n\epsilon} \arg\frac{\lambda^+(x)}{\lambda^-(x)}\,dx$ in an oscillatory region. Thus the total increment from $n = 1$ to $n = N$ (or $x = 0$ to $x = 1$), to leading order, is $\frac{1}{\epsilon}\int_0^1 \arg\frac{\lambda^+(x)}{\lambda^-(x)}\,dx$ where the integrand is zero when x is in an exponential region. The asymptotic condition for eigenvalues $z_k = e^{i\eta_k}$ is then

$$\frac{1}{\epsilon}\int_0^1 \arg\frac{\lambda^+(x, z_k)}{\lambda^-(x, z_k)}\,dx \sim 2\pi k \quad (\epsilon \to 0).$$

Using the expression (9) for the eigenvalues $\lambda^\pm(x; e^{i\eta})$, one computes $\arg\frac{\lambda^+(x;e^{i\eta})}{\lambda^-(x;e^{i\eta})}$ and finds that this condition becomes

$$\Psi(\eta_k) \sim \epsilon k \quad (\epsilon \to 0),$$

where the asymptotic spectral distribution Ψ is defined by

$$\Psi(\eta) = \frac{1}{\pi}\int_0^1 \arctan Re\frac{\sqrt{\sin^2\left(\eta + \frac{\phi'(x)}{2}\right) - q(x)^2}}{\cos\left(\eta + \frac{\phi'(x)}{2}\right)}\,dx.$$

To determine the limiting density of eigenvalues, we see that the number of eigenvalues in a η-interval on which Ψ is monotonic is given asymptotically by $1/\epsilon$ times the absolute value of the increment of Ψ over that interval. Thus we obtain the density

$$\rho(\eta) := |\Psi'(\eta)|$$

$$(12) \qquad = \frac{1}{\pi}\left|\int_0^1 Re\frac{\sin\left(\eta + \frac{\phi'(x)}{2}\right)}{\sqrt{\sin^2\left(\eta + \frac{\phi'(x)}{2}\right) - q(x)^2}}\,dx\right|, \qquad 0 \le \eta \le 2\pi.$$

This means that, for any subinterval $[\eta_1, \eta_2]$ of $[0, 2\pi]$,

$$\#[\eta_1, \eta_2] \sim \frac{1}{\epsilon}\int_{\eta_1}^{\eta_2} \rho(\eta)\,d\eta \quad (\epsilon \to 0),$$

where "$\#$" indicates the number of eigenvalues in the given interval.

One can confirm that the asymptotic analogs of the spectral properties in Part 2 of Proposition 1 do hold for this proposed density:

Asymptotic analogs of Proposition 1, Part 1.

a. The number of eigenvalues should be asymptotically equal to $2/\epsilon$. This is the statement that

$$\int_0^{2\pi} \rho(\eta) \, d\eta = 2.$$

When $\phi'(x)$ is taken to be constant, this is easily verified. In this case, $\Psi(\eta)$ is increasing (resp. decreasing) when $\sin(\eta + \frac{\phi'}{2})$ is positive (resp. negative), and one finds that its total variation is 2.

b. The asymptotic analog of the plus-minus parity is that $\rho(\eta) = \rho(\eta + \pi)$.

d. The Q_n being real corresponds to $\phi(x) \equiv 0$. In this case, $\rho(-\eta) = \rho(\eta)$, which is the analog of conjugate parity of eigenvalues. Because of (b), there is then a four-fold spectral symmetry.

e. Multiplying the Q_n all by $e^{in\chi}$ is asymptotically analogous to adding the constant χ to $\phi'(x)$. This does indeed shift the proposed density function by $-\chi/2$, as it should.

3.3. The asymptotic norming exponent. It can be shown that the norming constants have the following asymptotic behavior in the semiclassical limit:

$$\lim_{\epsilon \to 0} \epsilon \log G_{k_\epsilon} = \Im(\eta_*) \quad \text{if} \quad z_{k_\epsilon} \to e^{i\eta_*} \quad \text{as} \quad \epsilon \to 0,$$

where $\Im$ is a function defined on the support of the asymptotic spectral density and is determined by q and ϕ. In the case that these data give rise to exactly two turning points for values of η in the support of the asymptotic density, a candidate for the asymptotic norming exponent $\Im(\eta)$ has been proposed in [S]. One shows that the turning-point condition

$$q(x)^2 - \sin^2\left(\eta + \frac{\phi'(x)}{2}\right) = 0$$

is equivalent to

$$2\eta = \alpha(x) + 2k\pi \quad \text{or} \quad 2\eta = \beta(x) + 2k\pi \quad \text{for some } k$$

where

(13)
$$\begin{aligned} \alpha(x) &= 2\arcsin(q(x)) - \phi'(x), \\ \beta(x) &= 2\pi - 2\arcsin(q(x)) - \phi'(x). \end{aligned}$$

and $\beta(x) \geq \alpha(x)$, with equality only at 0 and 1.

The proposed form of the derivative of $\Im$ is

(14)
$$\frac{d\Im}{d\eta} = \left[\pm \int_0^{x_-(\eta)} \pm \int_{x_+(\eta)}^1 \right] Re \frac{\sin\left(\eta + \frac{\phi'(x)}{2}\right)}{\sqrt{-\sin^2\left(\eta + \frac{\phi'(x)}{2}\right) + q(x)^2}} \, dx,$$

which is defined in the support of the spectral density. The $\pm$ sign is chosen as illustrated in Example 2 of Section 4. One can compare with this formula the asymptotic analogs of the properties of the norming constants in Part 3 of Proposition 1.

Asymptotic analogs of Proposition 1, Part 2.

    d. The Q_n being real corresponds to $\phi(x) \equiv 0$. The symmetry of the norming constant about the angle $\pi/2$ corresponds the the anti-symmetry of $d\mathfrak{I}/d\eta$, which is confirmed in the proposed formula.

    e. Multiplying the Q_n all by $e^{in\chi}$ is asymptotically analogous to adding the constant χ to $\phi'(x)$, thus shifting the proposed asymptotic norming exponent by $-\chi/2$.

    f. The property $Q_{N-n} = \xi\bar{Q}_n$ corresponds to the symmetry of q and ϕ' about $x = 1/2$, and one shows that the candidate for $d\mathfrak{I}/d\eta$ is zero in this case. This corresponds to the converse of item (3f).

A natural candidate for $\mathfrak{I}(\eta)$ may be derived heuristically as follows:

$$\frac{1}{N}\log|F(e^{i\eta})| \sim \frac{1}{N}\log\prod_{n=0}^{N}|U_n(e^{i\eta})| \sim \frac{1}{N}\log\prod_{n=0}^{N}\max|\lambda_n^{\pm}(e^{i\eta})|$$

$$\sim \frac{1}{N}\sum_{n=0}^{N}\log\max|\lambda_n^{\pm}(e^{i\eta})|$$

$$\sim \int_0^1 \log\max|\lambda^{\pm}(x,e^{i\eta})|dx \;=?\; \mathfrak{I}(\eta).$$

One finds, indeed, that this integral coincides numerically with a limiting upper envelope of the functions $\frac{1}{N}\log|F(e^{i\eta})|$. However, $\frac{1}{N}\log|F(e^{i\eta})| = \frac{1}{N}\log G_k$, from $\eta = 0$ to $\eta = \pi$, has $N-1$ spikes emanating downward from this upper envelope, and the N points $(\eta_k, \frac{1}{N}\log G_k)$ lie at various places along these spikes. This is illustrated in Example 9. Thus, $\frac{1}{N}\log G_k$ is not given by $\int_0^1 \log\max|\lambda^{\pm}(x,e^{i\eta})|dx$. Recall that $\sum_{k=1}^{N}W_k = 1$ and

$$W_k = \frac{G_k}{\displaystyle\prod_{k'\neq k}|z_k^2 - z_{k'}^2|}.$$

So $\frac{1}{N}\log G_k < \frac{1}{N}\sum_{k'\neq k}\log|z_k^2 - z_{k'}^2|$ for each k. If one calculates numerically the asymptotic form of the right-hand side,

$$\frac{1}{N}\sum_{k'\neq k_N}\log|z_{k_N}^2 - z_{k'}^2| \sim \int_0^{\pi}\log|e^{2i\eta} - e^{2i\eta'}|\rho(\eta')d\eta' \qquad (N \to \infty)$$

$$\text{if} \quad z_{k_N} \longrightarrow e^{i\eta} \quad \text{as} \quad N \to \infty,$$

one finds that it also coincides with the limiting upper envelope of $\frac{1}{N}\log|F(e^{i\eta})|$ (even for values outside the support of ρ).

4. Numerical results. The numerical calculations in Examples 1 and 2 compare the proposed asymptotic spectral density $\rho(\eta)$ and norming exponent $\mathfrak{J}(\eta)$ defined in (12) and (14) with actual spectral data for various choices of q, ϕ, and N.

5. Asymptotics of the transfer matrix. In this section, we take a rigorous approach to determining the asymptotic behavior of the transfer matrix over an oscillatory region and over an exponential region and establish some asymptotics of the solution to the linear problem. Let $[a, b]$ be an oscillatory or exponential interval for data $q(x)$ and $\phi(x)$ and spectral value $e^{i\eta}$ whose distance from any turning point is bounded from below. Define $\underline{n} = \lceil a/\epsilon \rceil$ and $\overline{n} = \lfloor b/\epsilon \rfloor$, and let $[c_n^1, c_n^2]^t$ represent the vector $\breve{u}_n(e^{i\eta})$ in the eigenvector basis $\{p_n^{\pm}(e^{i\eta})\}$ for $\breve{U}_n(e^{i\eta})$.

THEOREM 2. *Given the notation above,*

1. *Let $[a, b]$ be an oscillatory interval. Then, for each sufficiently small $\epsilon > 0$, there exists a solution $\begin{bmatrix} c_n^1 & c_n^2 \end{bmatrix}^t$ such that, if $n\epsilon \in [a, b]$, then*

$$\arg \frac{c_n^1}{c_n^2} = \frac{1}{\epsilon} \int_a^{n\epsilon} \arg \frac{\lambda^+(y)}{\lambda^-(y)} dy + A(n\epsilon) + \varrho(\epsilon, n\epsilon),$$

$$c_n^{1,2} = \left(A^{\pm}(n\epsilon) + \varrho_{\pm}(\epsilon, n\epsilon) \right) \exp \left(\frac{1}{\epsilon} \int_a^{n\epsilon} \underline{\lambda}^{\pm}(y) dy \right),$$

in which $A(x)$ and $A^{\pm}(x)$ are continuous functions depending on q and ϕ and the choice of eigenvectors and

$$\varrho(\epsilon, x), \varrho_{\pm}(\epsilon, x) = o(1) \quad (\epsilon \to 0),$$

uniformly in x.

2. *Let $[a, b]$ be an exponential interval, and suppose that $0 < \lambda_n^- < \lambda_n^+$. Then, for each sufficiently small $\epsilon > 0$, there exists a solution $\begin{bmatrix} c_n^1 & c_n^2 \end{bmatrix}^t$ such that, if $n\epsilon \in [a, b]$, then*

$$c_n^1 = (B(n\epsilon) + \varrho_1(\epsilon, n\epsilon)) \exp \left(\frac{1}{\epsilon} \int_a^{n\epsilon} \underline{\lambda}^+(y) dy \right),$$

$$c_n^2 = \varrho_2(\epsilon, n\epsilon) \exp \left(\frac{1}{\epsilon} \int_a^{n\epsilon} \underline{\lambda}^+(y) dy \right),$$

in which B is determined by q and ϕ and the choice of eigenvectors and depends continuously on its arguments and

$$\varrho_{1,2}(\epsilon, x) = o(1) \quad (\epsilon \to 0),$$

uniformly in x.

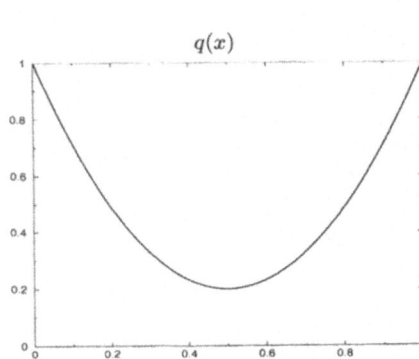

$q(x)$

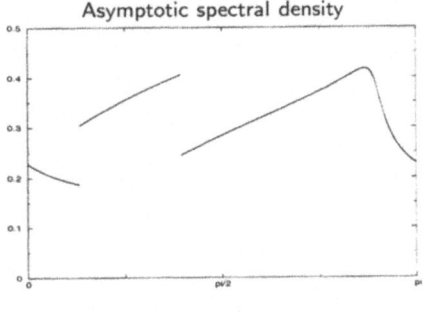

Asymptotic spectral density

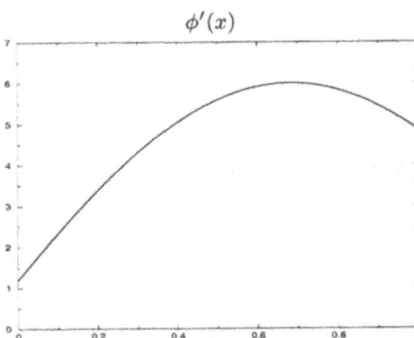

$\phi'(x)$

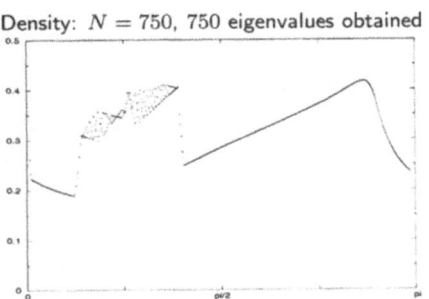

Density: $N = 750$, 750 eigenvalues obtained

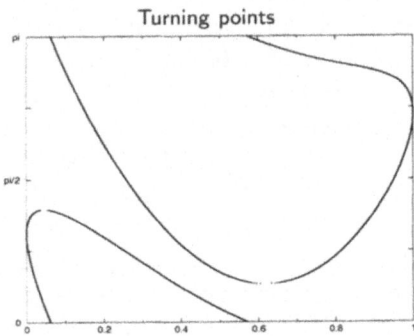

Turning points

Example 1. *This is an example in which ϕ' is not constant so that the spectral density has no symmetries. The value of the density changes abruptly at the values of η that separate regions with two turning points from those with four. The approximate density for $1/\epsilon = 750$ was obtained using 10 eigenvalues per density point. Two observations about the η-interval with four turning points: The three points where the upper and lower envelopes for the irregularly placed values of the approximate density come together coincide with the graph of the proposed asymptotic density. Using more eigenvalues per density point decreased the deviation from the asymptotic density.*

$d\mathfrak{I}/d\eta$ and diff. quot. of $\log(G_k)/N$ for $N = 16$.

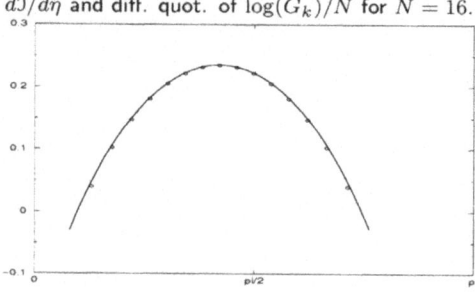

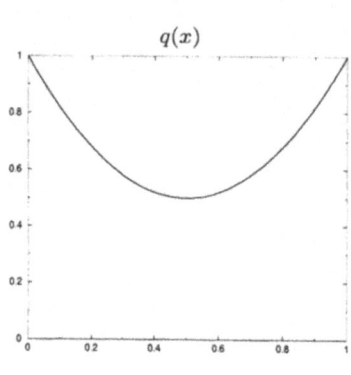

$q(x)$

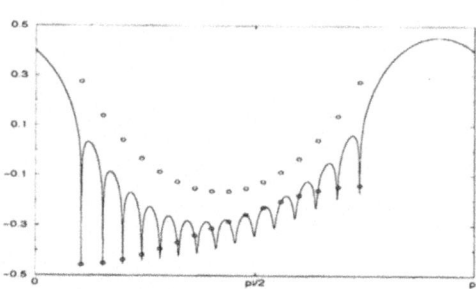

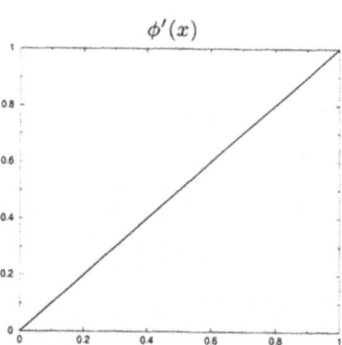

$\phi'(x)$

Example 2. $\mathfrak{I}'$ *is symmetric about* $\pi/2 -$ $1/4$. *This is because, if* q *and* ϕ *are symmetric about* $x = 1/2$, *then* $\mathfrak{I}'$ *is symmetric about* $\eta = \pi/2$, *and the shifting up of* ϕ' *by* $1/2$ *produces the shift to the left of the spectral data by half of that. The bottom graph above illustrates several things: The lower string of circles shows the values of* $\frac{1}{N} \log |F(e^{i\eta})|$ *(black line) evaluated at the eigenvalues* η_k *for* $N = 16$. *Such data are difficult to obtain for large values of* N *because of the sensitivity of* F *to changes in* η *on the spikes. For very large values of* N, *however, an upper envelope for* $\frac{1}{N} \log F(e^{i\eta})$ *can still be calculated, and the limiting values of this envelope as* $N \to \infty$ *is represented by the grey curve. Also coinciding with the grey curve are the two asymptotic quantities discussed in Subsection* 3.3—$\int_0^1 \log \max |\lambda^{\pm}(x, e^{i\eta})| dx$ *and* $\int_0^{\pi} \log |e^{2i\eta} - e^{2i\eta'}| \rho(\eta') d\eta'$. *The upper string of circles are the quantites* $\sum_{k' \neq k} \log |z_k^2 - z_{k'}^2|$ *plotted against* η_k.

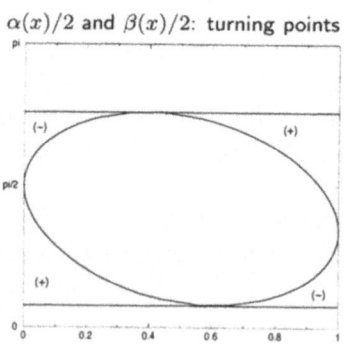

$\alpha(x)/2$ and $\beta(x)/2$: turning points

5.1. Preliminaries. Let $\tilde{U}_n$ be the matrix taking $\begin{bmatrix} c_n^1 & c_n^2 \end{bmatrix}^t$ to $\begin{bmatrix} c_{n+1}^1 & c_{n+1}^2 \end{bmatrix}^t$. Then $\tilde{U}_n = M_n \Lambda_n$ where $\Lambda_n = \mathrm{diag}(\lambda_n^+, \lambda_n^-)$ and M_n is the change-of-basis matrix from $\{\mathbf{p}_n^+, \mathbf{p}_n^-\}$ to $\{\mathbf{p}_{n+1}^+, \mathbf{p}_{n+1}^-\}$. Assuming three continuous derivatives of ϕ and two of q, and using the expansions (6), one computes that $M_n = I + \epsilon R_n$ where the entries r_n^{ij} of R_n have the property that, for some differentiable functions $\underline{r}^{ij}$ of x, $|r_n^{ij} - \underline{r}^{ij}(n\epsilon)| = \mathcal{O}(\epsilon)$ uniformly in $x \in [a, b]$. This means that $r_n^{ij} = \underline{r}^{ij\epsilon}(n\epsilon)$ for some functions $\underline{r}^{ij\epsilon}$ of x such that $\underline{r}^{ij\epsilon}(x) = \underline{r}^{ij}(x) + \mathcal{O}(\epsilon)$ as $\epsilon \to 0$ uniformly in x. We will study the asymptotic behavior of the transfer matrix T^ϵ taking $\begin{bmatrix} c_{\underline{n}}^1 & c_{\underline{n}}^2 \end{bmatrix}$ to $\begin{bmatrix} c_{\overline{n}+1}^1 & c_{\overline{n}+1}^2 \end{bmatrix}$:

$$T^\epsilon := \prod_{n=\underline{n}}^{\overline{n}} \tilde{U}_n = \prod_{n=\underline{n}}^{\overline{n}} (I + \epsilon R_n)\Lambda_n.$$

The multiplication is ordered, factors with a lower index being to the right of factors with a higher index. We will study the case in which $[a, b]$ is contained in an oscillatory x-region and the case in which it is contained in an exponential region. We begin the analysis by bringing this expression for T^ϵ into a form in which its structure and limiting behavior is more transparent. Expanding in powers of ϵ, T^ϵ takes the form

$$T^\epsilon = \sum_{\ell=0}^{L} \epsilon^\ell T_\ell$$

where $L = \overline{n} - \underline{n} + 1$ and

$$T_\ell := \sum_{\underline{n} \le n_1 < \dots < n_\ell \le \overline{n}} \left(\prod_{n=n_\ell+1}^{\overline{n}} \Lambda_n \right) R_{n_\ell} \left(\prod_{n=n_{\ell-1}+1}^{n_\ell} \Lambda_n \right) \cdots$$

$$\cdots R_{n_2} \left(\prod_{n=n_1+1}^{n_2} \Lambda_n \right) R_{n_1} \left(\prod_{n=\underline{n}}^{n_1} \Lambda_n \right)$$

and $T_0 := \prod_{n=\underline{n}}^{\overline{n}} \Lambda_n$. One can bring out a factor on the right, common to each T_ℓ, by using the following formula recursively: For any $i \le n' \le j$,

$$\left(\prod_{n=n'+1}^{j} \Lambda_n \right) R_{n'} \left(\prod_{n=i}^{n'} \Lambda_n \right) = \begin{bmatrix} r_{n'}^{11} & r_{n'}^{12} \displaystyle\prod_{n=n'+1}^{j} \frac{\lambda_n^+}{\lambda_n^-} \\ r_{n'}^{21} \displaystyle\prod_{n=n'+1}^{j} \frac{\lambda_n^-}{\lambda_n^+} & r_{n'}^{22} \end{bmatrix} \prod_{n=i}^{j} \Lambda_n.$$

Setting first (i, n', j) equal to $(n_{\ell-1}+1, n_\ell, \overline{n})$, then $(n_{\ell-2}+1, n_{\ell-1}, \overline{n})$, and so on up to $(\underline{n}, n_1, \overline{n})$, we arrive at the following expression for T_ℓ:

$$T_\ell = \left(\sum_{\underline{n} \leq n_1 < \ldots < n_\ell \leq \bar{n}} \hat{R}_{n_\ell} \cdots \hat{R}_{n_2} \hat{R}_{n_1} \right) \prod_{n=\underline{n}}^{\bar{n}} \Lambda_n,$$

in which

$$\hat{R}_{n'} := \begin{bmatrix} r_{n'}^{11} & r_{n'}^{12} \prod_{n=n'+1}^{\bar{n}} \frac{\lambda_n^+}{\lambda_n^-} \\ r_{n'}^{21} \prod_{n=n'+1}^{\bar{n}} \frac{\lambda_n^-}{\lambda_n^+} & r_{n'}^{22} \end{bmatrix}.$$

Using the notation

$$P_\ell := \sum_{\underline{n} \leq n_1 < \ldots < n_\ell \leq \bar{n}} \hat{R}_{n_\ell} \cdots \hat{R}_{n_1},$$

we can write

$$T_\ell = P_\ell \prod_{n=\underline{n}}^{\bar{n}} \Lambda_n$$

to obtain the form

$$T^\epsilon = \left(\sum_{\ell=0}^{L} \epsilon^\ell P_\ell \right) \prod_{n=\underline{n}}^{\bar{n}} \Lambda_n.$$

One computes the products $\hat{R}_{n_\ell} \cdots \hat{R}_{n_1}$ (the sums are over n):

$$\hat{R}_{n_2} \hat{R}_{n_1} =$$

$$\begin{bmatrix} r_{n_2}^{11} r_{n_1}^{11} + r_{n_2}^{12} r_{n_1}^{21} \prod_{n_1+1}^{n_2} \frac{\lambda_n^-}{\lambda_n^+} & r_{n_2}^{11} r_{n_1}^{12} \prod_{n_1+1}^{\bar{n}} \frac{\lambda_n^+}{\lambda_n^-} + r_{n_2}^{12} r_{n_1}^{22} \prod_{n_2+1}^{\bar{n}} \frac{\lambda_n^+}{\lambda_n^-} \\ r_{n_2}^{21} r_{n_1}^{11} \prod_{n_2+1}^{\bar{n}} \frac{\lambda_n^-}{\lambda_n^+} + r_{n_2}^{22} r_{n_1}^{21} \prod_{n_1+1}^{\bar{n}} \frac{\lambda_n^-}{\lambda_n^+} & r_{n_2}^{22} r_{n_1}^{22} + r_{n_2}^{21} r_{n_1}^{12} \prod_{n_1+1}^{n_2} \frac{\lambda_n^+}{\lambda_n^-} \end{bmatrix},$$

and the first column of $\hat{R}_{n_3} \hat{R}_{n_2} \hat{R}_{n_1}$ is

$$\begin{bmatrix} r_{n_3}^{11} r_{n_2}^{11} r_{n_1}^{11} + r_{n_3}^{12} r_{n_2}^{21} r_{n_1}^{11} \prod_{n_2+1}^{n_3} \frac{\lambda_n^-}{\lambda_n^+} + r_{n_3}^{11} r_{n_2}^{12} r_{n_1}^{21} \prod_{n_1+1}^{n_2} \frac{\lambda_n^-}{\lambda_n^+} + r_{n_3}^{12} r_{n_2}^{22} r_{n_1}^{21} \prod_{n_1+1}^{n_3} \frac{\lambda_n^-}{\lambda_n^+} \\ r_{n_3}^{21} r_{n_2}^{11} r_{n_1}^{11} \prod_{n_3+1}^{\bar{n}} \frac{\lambda_n^-}{\lambda_n^+} + r_{n_3}^{22} r_{n_2}^{21} r_{n_1}^{11} \prod_{n_2+1}^{\bar{n}} \frac{\lambda_n^-}{\lambda_n^+} + r_{n_3}^{22} r_{n_2}^{22} r_{n_1}^{21} \prod_{n_1+1}^{\bar{n}} \frac{\lambda_n^-}{\lambda_n^+} + r_{n_3}^{21} r_{n_2}^{12} r_{n_1}^{21} \prod_{n_1+1}^{n_2} \frac{\lambda_n^-}{\lambda_n^+} \prod_{n_3+1}^{\bar{n}} \frac{\lambda_n^-}{\lambda_n^+} \end{bmatrix}.$$

Inductively, we find that $\hat{R}_{n_\ell} \cdots \hat{R}_{n_1}$ includes the terms $r_{n_\ell}^{11} \cdots r_{n_1}^{11}$ and $r_{n_\ell}^{22} \cdots r_{n_1}^{22}$ in the upper left and lower right entries, respectively. The rest of the terms all contain factors that are products of the form $\prod_{n=n_1'}^{n_2'} \frac{\lambda_n^\pm}{\lambda_n^\mp}$. In the first column, λ_n^- always appears in the numerator, and in the second column, λ_n^+ always appears in the numerator. We find then that P_ℓ has the form

$$
P_\ell = \begin{bmatrix} \displaystyle\sum_{\underline{n} \le n_1 < \ldots < n_\ell \le \overline{n}} r_{n_\ell}^{11} \cdots r_{n_1}^{11} & 0 \\ 0 & \displaystyle\sum_{\underline{n} \le n_1 < \ldots < n_\ell \le \overline{n}} r_{n_\ell}^{22} \cdots r_{n_1}^{22} \end{bmatrix}
$$

(15)

$$
+ \sum_{\underline{n} \le n_1 < \ldots < n_\ell \le \overline{n}} \begin{bmatrix} (2^{\ell-1} - 1)\,\text{terms} & 2^{\ell-1}\,\text{terms} \\ 2^{\ell-1}\,\text{terms} & (2^{\ell-1} - 1)\,\text{terms} \end{bmatrix}
$$

(P_0 is the identity matrix) where the "terms" are as described above.

By induction on ℓ, one can prove the following Lemma on the structure of the first column of $\hat{R}_{n_\ell} \cdots \hat{R}_{n_1}$ and a similar lemma for the second column. $\hat{R}_{n_\ell} \cdots \hat{R}_{n_1}$ is assumed to be in simplified form in the sense that factors of the form $\frac{\lambda_n^+ \lambda_n^-}{\lambda_n^- \lambda_n^+}$ are removed.

LEMMA 3. *on the first column of* $\hat{R}_{n_\ell} \cdots \hat{R}_{n_1}$.

1. *The first entry contains the term* $r_{n_\ell}^{11} \cdots r_{n_1}^{11}$ *and* $2^{\ell-1} - 1$ *terms with factors of the form* $\displaystyle\prod_{n=m_1+1}^{m_2} \frac{\lambda_n^-}{\lambda_n^+}$ *for* $m_1, m_2 \in \{n_1, \ldots, n_\ell\}$ *(not the empty product). These factors have the following properties:*

   (a) *For any* n, *the factor* $\frac{\lambda_n^-}{\lambda_n^+}$ *occurs with multiplicity at most 1.*

   (b) *For one factor,* $m_2 = n_\ell$.

2. *The second entry contains* $2^{\ell-1}$ *terms with factors of the form* $\displaystyle\prod_{n=m_1+1}^{m_2} \frac{\lambda_n^-}{\lambda_n^+}$ *for* $m_1, m_2 \in \{n_1, \ldots, n_\ell, \overline{n}\}$ *(not the empty product). These factors have the following properties:*

   (a) *For any* n, *the factor* $\frac{\lambda_n^-}{\lambda_n^+}$ *occurs with multiplicity at most 1.*

   (b) *For one factor,* $m_2 = \overline{n}$.

5.2. Oscillatory region. Let us now consider the case in which $[a, b]$ is contained in an oscillatory region. The goal is to show that, as ϵ tends to zero, the transfer matrix is asymptotic to a diagonal matrix that depends only on q and ϕ, times $\prod_{n=\underline{n}}^{\overline{n}} \Lambda_n$. The task is to show that, by letting ϵ tend to zero, one can bring the expansion $\sum_{\ell=0}^{L} \epsilon^\ell P_\ell$ into any vicinity of a fixed diagonal matrix. Whereas P_0 is just the identity matrix, it is not obvious that the ϵ^1-term

$$
\epsilon P_1 = \epsilon \sum_{\underline{n} \le n_1 \le \overline{n}} \begin{bmatrix} r_{n_1}^{11} & r_{n_1}^{12} \displaystyle\prod_{n=n_1+1}^{\overline{n}} \frac{\lambda_n^+}{\lambda_n^-} \\ r_{n_1}^{21} \displaystyle\prod_{n=n_1+1}^{\overline{n}} \frac{\lambda_n^-}{\lambda_n^+} & r_{n_1}^{22} \end{bmatrix},
$$

for example, is tending to a diagonal form. One expects the diagonal entries to converge to the integrals $\int_a^b \underline{r}^{ii}(x)dx$ if the functions q and ϕ are sufficiently smooth. One can apply a naive formal argument to the other entries by replacing the product in, say, the $(1,2)$-entry by its asymptotic form $\exp\left(i\frac{1}{\epsilon}\int_a^x \underline{\theta}(x')dx'\right)$ and replacing the sum by an integral:

$$\epsilon \sum_{\underline{n}\leq n_1 \leq \overline{n}} r_{n_1}^{12} \prod_{n=n_1+1}^{\overline{n}} \frac{\lambda_n^+}{\lambda_n^-} \quad\longrightarrow\quad \int_a^b \underline{r}^{12}(x)\exp\left(i\frac{1}{\epsilon}\int_x^b \underline{\theta}(x')dx'\right)dx.$$

This formal limit does indeed tend to zero; however, converting what is essentially a Riemann sum into an integral is not so simple because of the fast oscillations in the integrand. Indeed, *the period of the oscillations is at the order of the mesh size ϵ.*

A similar but more complicated situation occurs in the higher-order terms in the expansion of T^ϵ. In an oscillatory region, the "terms" in expression (15) for P_ℓ are, for some index h, of the form

$$(16) \qquad r_{n_\ell}^* \cdots r_{n_h}^* \cdots r_{n_1}^*(u) \prod_{n=n_h+1}^{n_{h+1}} \frac{\lambda_n^\pm}{\lambda_n^\mp}$$

where n_{h+1} may be equal to $\overline{n}$ and n_{h-1} may be equal to $\underline{n}$, the asterisk $(*)$ represents any superscript from the set $\{11,12,21,22\}$, and u is a product of expressions that do not depend on n_h and are of the form $\prod_{n=n_1'}^{n_2'} \frac{\lambda_n^\pm}{\lambda_n^\mp}$ and are therefore unitary.

Again, one expects the quantities $\epsilon^\ell \sum_{\underline{n}\leq n_1<\ldots<n_\ell\leq\overline{n}} r_{n_1}^{ii}\cdots r_{n_\ell}^{ii}$ to converge to an ℓ-fold integral over the region $a \leq x_1 < \cdots < x_\ell \leq b$. Each of the other terms is oscillating in at least one of the variables n_h, and, extending the technique discussed above for the case when $\ell = 1$, one can show that each of these terms converges to zero.

Now, the number of these terms grows exponentially with ℓ, and, as ϵ decreases, the degree L of the expansion of T^ϵ in ϵ increases. One can solve these problems with the observation that the number of terms in a sum over $\underline{n} \leq n_1 < \ldots < n_\ell \leq \overline{n}$ is less than $\binom{N}{\ell}$ (recall that $\epsilon = 1/N$), and, upon multiplying by ϵ^ℓ, one can bound the whole expansion containing the "terms" of T^ϵ by a quantity that tends to zero as $\epsilon \to 0$. The details are in the proof of Proposition 5.

In formulating the lemma, the interval $[a,b]$ must be bounded away from any turning point so that the functions $\underline{r}^{ij}(x)$ are bounded and the function $e^{i\underline{\theta}(x)}$ is bounded away from the real axis. We make the following definitions:

- Let the number σ be such that $|1 - \exp(i\underline{\theta}(x))| > \sigma$ for $x \in [a,b]$.
- If ϕ'' is continuous on $[a,b]$, then $\frac{d\underline{\theta}}{dx}$ is bounded on $[a,b]$, so there exists a number κ such that $|\underline{\theta}(x_2) - \underline{\theta}(x_1)| \leq \kappa|x_2 - x_1|$ for all $x_1, x_2 \in [a,b]$.

- The difference quotients $\frac{\phi(x+\epsilon)-\phi(x)}{\epsilon}$ converge to $\phi'(x)$ uniformly on $[a, b]$ provided that ϕ'' is continuous. This implies the existence of a number τ such that $|\theta_n - \underline{\theta}(n\epsilon)| \leq \tau\epsilon$ whenever $n\epsilon \in [a, b]$.
- One can verify that the functions $\underline{r}^{ij}(x)$ have continuous derivatives on $[a, b]$, and this implies the existence of a number β such that $|\underline{r}^{ij}(x_2) - \underline{r}^{ij}(x_1)| \leq \beta|x_2 - x_1|$ for $x_1, x_2 \in [a, b]$.
- The existence of a number γ such that, for $i, j \in \{1, 2\}$, $|r_n^{ij} - \underline{r}^{ij}(n\epsilon)| < \gamma\epsilon$ has already been discussed.
- The continuity of the functions $\underline{r}^{ij}(x)$ and the previous bullet imply the existence of a number α such that, for $x, n\epsilon \in [a, b]$, $|\underline{r}^{ij}(x)| < \alpha$ and $|r_n^{ij}| < \alpha$.
- Define $S_i(a, b) = \int_a^b \underline{r}^{ii}(x)\, dx$, for $i = 1, 2$.

The first lemma obtains estimates on the oscillating terms in P_ℓ. One must understand why the sum over $\underline{n} \leq n_1 < \cdots < n_\ell \leq \overline{n}$ of any one of these terms, multiplied by ϵ^ℓ, tends to zero. Let Υ denote a general one of these quantities:

$$\Upsilon := \epsilon^\ell \sum_{\underline{n} \leq n_1 < \ldots < n_\ell \leq \overline{n}} r_{n_\ell}^* \cdots r_{n_h}^* \cdots r_{n_1}^*(u) \prod_{n=n_h+1}^{n_{h+1}} \frac{\lambda_n^{\pm}}{\lambda_n^{\mp}}.$$

LEMMA 4. *Let $[a, b]$ be an oscillatory interval with positive distance from the set of turning points, and let $\varrho > 0$ be given. Then $\epsilon' > 0$ can be chosen sufficiently small such that whenever $0 < \epsilon < \epsilon'$, $|\Upsilon| < \varrho\frac{(\alpha)^{\ell-1}}{(\ell-1)!}$ for any quantity of the type Υ.*

Proof. Denote by $\underline{\theta}(x)$ either $\arg\left(\frac{\lambda^+(x)}{\lambda^-(x)}\right)$ or $\arg\left(\frac{\lambda^-(x)}{\lambda^+(x)}\right)$, and by θ_n the corresponding quantity $\arg\left(\frac{\lambda_n^{\pm}}{\lambda_n^{\mp}}\right)$. Rewrite Υ as

$$\Upsilon = \epsilon^{\ell-1} \sum_{\underline{n} \leq n_1 < \ldots < \hat{n}_h < \ldots < n_\ell \leq \overline{n}} r_{n_\ell}^* \cdots \hat{r}_{n_h}^* \cdots r_{n_1}^*(u) \left[\epsilon \sum_{n_h=n_{h-1}+1}^{n_{h+1}-1} r_{n_h}^* \exp\left(i \sum_{n=n_h+1}^{n_{h+1}} \theta_n \right) \right].$$

The circumflex marks a removed factor or variable. Since $|r_{n_\ell}^* \cdots \hat{r}_{n_h}^* \cdots r_{n_1}^*(u)| < \alpha^{\ell-1}$ and there are no more than $\binom{N}{\ell-1} < \frac{N^{\ell-1}}{(\ell-1)!}$ terms in the outer sum, we see that

$$(17) \quad |\Upsilon| < \frac{\alpha^{\ell-1}}{(\ell-1)!} \max_{\underline{n} \leq n_1 < \ldots < \hat{n}_h < \ldots < n_\ell \leq \overline{n}} \left| \epsilon \sum_{n_h=n_{h-1}+1}^{n_{h+1}-1} r_{n_h}^* \exp\left(i \sum_{n=n_h+1}^{n_{h+1}} \theta_n \right) \right|.$$

Let us study a single quantity of the type

$$\Omega := \epsilon \sum_{n_h=n_{h-1}+1}^{n_{h+1}-1} r_{n_h}^* \exp\left(i \sum_{n=n_h+1}^{n_{h+1}} \theta_n \right).$$

Let $\bar{\varrho}$ be given such that $0 < \bar{\varrho} \leq \frac{1}{2}$ and $\bar{\varrho}^6 < \frac{1}{2}|b - x_*|$ for all turning points x_*, and assume that $0 < \epsilon < \bar{\varrho}^6$. Let M be a positive number such that $M^{-\frac{13}{2}} < \epsilon < M^{-6} < \bar{\varrho}^6$. The following procedure applies to any one of these quantities Ω. First, if $(n_{h+1} - n_{h-1} - 1)\epsilon < M^{-4}$, then it is clear that $|\Omega| < \alpha M^{-4}$. Otherwise, if $(n_{h+1} - n_{h-1} - 1)\epsilon > M^{-4}$, then we divide the interval $[(n_{h-1} + 1)\epsilon, n_{h+1}\epsilon)$ into disjoint subintervals $[\epsilon m_{k-1}, \epsilon m_k)$, $k = 1, ..., K$ where $m_0 = n_{h-1} + 1$ and $m_K = n_{h+1}$, such that, if we set $M_k = m_k - m_{k-1}$ for $k = 1, \ldots, K$, then $M^{-4} < M_k\epsilon < 2M^{-4}$. The first part of this inequality implies $K < M^4$. In summary, the conditions are

$$M^{-\frac{13}{2}} < \epsilon < M^{-6} < \bar{\varrho}^6,$$
$$M^{-4} < M_k\epsilon < 2M^{-4},$$
$$K < M^4.$$

Now break Ω into a sum

(18)
$$\Omega = \epsilon \sum_{k=1}^{K} \Omega_k,$$

in which

$$\Omega_k := \sum_{n_h = m_{k-1}}^{m_k - 1} r_{n_h}^* \exp\left(i \sum_{n=n_h+1}^{n_{h+1}} \theta_n\right).$$

If we define $r_{n_h}^k = r_{n_h}^* \exp\left(i \sum_{n=m_{k+1}}^{n_{h+1}} \theta_n\right)$, then we can rewrite Ω_k as

$$\Omega_k = \sum_{n_h = m_{k-1}}^{m_k - 1} r_{n_h}^k \exp\left(i \sum_{n=n_h+1}^{m_k} \theta_n\right)$$

and compare it with the "constant frequency and amplitude" quantity

$$\overline{\Omega}_k := \sum_{n_h = m_{k-1}}^{m_k - 1} r_{m_k}^k \exp[i(m_k - n_h)\underline{\theta}(m_k\epsilon)].$$

First, for any n_h such that $m_{k-1} \leq n_h < m_k$,

$$\exp\left(i \sum_{n=n_h+1}^{m_k} \theta_n\right) = \exp[i(m_k - n_h)\underline{\theta}(m_k\epsilon)] \exp\left(i \sum_{n=n_h+1}^{m_k} (\theta_n - \underline{\theta}(m_k\epsilon))\right)$$

and, from the definitions of κ and τ,

$$|\theta_n - \underline{\theta}(m_k\epsilon)| < M_k\epsilon\kappa + \epsilon\tau$$

whenever $m_{k-1} \leq n_h < m_k$, so we get the bound

$$\left|\sum_{n=n_h+1}^{m_k} (\theta_n - \underline{\theta}(m_k\epsilon))\right| < M_k(M_k\epsilon\kappa + \epsilon\tau).$$

Thus, using only $M_k \epsilon < 2M^{-4}$ and $M^{-\frac{13}{2}} < \epsilon$, we get

$$(19) \quad \left| \exp\left(i \sum_{n=n_h+1}^{m_k} \theta_n \right) - \exp[i(m_k - n_h)\underline{\theta}(m_k \epsilon)] \right| < M_k(M_k \epsilon \kappa + \epsilon \tau)$$

$$< 4\kappa M^{-\frac{3}{2}} + 2\tau M^{-4}$$

for $m_{k-1} \le n_h < m_k$. Second, by the definition of β,

$$(20) \qquad |r_{m_k}^k - r_{n_h}^k| < M_k \epsilon \beta + 2\epsilon \gamma < 2\beta M^{-4} + 2\gamma M^{-6}$$

whenever $m_{k-1} \le n_h < m_k$. Putting (19) and (20) together, we find that whenever $m_{k-1} \le n_h < m_k$,

$$\left| r_{n_h}^k \exp\left(i \sum_{n=n_h+1}^{m_k} \theta_j \right) - r_{m_k}^k \exp[i(m_k - n_h)\underline{\theta}(m_k \epsilon)] \right| < \bar{C} M^{-\frac{3}{2}}$$

for some constant $\bar{C}$ depending only on the functions q and ϕ. Using the same two inequalities as for the bound (19) and the fact that there are M_k elements in the sums Ω_k and $\bar{\Omega}_k$, this estimate implies

$$(21) \qquad |\Omega_k - \bar{\Omega}_k| < 2\bar{C}M.$$

To bound the quantities $\bar{\Omega}_k$, we write

$$\bar{\Omega}_k = r_{m_k}^k \sum_{n=1}^{M_k} \exp(in\underline{\theta}(m_k \epsilon)),$$

whence

$$(22) \qquad |\bar{\Omega}_k| \le \alpha \left| \frac{\exp(iM_k \underline{\theta}(m_k \epsilon)) - 1}{\exp(i\underline{\theta}(m_k \epsilon)) - 1} \right| < \alpha \frac{4}{\sigma}$$

where σ is as defined above. Combining (21) and (22) with our assumption that $M > 2$ yields

$$(23) \qquad |\Omega_k| < \bar{C}'' M.$$

Going back to (18) and using that $\epsilon K < M^{-2}$, we get

$$|\Omega| < CM^{-1}$$

where the constant C depends only on q and ϕ. Then going back to (17), we finally obtain the result

$$|\Upsilon| < \frac{\alpha^{\ell-1}}{(\ell-1)!} CM^{-1} < \frac{\alpha^{\ell-1}}{(\ell-1)!} C\bar{\varrho}.$$

The Lemma follows by taking

$$\bar{\varrho} = \min\left\{\frac{\varrho}{C}, \frac{1}{2}, 2^{-\frac{1}{6}}|b - x_*|^{\frac{1}{6}} : x_* \text{ a turning point}\right\}$$

and $\epsilon' = \bar{\varrho}^6$. $\qquad\Box$

PROPOSTION 5. *Let $[a, b]$ be an oscillatory interval with positive distance from the set of turning points, and let $\varrho > 0$ be given. Then there exists $\epsilon' > 0$ sufficiently small such that whenever $0 < \epsilon < \epsilon'$,*

$$T^\epsilon = \begin{bmatrix} e^{S_1(a,b)} + \varrho^{11} & \varrho^{12} \\ \\ \varrho^{21} & e^{S_2(a,b)} + \varrho^{22} \end{bmatrix} \begin{bmatrix} \prod_{n=\underline{n}}^{\overline{n}} \lambda_n^+ & 0 \\ \\ 0 & \prod_{n=\underline{n}}^{\overline{n}} \lambda_n^- \end{bmatrix}$$

for some complex numbers ϱ^{ij} such that $|\varrho^{ij}| < \varrho$ for $i, j = 1, 2$.

REMARK. The convergence of the left-hand factor to a diagonal form as ϵ tends to zero is not uniform as a or b nears a turning point. However, it is uniform over all x-intervals whose distance from any turning point is bounded below by some positive number. Notice that $S_{ii}(a, b)$ depend on the choice of eigenvectors.

Part (1) of Theorem 2 can be deduced from Proposition 5, though the details are not presented here.

Proof. Consider first ϵ^ℓ times the diagonal matrix in expression (15) for P_ℓ. The quantities $\epsilon^\ell \sum_{\underline{n} \leq n_1 < \ldots < n_\ell \leq \overline{n}} r_{n_1}^{ii} \cdots r_{n_\ell}^{ii}$ are essentially Riemann sums for the integrals

$$\int \cdots \int_{\mathcal{R}_\ell} \underline{r}^{ii}(x_1) \cdots \underline{r}^{ii}(x_\ell) \, dx_1 \cdots dx_\ell$$

in which the integration is over the subregion $\mathcal{R}_\ell$ of $[0, 1]^\ell$ described by the inequalities $a < x_1 < \ldots < x_\ell < b$. These integrals are in fact equal to

$$\frac{1}{\ell!} \int_a^b \cdots \int_a^b \underline{r}^{ii}(x_1) \cdots \underline{r}^{ii}(x_\ell) \, dx_1 \cdots dx_\ell = \frac{1}{\ell!}\left(\int_a^b \underline{r}^{ii}(x) \, dx\right)^\ell$$

$$= \frac{1}{\ell!}(S_i(a, b))^\ell.$$

The sum over all ℓ should then converge to $e^{S_i(a,b)}$. This can be made rigorous, but the details are omitted.

Regarding the second summand of $\epsilon^\ell P_\ell$ in equation (15), we see that any one of its entries contains no more that $2^{\ell-1}$ terms of the type Υ, and so by Lemma 4, given $\bar{\varrho} > 0$, the sum over all these terms can be made to be less in modulus than $\bar{\varrho}\frac{(2\alpha)^{\ell-1}}{(\ell-1)!}$ for each ℓ by taking ϵ sufficiently small, and thus, in the sum over all ℓ, this second summand contributes less than $\bar{\varrho}\exp(2\alpha)$ in modulus to any entry of the matrix $\sum_{\ell=0}^{L} \epsilon^\ell P_\ell$. $\qquad\Box$

5.3. Exponential region. We now turn to an exponential region $[a, b]$. Let us suppose, for the sake of the argument, that $0 < \lambda^-(x) < \lambda^+(x)$ on this interval. Referring to the form of P_ℓ computed on page 251, we observe that each entry of its second column consists of terms containing a factor of the form $\prod_{n=n_\ell}^{\bar{n}} \frac{\lambda_n^-}{\lambda_n^+}$, which are summed over $\underline{n} \le n_1 < \dots < n_\ell \le \bar{n}$, and one expects these to converge to zero as $\epsilon \to 0$ since $\frac{\lambda_n^-}{\lambda_n^+} < 1$. The first column contains a sum over $\underline{n} \le n_1 < \dots < n_\ell \le \bar{n}$ of terms of the form $\prod_{n=n_h}^{n_{h+1}-1} \frac{\lambda_n^+}{\lambda_n^-}$, but n_{h+1} is never equal to $\bar{n}$ except in the term containing the product $r_{n_1}^{ii} \cdots r_{n_\ell}^{ii}$, which has $\prod_{n=\underline{n}}^{\bar{n}} \frac{\lambda_n^+}{\lambda_n^-}$ as a factor. This suggests that, although both entries of the first column diverge as $\epsilon \to 0$, the upper left entry will dominate. This can be proved rigorously as long as the interval $[a, b]$ is bounded away from any turning point. Suppose that, for some fixed value of s with $0 < s < 1$,

$$\frac{\lambda^-(x)}{\lambda^+(x)} < s \qquad \text{for} \quad x \in [a, b],$$

and let α, β, γ, and $S_i(a, b)$, be defined as before (α, β, and γ depend on s). The following proposition makes precise what is meant by the dominance of the $(1, 1)$-entry of T^ϵ.

PROPOSTION 6. *Let $[a, b]$ be an exponential interval with positive distance from the set of turning points, and assume that $0 < \lambda^-(x) < \lambda^+(x)$ for every x in $[a, b]$. Then, for any $\varrho > 0$, $\epsilon' > 0$ can be chosen sufficiently small such that*

$$T^\epsilon = \begin{bmatrix} e^{S_1(a,b)} + \varrho_{11} & \varrho_{12} \\ \\ \varrho_{21} & \varrho_{22} \end{bmatrix} \prod_{n=\underline{n}}^{\bar{n}} \lambda_n^+$$

whenever $0 < \epsilon < \epsilon'$, for certain numbers ϱ_{ij} with modulus less than ϱ.
Proof is omitted.

REMARK. The convergence of the left-hand matrix as ϵ tends to zero is not uniform as a or b nears a turning point since a suitable value of s approaches 1 near a turning point. However, it is uniform over all exponential x-intervals whose distance from any turning point is bounded below by some positive number.

Part (2) of Theorem 2 follows from Proposition 6.

Acknowledgment. I am grateful to have had the valuable input of Nicholas Ercolani and Kenneth T-R. McLaughlin on this research.

REFERENCES

[B]   J.C. BRONSKI, *Semiclassical eigenvalue distribution of the Zakharov-Shabat eigenvalue problem*, Physica D, **97** (1996), 376–397.

[C]   O. COSTIN AND R. COSTIN, *Rigorous WKB for Finite-Order Linear Recurrence Relations with Smooth Coefficients*, SIAM J. Math. Anal., Vol. **27**, no. 1 (1996).

[DM]   P. DEIFT AND K.T.-R. MCLAUGHLIN, *A Continuum Limit of the Toda Lattice*, Mem. Amer. Math. Soc., **131** (1998), no. 624.

[GS]   J. GERONIMO AND D. SMITH, *WKB Analysis of Second Order Difference Equations and Applications*, Journal of Approximation Theory, **69**(3) (1992), 269–301.

[S]   S.P. SHIPMAN, *Modulated Waves in a Semiclassical Continuum Limit of an Integrable NLS Chain*, Comm. Pure & Appl. Math., Vol. **LIII**, 0243–0279 (2000).

[V]   V. VEKSLERCHIK, *Finite Nonlinear Schrödinger Chain*, Phys. Lett. A, **174** (1993), 285–288.

BIFURCATION ANALYSIS OF CYLINDRICAL COUETTE FLOW WITH EVAPORATION AND CONDENSATION BY THE BOLTZMANN EQUATION

YOSHIO SONE* AND TOSHIYUKI DOI†

Abstract. Time-independent behavior of a gas between two coaxial circular cylinders made of the condensed phase of the gas, where evaporation or condensation takes place, is considered when the cylinders are rotating around their common axis. The problem is studied analytically on the basis of the Boltzmann equation, when the speeds of rotation of the cylinders and the Knudsen number of the system are small. The explicit solution is obtained when the flow field is axially symmetric and uniform, and the bifurcation of solution is found to occur even in this simple field.

Key words. bifurcation, Boltzmann equation, kinetic theory, evaporation, condensation, Couette flow.

AMS(MOS) subject classifications. 76P05, 70K50, 82B40, 34E15, 76T10.

1. Introduction. Recently, the first author, in collaboration with Sugimoto and Aoki, studied flows of a rarefied gas between coaxial circular cylinders made of the condensed phase of the gas mainly numerically and found a new type of bifurcation of flow, different from well-known Taylor-Couette type, that occurs when the flow field is axially symmetric and uniform (or axially and circumferentially uniform) [1]. The parameters range studied, however, is rather limited because of numerical analysis. Bifurcation in the case where only the inner cylinder is rotating with high speed is found. In the present paper, we will investigate the problem analytically when the speeds of rotation of the cylinders and the Knudsen number of the system are small and try to clarify the comprehensive feature of the flow, especially the structure of the solution. Another purpose of the study is to present an example of bifurcation in a mathematically simple system and to promote the mathematical study of bifurcation of the Boltzmann system.

For problems with evaporation and condensation on a boundary in general, analysis on the basis of kinetic theory is inevitable even in the limit that the Knudsen number tends to zero (or the continuum limit) because of a nonequilibrium layer on the boundary. Furthermore, the present special problem provides an important example of the ghost effect (or incompleteness of the classical continuum gas dynamics), which will be explained at the last paragraph in Section 7.

---

*230-133 Iwakura-Nagatani-cho Sakyo-ku, Kyoto 606-0026, Japan (yosone@ip.media.kyoto-u.ac.jp). The author is partially supported by IMA for the expense of the travel for IMA Workshop.

†Department of Applied Mathematics and Physics, Tottori University, Tottori 680-8552, Japan (doi@damp.tottori-u.ac.jp).

This article is the outline of the talk given at the IMA Workshop, Workshop 8: Simulation of Transport in Transition Regimes (May 2000). The full paper will be published elsewhere [2].

2. Problem. Consider a gas between two rotating coaxial cylinders made of the condensed phase of the gas. Let the radius, temperature, and circumferential velocity of the inner cylinder be, respectively, L_A, T_A, and $V_{\theta A}$, and let the corresponding quantities of the outer cylinder be L_B, T_B, and $V_{\theta B}$. The saturation gas pressure at temperature T_A (T_B) is denoted by p_{SA} (p_{SB}). We assume that (i) the behavior of the gas is described by the Boltzmann equation; (ii) the velocity distribution function of the gas molecules leaving a condensed phase is the corresponding part of the Maxwell distribution with the velocity and temperature of the condensed phase and with the pressure of saturated gas at the temperature of the condensed phase (complete condensation condition); (iii) the flow field is axially symmetric and uniform (or axially and circumferentially uniform); and (iv) the speeds of rotation of the cylinders and the Knudsen number of the system are small.

The variables in the system being reduced to nondimensional variables appropriately with the aid of boundary data, the system is found to be determined by the following six parameters:

$$(2.1) \qquad \hat{r}_B = \frac{L_B}{L_A}, \quad \frac{V_{\theta A}}{\sqrt{2RT_A}}, \quad \frac{V_{\theta B}}{\sqrt{2RT_A}}, \quad \frac{T_B}{T_A}, \quad \frac{p_{SB}}{p_{SA}}, \quad k,$$

where R is the specific gas constant (or the Boltzmann constant divided by the mass of a molecule) and k is defined by

$$(2.2) \qquad k = \frac{\sqrt{\pi}}{2} \frac{l_A}{L_A},$$

where l_A is the mean free path of the gas molecules at the saturated equilibrium state at rest at temperature T_A. In the main text of this article, taking a small quantity ε, we investigate the asymptotic solution of the time-independent boundary-value problem of the Boltzmann equation for the problem mentioned above when the parameters are limited to the following conditions:

$$(2.3) \quad \begin{array}{ccc} \hat{r}_B - 1 = O(1), & \dfrac{V_{\theta A}}{\sqrt{2RT_A}} = \varepsilon u_{\theta A1}, & \dfrac{V_{\theta B}}{\sqrt{2RT_A}} = \varepsilon u_{\theta B1}, \\[3mm] \dfrac{T_B}{T_A} - 1 = \varepsilon^2 \tau_{B2}, & \dfrac{p_{SB}}{p_{SA}} - 1 = \varepsilon^2 P_{SB2}, & k = \varepsilon^m, \end{array}$$

where $u_{\theta A1}$, $u_{\theta B1}$, τ_{B2}, and P_{SB2} are quantities of the order of unity and $m \geq 3$. The other cases, where $p_{SB}/p_{SA} - 1$ is of the order of ε instead of ε^2, or $m = 1$ or 2, will be discussed briefly in Section 6.

For analysis in the above range of the parameters, the nondimensional Boltzmann equation and the boundary condition for a perturbed velocity distribution function are convenient. The Boltzmann equation is given by

$$(2.4) \qquad \zeta_r \frac{\partial \phi}{\partial \hat{r}} + \frac{\zeta_\theta^2}{\hat{r}} \frac{\partial \phi}{\partial \zeta_r} - \frac{\zeta_r \zeta_\theta}{\hat{r}} \frac{\partial \phi}{\partial \zeta_\theta} = \frac{1}{\varepsilon^m} [\mathcal{L}(\phi) + \hat{J}(\phi, \phi)],$$

where the radial distance r from the axis of the cylinders, the molecular velocity $(\xi_r, \xi_\theta, \xi_z)$ in the cylindrical coordinate system (r, θ, z) with the axis of the cylinders being taken as z axis, and the velocity distribution function $f(r, \xi_r, \xi_\theta, \xi_z)$ are expressed, respectively, by

$$\begin{aligned} r &= L_A \hat{r}, \quad (\xi_r, \xi_\theta, \xi_z) = \sqrt{2RT_A}(\zeta_r, \zeta_\theta, \zeta_z), \\ (2.5) \qquad f &= 2p_{SA}(2RT_A)^{-5/2} E(\zeta)[1 + \phi(\hat{r}, \zeta_r, \zeta_\theta, \zeta_z)], \\ E(\zeta) &= \pi^{-3/2} \exp(-\zeta^2), \quad \zeta = (\zeta_r^2 + \zeta_\theta^2 + \zeta_z^2)^{1/2}, \end{aligned}$$

and $\mathcal{L}(\phi)$ and $\hat{J}(\phi, \phi)$ are, respectively, nondimensional linearized and original collision integrals, which are not shown explicitly. The boundary conditions on the two cylinders are

$$(2.6a) \qquad \phi = \phi_e(0, 0, \varepsilon u_{\theta A1}) \quad (\zeta_r > 0) \text{ at } \hat{r} = 1,$$

$$(2.6b) \qquad \phi = \phi_e(\varepsilon^2 P_{SB2}, \varepsilon^2 \tau_{B2}, \varepsilon u_{\theta B1}) \quad (\zeta_r < 0) \text{ at } \hat{r} = \hat{r}_B,$$

where ϕ_e is the perturbed part of a local Maxwellian from the Maxwellian at rest with T_A and p_{SA}, that is,

$$E(\zeta)[1 + \phi_e(a, b, c)] = \frac{1+a}{\pi^{3/2}(1+b)^{5/2}} \exp\left(-\frac{\zeta_r^2 + (\zeta_\theta - c)^2 + \zeta_z^2}{1+b}\right).$$

The macroscopic variables: the density $(p_{SA}/RT_A)(1+\omega)$, flow velocity $\sqrt{2RT_A}(u_r, u_\theta, 0)$, pressure $p_{SA}(1 + P)$, and temperature $T_A(1 + \tau)$ of the gas are given by the moments of ϕ in the following form:

$$(2.7a) \qquad \omega = \int \phi E(\zeta) d\zeta,$$

$$(2.7b) \qquad (u_r, u_\theta) = \frac{1}{1+\omega} \int (\zeta_r, \zeta_\theta) \phi E(\zeta) d\zeta,$$

$$(2.7c) \qquad \frac{3}{2}\tau = \frac{1}{1+\omega} \int \left(\zeta^2 - \frac{3}{2}\right) \phi E(\zeta) d\zeta - u_r^2 - u_\theta^2,$$

$$(2.7d) \qquad 1 + P = (1 + \omega)(1 + \tau),$$

$$d\zeta = d\zeta_r d\zeta_\theta d\zeta_z.$$

3. Solution type I.

3.1. Analysis. The solution that describes the overall behavior of the gas is obtained in a power series of ε under the assumption that $\partial\phi/\partial\hat{r} = O(\phi)$ (an extended Hilbert expansion):

$$(3.1) \qquad \phi_H = \phi_{H1}\varepsilon + \phi_{H2}\varepsilon^2 + \cdots,$$

where the subscript H is attached to discriminate the solution of this class (Hilbert solution). Corresponding to this expansion, the moments of ϕ_H, the macroscopic variables: the density, flow velocity, pressure, and temperature, are also expanded in a power series of ε:

$$h_H = h_{H1}\varepsilon + h_{H2}\varepsilon^2 + \cdots,$$

where h represents ω, u_r, etc. Substitution of the series into the Boltzmann equation (2.4) leads to a series of integral equations of the form:

$$(3.2) \qquad \mathcal{L}(\phi_{Hs}) = \mathrm{ihm}_s,$$

where ihm_s is their inhomogeneous term expressed by lower-order quantities. The corresponding homogeneous equation $\mathcal{L}(\phi_{Hs}) = 0$ has five independent solutions 1, ζ_r, ζ_θ, ζ_z, ζ^2, but ζ_z is excluded in the present problem from the symmetry request. Up to the order $s = m$, it consists only of $\hat{J}(\phi_{Hs-n}, \phi_{Hn})$ terms, and thus, ϕ_{Hs} is the corresponding term of the expansion of the local Maxwellian. For $s \geq m + 1$, additional inhomogeneous terms, which consist of the derivative terms of Eq. (2.4), enter. The newly entered derivative terms must satisfy the solvability condition:

$$(3.3) \qquad \int (1, \zeta_r, \zeta_\theta, \zeta^2)\left(\zeta_r \frac{\partial \phi_{Hs-m}}{\partial \hat{r}} + \frac{\zeta_\theta^2}{\hat{r}} \frac{\partial \phi_{Hs-m}}{\partial \zeta_r} - \frac{\zeta_r \zeta_\theta}{\hat{r}} \frac{\partial \phi_{Hs-m}}{\partial \zeta_\theta} \right) E(\zeta)\mathrm{d}\zeta = 0.$$

This condition gives the differential equations that determine the undetermined coefficients of 1, ζ_r, ζ_θ, and ζ^2 in the solution, introduced by the solution of the corresponding homogeneous equation. The coefficients are related to the component functions of the expansion of the macroscopic variables ω_H, u_{rH}, $u_{\theta H}$, P_H, and τ_H. The coefficients (or macroscopic variables) being determined from the differential equations (the boundary condition being put aside for a moment) from the lowest order successively, the series of solutions of the integral equations, the component functions ϕ_{Hs}, is determined in the same way. The expansion can be carried out consistently with a special assumption $u_{rH1} = P_{H1} = 0$, which is consistent with the boundary conditions (2.6a) and (2.6b) to be studied later. (This is derived if we start from $u_{rH1} \neq 0$ and $P_{H1} \neq 0$, but the process is omitted for shortness of description.) The differential equations for the leading order are

$$\text{(3.4a)} \qquad \frac{d\, u_{rH2}\hat{r}}{d\,\hat{r}} = 0,$$

$$\text{(3.4b)} \qquad \frac{u_{\theta H1}^2}{\hat{r}} = \frac{1}{2}\frac{d\, P_{H2}}{d\,\hat{r}},$$

$$\text{(3.4c)} \qquad u_{rH2}\frac{d\, u_{\theta H1}\hat{r}}{d\,\hat{r}} = 0,$$

$$\text{(3.4d)} \qquad u_{rH2}\frac{d\, \tau_{H1}}{d\,\hat{r}} = 0.$$

The Hilbert solution ϕ_H, however, does not have freedom enough to be matched to the kinetic boundary conditions (2.6a) and (2.6b), because the Hilbert solution has a special form of molecular velocity $(\zeta_r, \zeta_\theta, \zeta_z)$ as the solution of the integral equation (3.2). Thus, we introduce the correction to ϕ_H in two steps. First, we consider a solution that varies sharply in the neighborhood of the cylinders with the length scale of variation of the variable of the order of $\varepsilon^{m-2}L_A$, that is, $\partial\phi/\partial\hat{r} = O(\varepsilon^{-m+2}\phi)$ in the neighborhood of the cylinders and that is continuously transformed into the Hilbert solution as going away from the cylinders. With these assumptions, the solution of Eq. (2.4) is obtained in a power series of ε. The process of analysis is similar to that of the Hilbert expansion except that the derivative term in the integral equation (3.2) is upgraded by $m - 2$. The solution is discriminated by the subscript V. We start under the assumption $u_{rV1} = P_{V1} = 0$ by the same reason as in the analysis of the Hilbert solution. The differential equations for the macroscopic variables for the leading order are

$$\text{(3.5a)} \qquad \frac{d\, u_{rV2}}{d\,y} = 0,$$

$$\text{(3.5b)} \qquad \frac{d\, P_{V2}}{d\,y} = 0,$$

$$\text{(3.5c)} \qquad u_{rV2}\frac{d\, u_{\theta V1}}{d\,y} = \frac{\gamma_1}{2}\frac{d^2 u_{\theta V1}}{d\,y^2},$$

$$\text{(3.5d)} \qquad u_{rV2}\frac{d\, \tau_{V1}}{d\,y} = \frac{\gamma_2}{2}\frac{d^2 \tau_{V1}}{d\,y^2},$$

$$y = (r - r_L)/\varepsilon^{m-2}L_A = (\hat{r} - \hat{r}_L)/\varepsilon^{m-2},$$

where r_L and $\hat{r}_L$ are, respectively, L_A and 1 for the solution V on the inner cylinder, or L_B and $\hat{r}_B$ for the solution V on the outer cylinder, and γ_1 and γ_2 are constants that depend on molecular models, e.g.,

$$\gamma_1 = 1.270042427, \quad \gamma_2 = 1.922284066 \quad \text{(a hard-sphere gas)},$$
$$\gamma_1 = \gamma_2 = 1 \quad \text{(BKW model [3, 4])}.$$

Equations (3.5a)–(3.5d) are simplified by the condition that the equations are applied in the region $y = O(1)$ in addition to the above assumptions.

The connection of the Hilbert solution and the solution V is carried out by the conditions: As $y \to \infty$ on the inner cylinder or as $y \to -\infty$ on the outer cylinder,

$$(3.6) \qquad h_{V1} \sim (h_{H1})_L, \quad h_{V2} \sim (h_{H2})_L + y \left(\frac{d\,h_{H1}}{d\hat{r}} \right)_L \delta_3(m),$$

where h represents P, u_r, u_θ, and τ; $\delta_3(m) = 1$ $(m = 3)$ and $\delta_3(m) = 0$ $(m \neq 3)$; and the quantities in the parentheses with the subscript L are evaluated on the inner or outer cylinder depending on $L = A$ or $L = B$.

On an evaporating cylinder [on the inner cylinder for $u_{rV2} > 0$ (or $u_{rH2} > 0$ by Eq. (3.6)) or on the outer cylinder for $u_{rV2} < 0$ (or $u_{rH2} < 0$)], the solutions $u_{\theta V1}$ and τ_{V1} in Eqs. (3.5c) and (3.5d) diverge exponentially as $y \to \infty$ on the inner cylinder or as $y \to -\infty$ on the outer cylinder. The solution that can be connected continuously to the Hilbert solution is possible only on the condensing cylinder. Thus, we call the solution V the suction boundary-layer solution.

The suction boundary-layer solution has more freedom to be adjusted to the boundary conditions than the Hilbert solution because Eqs. (3.5c) and (3.5d) are of the second degree in contrast to Eqs. (3.4c) and (3.4d). At the level of order ε, the combination of ϕ_H and ϕ_V is made to satisfy the boundary conditions (2.6a) and (2.6b) by matching the boundary value of $u_{\theta H1}$ or $u_{\theta V1}$ to $u_{\theta A1}$ at $\hat{r} = 1$ and to $u_{\theta B1}$ at $\hat{r} = \hat{r}_B$ and that of τ_{H1} or τ_{V1} to zero at $\hat{r} = 1$ and $\hat{r}_B$, because the velocity distribution function is Maxwellian with $u_{rH1} = u_{rV1} = P_{H1} = P_{V1} = 0$. However, the velocity distribution function ϕ_V has a special form of the molecular velocity as the Hilbert solution and still does not have freedom enough to be matched to the kinetic boundary conditions (2.6a) and (2.6b) at the higher orders. Thus we introduce the Knudsen-layer correction. That is, the solution of the problem is put in the form:

$$(3.7) \qquad\qquad \phi = \phi_G + \phi_K,$$

where $\phi_G = \phi_H$ or ϕ_V depending on the correction on the evaporating or condensing cylinder. The correction ϕ_K is assumed to vary appreciably over the distance of the order of the mean free path $[\partial \phi_K / \partial \hat{r} = O(\varepsilon^{-m} \phi_K)]$ and to decay rapidly away from the cylinders. The correction ϕ_K is also expanded in a power series of ε:

$$(3.8) \qquad\qquad \phi_K = \phi_{K2}\varepsilon^2 + \phi_{K3}\varepsilon^3 + \cdots,$$

where the series starts from the second order of ε because the boundary conditions are satisfied only by ϕ_H and ϕ_V at the first order.

The equation for ϕ_{K2} is given by

$$(3.9) \qquad\qquad \zeta_r \frac{\partial \phi_{K2}}{\partial \eta} = \mathcal{L}(\phi_{K2}),$$

where $\eta = (\hat{r} - 1)/\varepsilon^m$ on the inner cylinder and $\eta = (\hat{r} - \hat{r}_B)/\varepsilon^m$ on the outer cylinder. The boundary condition for ϕ_{K2} is given as follows: On the inner cylinder ($\hat{r} = 1$),

$$\phi_{K2} = -P_{G2} - 2\zeta_\theta u_{\theta G2} - \left(\zeta^2 - \frac{5}{2}\right)\tau_{G2} - 2\zeta_r u_{rG2} \quad (\zeta_r > 0) \quad \text{at } \eta = 0,$$

$$\phi_{K2} \to 0 \quad \text{as } \eta \to \infty,$$

and on the outer cylinder ($\hat{r} = \hat{r}_B$),

$$\phi_{K2} = P_{SB2} - P_{G2} - 2\zeta_\theta u_{\theta G2} + \left(\zeta^2 - \frac{5}{2}\right)(\tau_{B2} - \tau_{G2}) - 2\zeta_r u_{rG2}$$

$$(\zeta_r < 0) \quad \text{at } \eta = 0,$$

$$\phi_{K2} \to 0 \quad \text{as } \eta \to -\infty,$$

where the subscript G represents H on the evaporating cylinder and V on the condensing cylinder.

This is the half-space boundary-value problem of the linearized one-dimensional Boltzmann equation studied mathematically by various authors [5, 6, 7]. According to them, the solution ϕ_{K2} exists uniquely when and only when the boundary values of P_{G2}, $u_{\theta G2}$, τ_{G2}, and u_{rG2} satisfy some relations. The conditions on the boundary values of $u_{\theta H1}$ or $u_{\theta V1}$ and τ_{H1} or τ_{V1} mentioned before and one of the relations just mentioned give the boundary conditions for the equations (3.4a)–(3.4d) or (3.5a)–(3.5d) on the cylinders. That is,

(3.10a) $u_{\theta G1} = u_{\theta A1}, \quad \tau_{G1} = 0, \quad P_{G2} = C_4^* u_{rG2} \quad$ at $\hat{r} = 1$,

(3.10b) $u_{\theta G1} = u_{\theta B1}, \quad \tau_{G1} = 0, \quad P_{G2} = P_{SB2} - C_4^* u_{rG2} \quad$ at $\hat{r} = \hat{r}_B$,

where C_4^* is a constant that depends on molecular model, e.g.,

$$C_4^* = -2.1412 \text{ (hard-sphere)}, \quad C_4^* = -2.13204 \text{ (BKW)}.$$

Now the material enough to obtain the asymptotic solution at the leading order as $\varepsilon \to 0$ is prepared. We only have to solve Eqs. (3.4a)–(3.4d) and Eqs. (3.5a)–(3.5d) under the connection condition (3.6) and the boundary conditions (3.10a) and (3.10b).

The correction to the Hilbert solution being carried out in two steps, the suction boundary layer solution and the Knudsen-layer correction, the equation of the Knudsen layer, for which the discussion at the level of the velocity distribution function is required, is reduced to a linear equation.

3.2. Solution. At this stage, we do not know on which cylinder condensation is taking place. Thus, we will obtain the solutions of the forementioned system of equations and boundary conditions for both cases and examine their consistency with the assumption. The two solutions are given as follows:

(i) Case $u_{rH2} > 0$ (condensation is taking place on the outer cylinder).

$$(3.11) \quad u_{\theta H1} = \frac{u_{\theta A1}}{\hat{r}}, \quad \tau_{H1} = 0, \quad u_{rH2} = \frac{c_{r2}}{\hat{r}}, \quad P_{H2} = c_{p2} - \left(\frac{u_{\theta A1}}{\hat{r}}\right)^2,$$

$$(3.12)$$
$$u_{\theta V1} = \frac{u_{\theta A1}}{\hat{r}_B} + \left(u_{\theta B1} - \frac{u_{\theta A1}}{\hat{r}_B}\right) \exp\left(\frac{2c_{r2}y}{\gamma_1 \hat{r}_B}\right), \quad \tau_{V1} = 0,$$

$$u_{rV2} = \frac{c_{r2}}{\hat{r}_B}, \quad P_{V2} = c_{p2} - \left(\frac{u_{\theta A1}}{\hat{r}_B}\right)^2,$$

where

$$c_{r2} = \frac{-1}{C_4^*}\left[-\frac{\hat{r}_B}{\hat{r}_B + 1}P_{SB2} + \frac{\hat{r}_B - 1}{\hat{r}_B}u_{\theta A1}^2\right],$$

$$c_{p2} = \frac{1}{\hat{r}_B}u_{\theta A1}^2 + \frac{\hat{r}_B}{\hat{r}_B + 1}P_{SB2}.$$

The solution satisfies the condition $u_{rH2} > 0$, or exists when $P_{SB2} < (\hat{r}_B^2 - 1)(u_{\theta A1}/\hat{r}_B)^2$ because $C_4^* < 0$. This solution will be called solution I_+.

(ii) Case $u_{rH2} < 0$ (condensation is taking place on the inner cylinder).

$$(3.13)$$
$$u_{\theta H1} = \frac{u_{\theta B1}\hat{r}_B}{\hat{r}}, \quad \tau_{H1} = 0,$$

$$u_{rH2} = \frac{c_{r2}}{\hat{r}}, \quad P_{H2} = c_{p2} - \left(\frac{u_{\theta B1}\hat{r}_B}{\hat{r}}\right)^2,$$

$$(3.14)$$
$$u_{\theta V1} = u_{\theta B1}\hat{r}_B + (u_{\theta A1} - u_{\theta B1}\hat{r}_B)\exp\left(\frac{2c_{r2}y}{\gamma_1}\right), \quad \tau_{V1} = 0,$$

$$u_{rV2} = c_{r2}, \quad P_{V2} = c_{p2} - (u_{\theta B1}\hat{r}_B)^2,$$

where

$$c_{r2} = \frac{-1}{C_4^*}\left[-\frac{\hat{r}_B}{\hat{r}_B + 1}P_{SB2} + \hat{r}_B(\hat{r}_B - 1)u_{\theta B1}^2\right],$$

$$c_{p2} = \hat{r}_B u_{\theta B1}^2 + \frac{\hat{r}_B}{\hat{r}_B + 1}P_{SB2}.$$

The solution satisfies the condition $u_{rH2} < 0$, or exists when $P_{SB2} > (\hat{r}_B^2 - 1)u_{\theta B1}^2$. This solution will be called solution I_-.

Combining the results, we find that

(A) the two solutions, I_+ and I_-, exist when

$$(3.15) \quad (u_{\theta B1}\hat{r}_B)^2 < \frac{\hat{r}_B^2}{\hat{r}_B^2 - 1}P_{SB2} < u_{\theta A1}^2;$$

(B) no solution exists when

$$(3.16) \qquad u_{\theta A1}^2 < \frac{\hat{r}_B^2}{\hat{r}_B^2 - 1} P_{SB2} < (u_{\theta B1}\hat{r}_B)^2;$$

(C) one of the solutions, I_+ or I_-, exists in the other cases.

As the parameter P_{SB2} approaches $(\hat{r}_B^2 - 1)u_{\theta A1}^2/\hat{r}_B^2$ from below, c_{r2} (> 0) tends to zero and finally solution I_+ ceases to exist, and on the other hand, as P_{SB2} approaches $(\hat{r}_B^2 - 1)u_{\theta B1}^2$ from above, c_{r2} (< 0) tends to zero and finally solution I_- ceases to exist. We will call the point $P_{SB2} = (\hat{r}_B^2 - 1)u_{\theta A1}^2/\hat{r}_B^2$ critical point C_+ and the point $P_{SB2} = (\hat{r}_B^2 - 1)u_{\theta B1}^2$ critical point C_-. The bifurcation diagrams in terms of c_{r2} versus P_{SB2} are shown in Fig. 1. Hereafter, we call the two solutions I_+ and I_- combined solution I.

In the solution obtained in this section, solution I, the effect of viscosity and thermal conductivity (γ_1 and γ_2) is confined in the suction boundary layer and is a secondary one in most of the flow field.

4. Solution type II. The two solutions I_+ and I_- are disconnected unless the two critical points C_+ and C_- coincide, or unless $u_{\theta A1}^2 = u_{\theta B1}^2\hat{r}_B^2$. In the present section, we present another type of solution (say, solution II) connecting the two solutions. The Hilbert expansion of the Boltzmann equation can be carried out consistently under the assumption that $u_{rHs} = 0$ ($s = 1, \cdots, m - 1$) as well as $P_{H1} = 0$, and a series of differential equations that determine the variables $(u_{\theta Hs}, \tau_{Hs}, P_{Hs+1}, u_{rHs+m-1})$ as a set is obtained from the solvability conditions of the linear integral equations for the component functions of the velocity distribution function in a power series expansion in ε. The set of equations for the leading order is given as follows:

$$(4.1a) \qquad \frac{d\, u_{rHm}\hat{r}}{d\,\hat{r}} = 0,$$

$$(4.1b) \qquad \frac{u_{\theta H1}^2}{\hat{r}} = \frac{1}{2}\frac{d\, P_{H2}}{d\,\hat{r}},$$

$$(4.1c) \qquad \frac{u_{rHm}}{\hat{r}}\frac{d\, u_{\theta H1}\hat{r}}{d\,\hat{r}} = \frac{\gamma_1}{2}\left[\frac{1}{\hat{r}}\frac{d}{d\,\hat{r}}\left(\hat{r}\frac{d\, u_{\theta H1}}{d\,\hat{r}}\right) - \frac{u_{\theta H1}}{\hat{r}^2}\right],$$

$$(4.1d) \qquad u_{rHm}\frac{d\,\tau_{H1}}{d\,\hat{r}} = \frac{\gamma_2}{2}\frac{1}{\hat{r}}\frac{d}{d\,\hat{r}}\left(\hat{r}\frac{d\,\tau_{H1}}{d\,\hat{r}}\right).$$

Equations (4.1c) and (4.1d) for $u_{\theta H1}$ and τ_{H1} are of the second degree in contrast to Eqs. (3.4c) and (3.4d). Thus, the solution of the original boundary-value problem is expressed with this solution and its Knudsen-layer correction. The suction boundary layer is unnecessary (or absent). The boundary conditions for these equations are obtained by analysis of

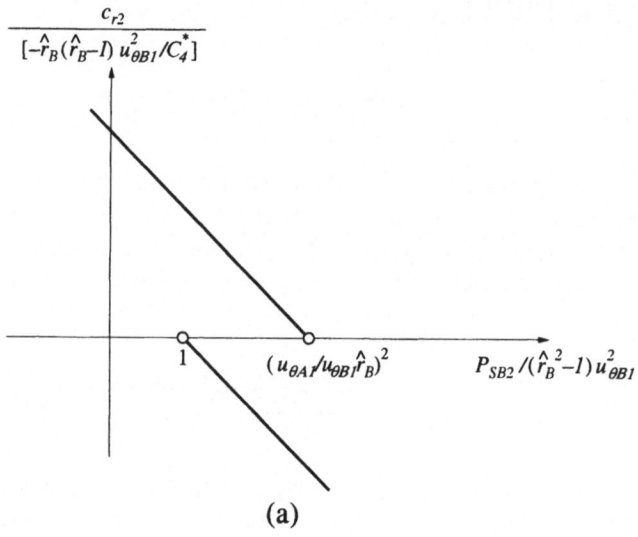

(a)

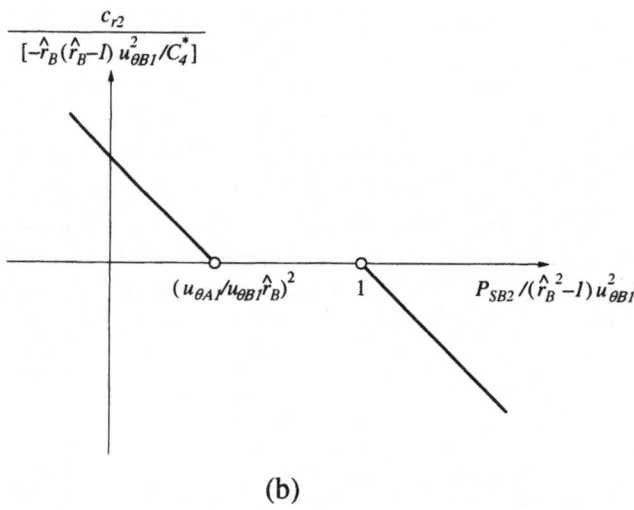

(b)

FIG. 1. *Bifurcation diagram of the solution obtained in Section 3 (c_{r2} versus P_{SB2}).*
(a) $u_{\theta A1}^2 > (u_{\theta B1}\hat{r}_B)^2$ and (b) $u_{\theta A1}^2 < (u_{\theta B1}\hat{r}_B)^2$. The slope is -1 irrespective of the
parameters.

the Knudsen layer. The boundary conditions for the set of equations
(4.1a)–(4.1d) are as follows:

(4.2a) $u_{\theta H1} = u_{\theta A1},\ \tau_{H1} = 0,\ P_{H2} = 0$ at $\hat{r} = 1$,

(4.2b) $u_{\theta H1} = u_{\theta B1},\ \tau_{H1} = 0,\ P_{H2} = P_{SB2}$ at $\hat{r} = \hat{r}_B$.

The solution of this boundary-value problem, Eqs. (4.1a)–(4.1d) with (4.2a) and (4.2b), is easily obtained in the following form:

(4.3)
$$u_{\theta H1} = \frac{a}{\hat{r}} + b\hat{r}^{1+\alpha}, \quad \tau_{H1} = 0, \quad u_{rHm} = \frac{\gamma_1 \alpha}{2\hat{r}},$$

$$P_{H2} = P_{SB2} + a^2 \left(\frac{1}{\hat{r}_B^2} - \frac{1}{\hat{r}^2} \right)$$

$$+ \frac{b^2}{\alpha+1} (\hat{r}^{2(1+\alpha)} - \hat{r}_B^{2(1+\alpha)}) + \frac{4ab}{\alpha} (\hat{r}^\alpha - \hat{r}_B^\alpha),$$

where

$$a = \frac{\hat{r}_B(u_{\theta A1}\hat{r}_B^{1+\alpha} - u_{\theta B1})}{\hat{r}_B^{2+\alpha} - 1}, \quad b = \frac{(u_{\theta B1}\hat{r}_B - u_{\theta A1})}{\hat{r}_B^{2+\alpha} - 1},$$

and α is the solution of the following transcendental equation:

(4.4)
$$\frac{P_{SB2}}{(\hat{r}_B^2 - 1)u_{\theta B1}^2} = \frac{(\hat{r}_B^{2+\alpha} M_u - 1)^2}{(\hat{r}_B^{2+\alpha} - 1)^2} + \frac{\hat{r}_B^2(\hat{r}_B^{2(1+\alpha)} - 1)(M_u - 1)^2}{(\alpha+1)(\hat{r}_B^2 - 1)(\hat{r}_B^{2+\alpha} - 1)^2}$$

$$- \frac{4\hat{r}_B^2(\hat{r}_B^\alpha - 1)(M_u - 1)(\hat{r}_B^{2+\alpha} M_u - 1)}{\alpha(\hat{r}_B^2 - 1)(\hat{r}_B^{2+\alpha} - 1)^2},$$

$$M_u = \frac{u_{\theta A1}}{u_{\theta B1}\hat{r}_B}.$$

The $P_{SB2}/(\hat{r}_B^2 - 1)u_{\theta B1}^2$ is a continuous function of α, and it approaches 1 as $\alpha \to -\infty$ and $(u_{\theta A1}/u_{\theta B1}\hat{r}_B)^2$ as $\alpha \to \infty$. Thus, Eq. (4.4) for α has, at least, one solution when $P_{SB2}/(\hat{r}_B^2 - 1)u_{\theta B1}^2$ lies in the range

(4.5)
$$\left[\frac{P_{SB2}}{(\hat{r}_B^2 - 1)u_{\theta B1}^2} - 1 \right] \left[\frac{P_{SB2}}{(\hat{r}_B^2 - 1)u_{\theta B1}^2} - \left(\frac{u_{\theta A1}}{u_{\theta B1}\hat{r}_B} \right)^2 \right] < 0.$$

The point $\alpha = \infty$ corresponds to critical point C_+, and the point $\alpha = -\infty$ corresponds to critical point C_-. Therefore, there is another solution that connects the two critical points C_+ and C_-.

The behavior of Eq. (4.4) is examined numerically with the aid of its asymptotic behavior as $\alpha \to \pm\infty$. When the two cylinders are rotating in the same direction (or $u_{\theta A1}u_{\theta B1} > 0$), $P_{SB2}/(\hat{r}_B^2 - 1)u_{\theta B1}^2$ is a monotonic function of α, monotonic increasing for $(u_{\theta A1}/u_{\theta B1}\hat{r}_B)^2 > 1$ or decreasing for $(u_{\theta A1}/u_{\theta B1}\hat{r}_B)^2 < 1$. On the other hand when they are rotating in opposite directions (or $u_{\theta A1}u_{\theta B1} < 0$), it is not necessarily monotonic and may have three solutions in some range of Eq. (4.5) and furthermore may have two solutions in some range outside Eq. (4.5). Several examples of the profile $P_{SB2}/(\hat{r}_B^2 - 1)u_{\theta B1}^2$ versus α are shown in Figs. 2, 3, and 4. The bifurcation diagram of solution II for $u_{\theta A1}u_{\theta B1} < 0$ derived from these figures is classified into six types, which are shown in Fig. 5.

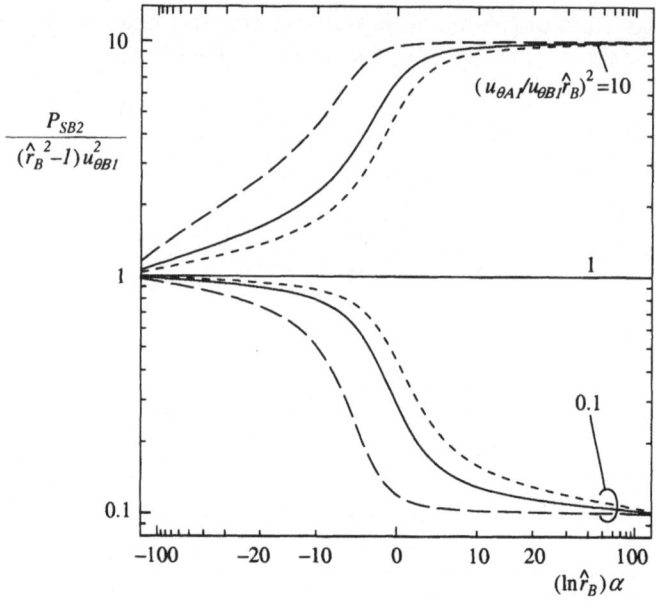

FIG. 2. $P_{SB2}/(\hat{r}_B^2 - 1)u_{\theta B1}^2$ *versus* α *[Eq. (4.4)] I: The two cylinders are rotating in the same direction* $(u_{\theta A1}u_{\theta B1} > 0)$. *The solid lines* ——— *indicate the case* $\hat{r}_B = 2$, *the long dashed lines* — — —: $\hat{r}_B = 10$, *and the short dashed lines* - - -: $\hat{r}_B = 1.1$.

5. Bifurcation diagram and transition solution.

Each solution can be identified by the value of c_{r2} (or $u_{rH2}\hat{r}$) for solution I and that of α (or $u_{rHm}\hat{r}$) for solution II. Summarizing the results for solutions I and II obtained in Sections 3 and 4, we show the diagram $\lim_{\varepsilon \to 0} u_{rH}\hat{r}/\varepsilon^2$ versus P_{SB2} for $u_{\theta A1}u_{\theta B1} > 0$ in Fig. 6 and for $u_{\theta A1}u_{\theta B1} < 0$ in Fig. 7. In these figures, $\lim_{\varepsilon \to 0} u_{rH}\hat{r}/\varepsilon^2$ vanishes for all the solutions of type II. Thus, the microstructure is shown in these figures. When both cylinders are rotating in the same direction $(u_{\theta A1}u_{\theta B1} > 0)$, the situation is simple: (a) If $(u_{\theta A1}/u_{\theta B1}\hat{r}_B)^2 > 1$, three solutions, I$_+$, I$_-$, and II, exist in the range $1 < P_{SB2}/(\hat{r}_B^2 - 1)u_{\theta B1}^2 < (u_{\theta A1}/u_{\theta B1}\hat{r}_B)^2$, and the solution is unique in the other range; (b) if $(u_{\theta A1}/u_{\theta B1}\hat{r}_B)^2 < 1$, the solution is unique in the whole range. When the two cylinders are rotating in opposite directions $(u_{\theta A1}u_{\theta B1} < 0)$, the situation is classified into the six cases shown in Fig. 7, depending on the values of $(u_{\theta A1}/u_{\theta B1}\hat{r}_B)^2$ and $\hat{r}_B$.

Up to now, we have excluded the discussion at the critical points C_+ and C_-. The two solutions I$_+$ and II or I$_-$ and II are not confirmed to be transformed into each other continuously because of the difference of the order of the radial velocity. In the foregoing analyses, the asymptotic solution as $\varepsilon \to 0$ is sought for a fixed set of the parameters $\hat{r}_B$, $u_{\theta A1}$, $u_{\theta B1}$, τ_{B2}, and P_{SB2}. The parameters may be arbitrarily close to a critical point as far as they are fixed in the process as $\varepsilon \to 0$. Therefore the bifurcation

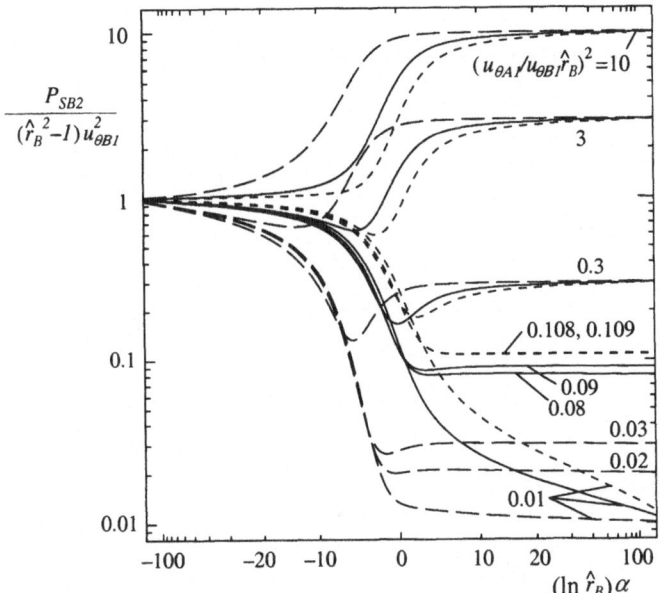

FIG. 3. $P_{SB2}/(\hat{r}_B^2 - 1)u_{\theta B1}^2$ versus α [Eq. (4.4)] II: The two cylinders are rotating in opposite directions ($u_{\theta A1}u_{\theta B1} < 0$). The solid lines —— indicate the case $\hat{r}_B = 2$, the long dashed lines — — —: $\hat{r}_B = 10$, and the short dashed lines - - -: $\hat{r}_B = 1.1$. The profile for $[(u_{\theta A1}/u_{\theta B1}\hat{r}_B)^2 = 0.109, \hat{r}_B = 1.1]$ and that for $[(u_{\theta A1}/u_{\theta B1}\hat{r}_B)^2 = 0.03,$ $\hat{r}_B = 10]$ show a similar oscillatory behavior, with the minimum at a finite α, to that for $[(u_{\theta A1}/u_{\theta B1}\hat{r}_B)^2 = 0.09, \hat{r}_B = 2]$. The profile for $[(u_{\theta A1}/u_{\theta B1}\hat{r}_B)^2 = 0.108, \hat{r}_B = 1.1]$ and that for $[(u_{\theta A1}/u_{\theta B1}\hat{r}_B)^2 = 0.02, \hat{r}_B = 10]$ show a similar oscillatory behavior to that for $[(u_{\theta A1}/u_{\theta B1}\hat{r}_B)^2 = 0.08, \hat{r}_B = 2]$. By extensive examination for various cases, $u_{\theta A1} = -3u_{\theta B1}\hat{r}_B$ and $u_{\theta A1} = -u_{\theta B1}\hat{r}_B/3$ are found to be the boundaries of two different types of profile. This is consistent with the asymptotic behavior for $\alpha = \pm\infty$. See Fig. 4.

diagram as $\varepsilon \to 0$ is not affected. Thus, in the present paper, we examine the transition solution only for the case $m = 4$.

The analysis of the transition solution is carried out in the neighborhood of the critical point C_+ or C_- variable with ε. That is, putting

$$P_{SB2} - (\hat{r}_B^2 - 1)(u_{\theta A1}/\hat{r}_B)^2 = \Delta_{(C+)}\varepsilon \quad \text{(around } C_+\text{)},$$

$$\text{or} \quad P_{SB2} - (\hat{r}_B^2 - 1)u_{\theta B1}^2 = \Delta_{(C-)}\varepsilon \quad \text{(around } C_-\text{)},$$

we carry out analysis similar to that in Section 3, which can be done consistently to higher orders. Brief summary of the results is as follows:

(i) In the neighborhood of critical point C_+:

    (a) When $(3u_{\theta A1} + u_{\theta B1}\hat{r}_B)(u_{\theta A1} - u_{\theta B1}\hat{r}_B) < 0$, the solution with positive u_{rH} exists uniquely for all values of $\Delta_{(C+)}$. This solution is continuously transformed into solution I_+ in the intermediate region $O(\varepsilon) \ll -\Delta_{(C+)}\varepsilon \ll O(1)$ and into solution II in a region $O(\varepsilon) \ll \Delta_{(C+)}\varepsilon \ll O(1)$.

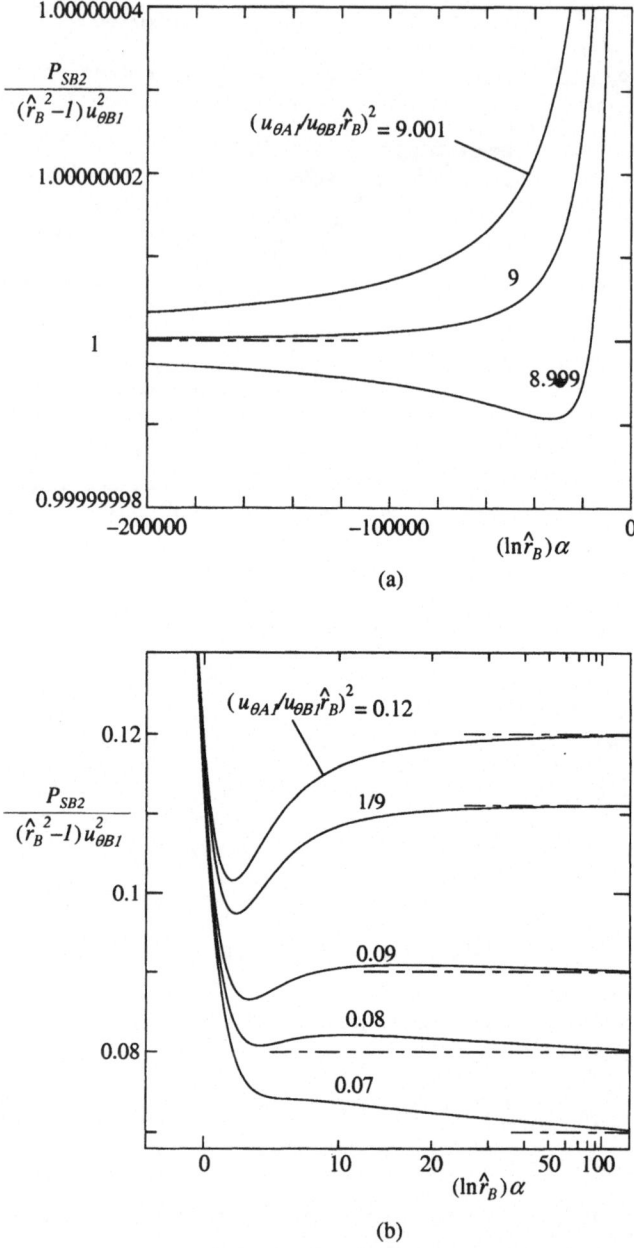

FIG. 4. $P_{SB2}/(\hat{r}_B^2 - 1)u_{\theta B1}^2$ versus α [Eq. (4.4)] III: Magnified local profiles $(u_{\theta A1}u_{\theta B1} < 0)$. (a) Profiles $(\hat{r}_B = 2)$ around $u_{\theta A1} = -3u_{\theta B1}\hat{r}_B$, and (b) profiles $(\hat{r}_B = 2)$ around $u_{\theta A1} = -u_{\theta B1}\hat{r}_B/3$. In panel (a), the chain line $--$ indicates the line $P_{SB2}/(\hat{r}_B^2 - 1)u_{\theta B1}^2 = 1$. In panel (b), the chain lines $---$ indicate the asymptotes as $\alpha \to \infty$ [or $P_{SB2}/(\hat{r}_B^2 - 1)u_{\theta B1}^2 = (u_{\theta A1}/u_{\theta B1}\hat{r}_B)^2$].

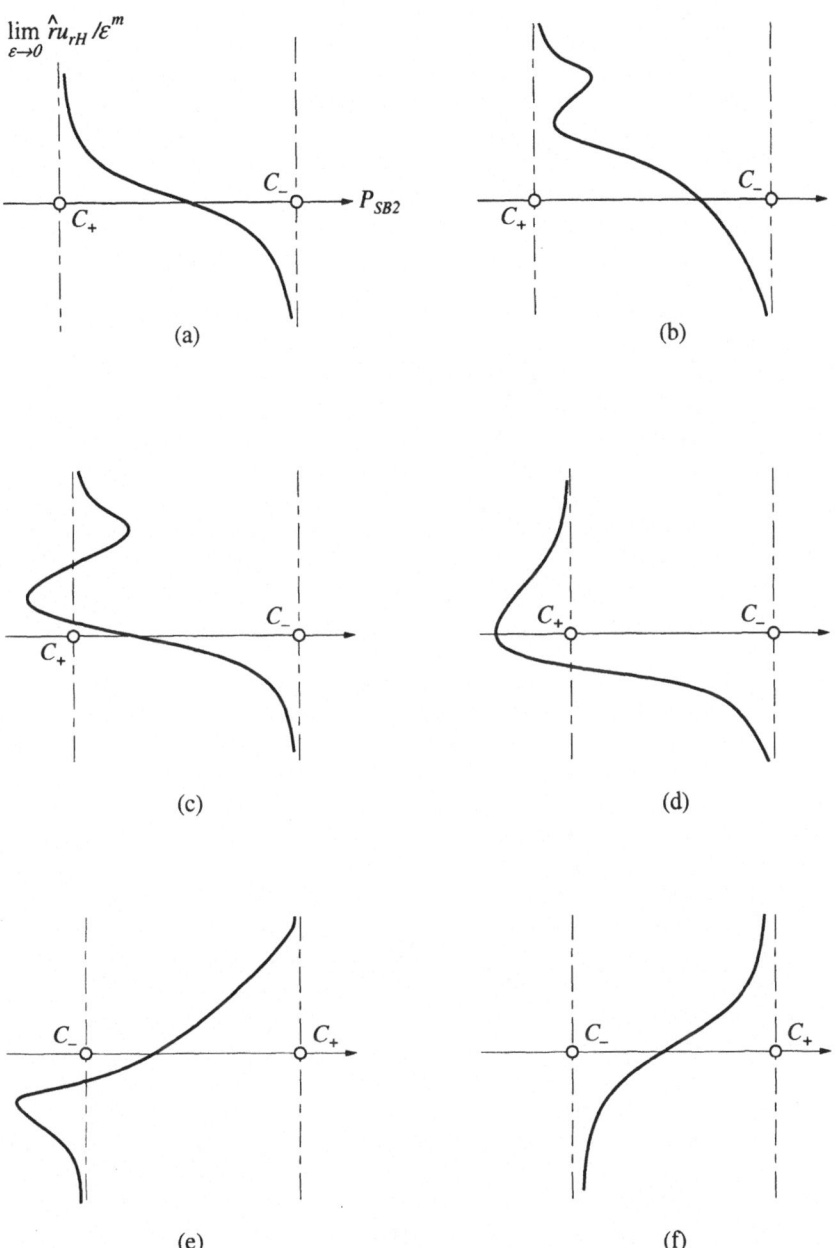

FIG. 5. *Bifurcation diagram of solution II (or schematic profiles* $\lim_{\varepsilon \to 0} \hat{r} u_{rH}/\varepsilon^m$ *versus* P_{SB2}), *when the two cylinders are rotating in opposite directions* ($u_{\theta A1} u_{\theta B1} < 0$). (a) $(u_{\theta A1}/u_{\theta B1}\hat{r}_B)^2 \leq 1/9c_1$ ($c_1 > 1$), (b) $1/9c_1 < (u_{\theta A1}/u_{\theta B1}\hat{r}_B)^2 < 1/9c_2$ ($c_1 > c_2 > 1$), (c) $1/9c_2 < (u_{\theta A1}/u_{\theta B1}\hat{r}_B)^2 < 1/9$, (d) $1/9 \leq (u_{\theta A1}/u_{\theta B1}\hat{r}_B)^2 < 1$, (e) $1 < (u_{\theta A1}/u_{\theta B1}\hat{r}_B)^2 < 9$, *and* (f) $(u_{\theta A1}/u_{\theta B1}\hat{r}_B)^2 \geq 9$, *where* c_1 *and* c_2 *depend on* $\hat{r}_B$.

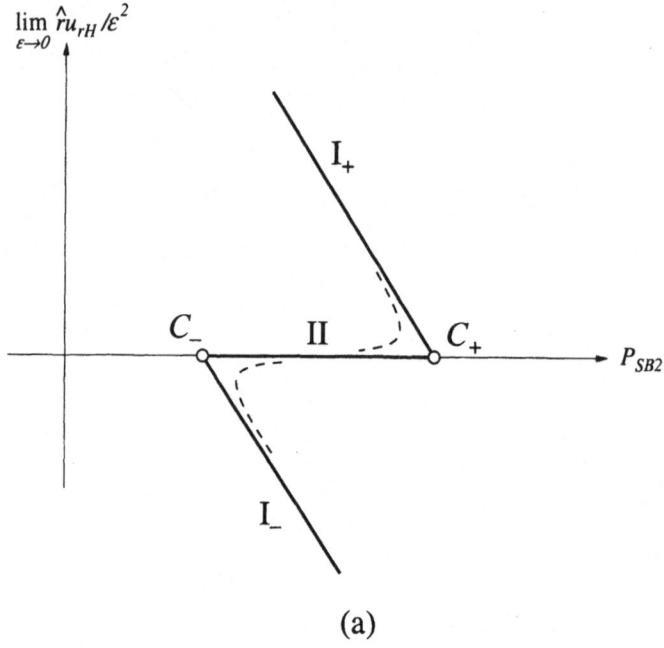

(a)

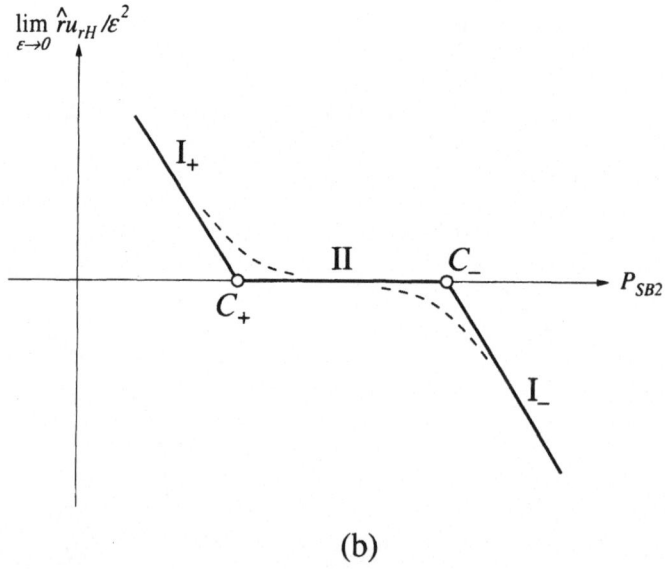

(b)

FIG. 6. *Bifurcation diagram (or schematic profiles* $\lim_{\varepsilon \to 0} \hat{r} u_{rH}/\varepsilon^2$ *versus* P_{SB2}) I: The two cylinders are rotating in the same direction $(u_{\theta A1} u_{\theta B1} > 0)$. (a) $u_{\theta A1}^2 > (u_{\theta B1} \hat{r}_B)^2$ and (b) $u_{\theta A1}^2 < (u_{\theta B1} \hat{r}_B)^2$. The solution is indicated by thick solid lines. The connection of the two solutions I and II by a transition solution is shown by dashed lines.*

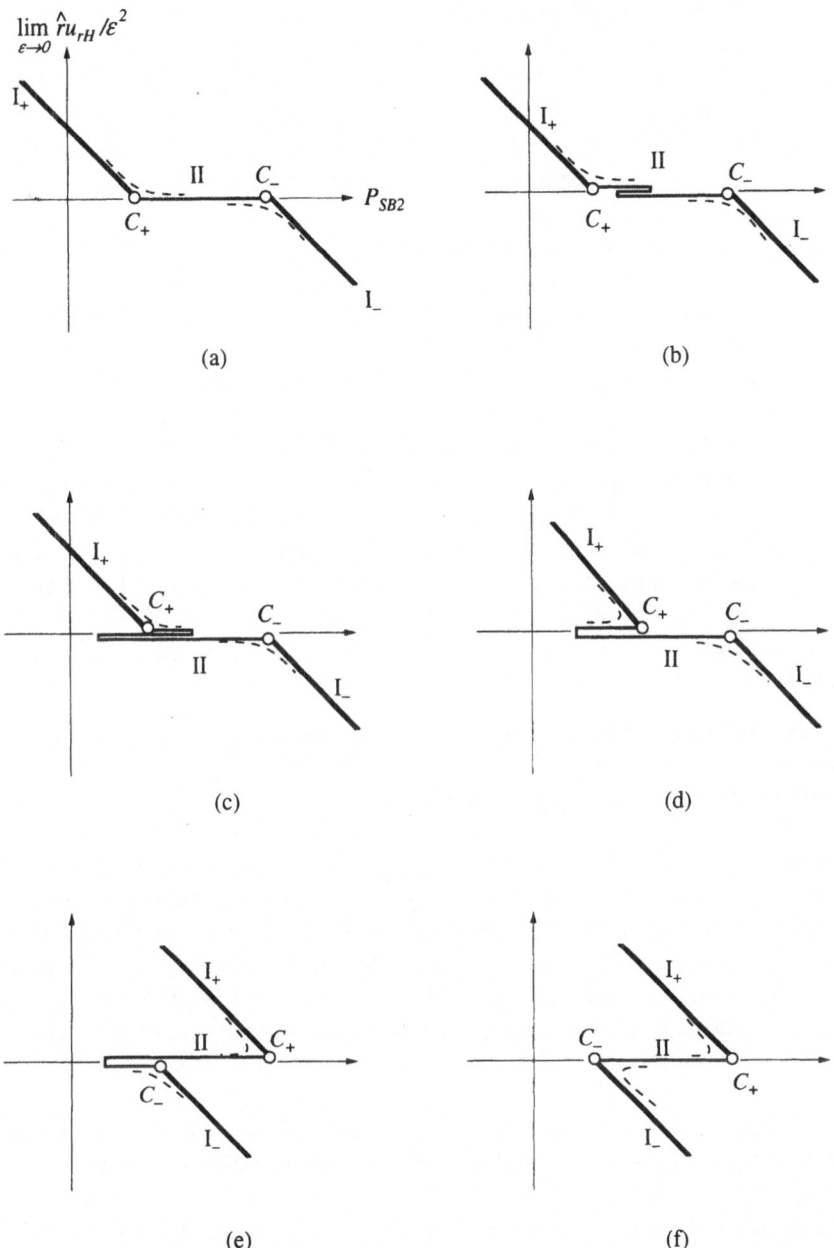

FIG. 7. *Bifurcation diagram (or schematic profiles* $\lim_{\varepsilon \to 0} \hat{r}u_{rH}/\varepsilon^2$ *versus* P_{SB2})
II: The two cylinders are rotating in opposite directions $(u_{\theta A1}u_{\theta B1} < 0)$. *(a)*
$(u_{\theta A1}/u_{\theta B1}\hat{r}_B)^2 \leq 1/9c_1$, *(b)* $1/9c_1 < (u_{\theta A1}/u_{\theta B1}\hat{r}_B)^2 < 1/9c_2$, *(c)* $1/9c_2 <$
$(u_{\theta A1}/u_{\theta B1}\hat{r}_B)^2 < 1/9$, *(d)* $1/9 \leq (u_{\theta A1}/u_{\theta B1}\hat{r}_B)^2 < 1$, *(e)* $1 < (u_{\theta A1}/u_{\theta B1}\hat{r}_B)^2 < 9$,
and (f) $(u_{\theta A1}/u_{\theta B1}\hat{r}_B)^2 \geq 9$, *where* c_1 *and* c_2 *depend on* $\hat{r}_B$. *The solution is indicated
by thick or medium solid lines. The connection of the two solutions I and II by a
transition solution is shown by dashed lines.*

(b) When $(3u_{\theta A1} + u_{\theta B1}\hat{r}_B)(u_{\theta A1} - u_{\theta B1}\hat{r}_B) > 0$, two solutions with positive u_{rH} exist for $\Delta_{(C+)}$ that is smaller than some negative value, and they are continuously joined at the threshold value of $\Delta_{(C+)}$. No solution exists for the other values of $\Delta_{(C+)}$. One of the solution with larger u_{rH} is transformed into solution I_+ and the other solution into II in an intermediate region $O(\varepsilon) \ll -\Delta_{(C+)}\varepsilon \ll O(1)$.

(ii) In the neighborhood of critical point C_-:

(a) When $(u_{\theta A1} + 3u_{\theta B1}\hat{r}_B)(u_{\theta A1} - u_{\theta B1}\hat{r}_B) < 0$, the solution with negative u_{rH} exists uniquely for all values of $\Delta_{(C-)}$. This solution is continuously transformed into solution I_- in the intermediate region $O(\varepsilon) \ll \Delta_{(C-)}\varepsilon \ll O(1)$ and into solution II in a region $O(\varepsilon) \ll -\Delta_{(C-)}\varepsilon \ll O(1)$.

(b) When $(u_{\theta A1} + 3u_{\theta B1}\hat{r}_B)(u_{\theta A1} - u_{\theta B1}\hat{r}_B) > 0$, two solutions with negative u_{rH} exist for $\Delta_{(C-)}$ that is larger than some positive value, and they are continuously joined at the threshold value of $\Delta_{(C-)}$. No solution exists for the other values of $\Delta_{(C-)}$. These solutions are continuously transformed into solution I_- or II in a region $O(\varepsilon) \ll \Delta_{(C-)}\varepsilon \ll O(1)$.

The behavior of the transition solutions in the bifurcation diagrams in Figs. 6 and 7 is shown by dashed lines.

6. Supplementary discussions. In this section, we briefly discuss the cases where $p_{SB}/p_{SA} - 1$ is of the order of ε, or $m = 1$ or 2, which are excluded in the foregoing analysis [8].

When $p_{SB}/p_{SA} - 1$ is of the order of ε, according to the asymptotic theory in [9], [10], and [11] (see also [12]), the radial velocity of the order of ε is uniquely determined by $p_{SB}/p_{SA} - 1$, and the tangential velocity is also determined uniquely and independently of the higher order radial velocity. The general asymptotic theory for the case $m = 1$ is developed in [10]. When $m = 1$ and $p_{SB}/p_{SA} - 1 = O(\varepsilon^2)$, the radial velocity is of the order ε^2, and the leading tangential velocity is uniquely determined by the Stokes set of equations without convection terms (no radial velocity enters in this set).

When $m = 2$ and $p_{SB}/p_{SA} - 1 = P_{SB2}\varepsilon^2$ $[P_{SB2} = O(1)]$, the asymptotic analysis for small ε leads to the sets of equations similar to those obtained in Section 4. That is, $u_{rH1} = P_{H1} = 0$, and the equations governing $u_{\theta H1}, \tau_{H1}, u_{rH2},$ and P_{H2} are Eqs. (4.1a)–(4.1d) with $m = 2$. The boundary conditions are

$$u_{\theta H1} = u_{\theta A1}, \quad \tau_{H1} = 0, \quad P_{H2} = C_4^* u_{rH2} \quad \text{at } \hat{r} = 1,$$

$$u_{\theta H1} = u_{\theta B1}, \quad \tau_{H1} = 0, \quad P_{H2} = P_{SB2} - C_4^* u_{rH2} \quad \text{at } \hat{r} = \hat{r}_B.$$

The equations are of type II, but the boundary conditions are of type I. Thus, the solution has mixed features of solutions I and II: The solution is of type II in the sense that the viscous effect is not confined in a suction

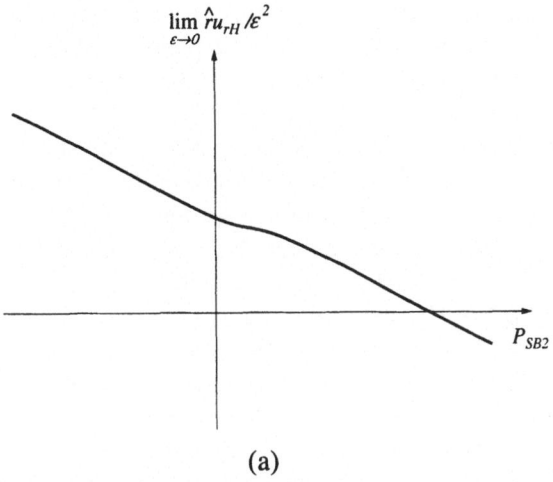

(a)

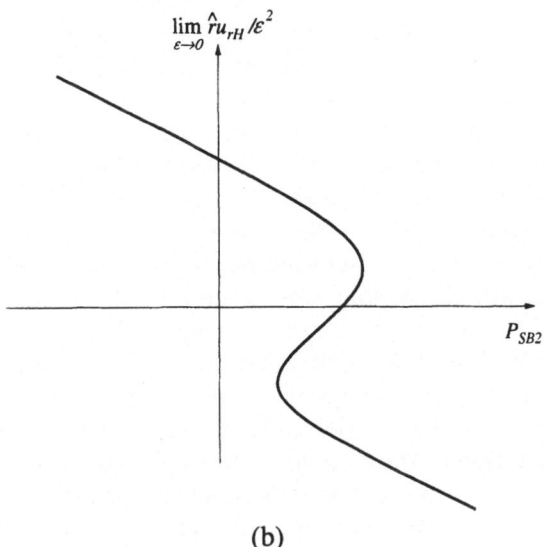

(b)

FIG. 8. *Bifurcation diagram (or schematic profiles* $\lim_{\varepsilon \to 0} \hat{r}u_{rH}/\varepsilon^2$ *versus* P_{SB2}) *for the case* $m = 2$ *(Section 6). (a) Unique solution and (b) bifurcation.*

boundary layer but extends over the whole domain, but is of type I in the sense that $\lim_{\varepsilon \to 0} \hat{r}u_{rH}/\varepsilon^2$ diverges to ∞ or $-\infty$ as P_{SB2} tends to $-\infty$ or ∞. The bifurcation diagram ($\lim_{\varepsilon \to 0} \hat{r}u_{rH}/\varepsilon^2$ versus P_{SB2}) has two types as shown in Figs. 8(a) and 8(b) in the latter of which three solutions are possible.

7. Concluding remarks. We considered a gas between two coaxial circular cylinders made of the condensed phase of the gas, where evaporation or condensation takes place. The time-independent behavior of the gas was investigated analytically with special interest in bifurcation of the flow, when the speeds of rotation of the cylinders and the Knudsen number were small. Bifurcation of solution occurs when the Knudsen number Kn is of the second order or the higher of the speeds $O(\varepsilon)$ of the circumferential velocity of the cylinders (Kn $= \varepsilon^m$, $m \geq 2$) even in the simple case where the flow field is axially symmetric and uniform. The solution including its existence range of the parameters is generally obtained in an analytic form. When $m \geq 3$, the solutions are classified into two types, I and II, (Sections 3 and 4) according to the order of the magnitude of the radial flow velocity compared with the speeds of rotation of the cylinders (or whether the effect of viscosity is confined in a thin layer or extends over the whole field). The type-I solutions are subclassified into two classes, I_+ and I_-, of positive or negative radial velocity. In some range of the parameters explicitly given, the two subclasses of solutions coexist and in some other range explicitly given, no solution I (neither I_+ nor I_-) exists. In these ranges, there exists at least one solution of type II. Two or three solutions of type II also exist in some range outside or inside these ranges when the two cylinders are rotating in opposite directions. As is clear from the formulas, the formulas of classification and range in the bifurcation diagrams are independent of molecular models (or γ_1, γ_2, and C_4^*) as far as $C_4^* < 0$. (The difference appears only in the scale of the ordinate of the bifurcation diagrams.) When $m = 2$, the solution has mixed features of the two types and bifurcation with three solutions occurs. In this case the formulas of classification and range depend on molecular models (or γ_1 and C_4^*). The comprehensive feature of the bifurcation of solution (a new type as well as the one first pointed out in our previous paper [1]) is clarified.

The feature of the bifurcation is independent of m (the relative size of the speeds of rotation of the cylinders and the Knudsen number) if $m \geq 3$. The limiting case $m \to \infty$ corresponds to a flow with small Mach numbers in the continuum limit. In solution II, the radial flow velocity u_r vanishes in this limit. That is, if we live in the world of the continuum limit, there is no radial velocity. However, if we solve the Navier-Stokes set of equations for small Mach numbers neglecting the radial velocity, we obtain only the solution II with $\alpha = 0$ in Section 4. The solutions II with the other α, which are also the solutions in the continuum limit, cannot be obtained in the framework of the classical gas dynamics. Something not existing in the world of the continuum limit affects a finite effect on the behavior of a gas in that world, which is the ghost effect discussed in [13] and [14] and shows the incompleteness of the classical gas dynamics.

REFERENCES

[1] Y. SONE, H. SUGIMOTO, AND K. AOKI, Cylindrical Couette flows of a rarefied gas with evaporation and condensation: Reversal and bifurcation of flows, *Phys. Fluids* (1999), **11**: 476–490.

[2] Y. SONE AND T. DOI, Analytical study of bifurcation of a flow of a gas between coaxial circular cylinders with evaporation and condensation, *Phys. Fluids* (2000), **12**: 2639–2660.

[3] P. WELANDER, On the temperature jump in a rarefied gas, *Ark. Fys.* (1954), **7**: 507–553.

[4] M.N. KOGAN, On the equation of motion of a rarefied gas, *Appl. Math. Mech.* (1958), **22**: 597–607.

[5] C. BARDOS, R.E. CAFLISCH, AND B. NICOLAENKO, The Milne and Kramers problems for the Boltzmann equation of a hard sphere gas, *Commun. Pure Appl. Math.* (1986), **39**: 323–352.

[6] C. CERCIGNANI, Half-space problem in the kinetic theory of gases, Trends in Applications of Pure Mathematics to Mechanics, E. Kröner and K. Kirchgässner (eds.), Springer-Verlag, Berlin, 1986, pp. 35–50.

[7] F. GOLSE AND F. POUPAUD, Stationary solutions of the linearized Boltzmann equation in a half-space, *Math. Methods Appl. Sci.* (1989), **11**: 483–502.

[8] The saturation pressure of a gas is a rapidly increasing function of its temperature. Thus, we consider only the case $T_B/T_A - 1 = O(p_{SB}/p_{SA} - 1)$.

[9] Y. ONISHI AND Y. SONE, Kinetic theory of slightly strong evaporation and condensation–Hydrodynamic equation and slip boundary condition for finite Reynolds number–, *J. Phys. Soc. Jpn* (1979), **47**: 1676–1685.

[10] Y. SONE AND K. AOKI, Steady gas flows past bodies at small Knudsen numbers–Boltzmann and hydrodynamic systems, *Transp. Theory Stat. Phys.* (1987), **16**: 189–199.

[11] K. AOKI AND Y. SONE, Gas flows around the condensed phase with strong evaporation or condensation–Fluid dynamic equation and its boundary condition on the interface and their application, Advances in Kinetic Theory and Continuum Mechanics, R. Gatignol and Soubbaramayer (eds.), Springer-Verlag, Berlin, 1991, pp. 43–54.

[12] Y. SONE, Theoretical and Numerical Analyses of the Boltzmann Equation–Theory and Analysis of Rarefied Gas Flows–Part I, Lecture Notes, Department of Aeronautics and Astronautics, Graduate School of Engineering, Kyoto University, Kyoto, 1998 (http://www.users.kudpc.kyoto-u.ac.jp/~a50077).

[13] Y. SONE, K. AOKI, S. TAKATA, H. SUGIMOTO, AND A.V. BOBYLEV, Inappropriateness of the heat-conduction equation for description of a temperature field of a stationary gas in the continuum limit: Examination by asymptotic analysis and numerical computation of the Boltzmann equation, *Phys. Fluids* (1996), **8**: 628–638; Erratum, *Phys. Fluids* (1996), **8**: 841.

[14] Y. SONE, Continuum gas dynamics in the light of kinetic theory and new features of rarefied gas flows, Rarefied Gas Dynamics, Ching Shen (ed.), Peking University Press, Beijing, 1997, pp. 3–24.

MAGNETIC INSTABILITY IN A COLLISIONLESS PLASMA*

WALTER A. STRAUSS[†]

1. Introduction. We consider the Vlasov Maxwell equations. Consider an equilibrium $f = \mu(v)$ with $E = B = 0$. Consider general electromagnetic perturbations $f = \mu + \delta f, E = \delta E, B = \delta B$. Is the equilibrium stable or not?

In this lecture I report on joint work with Yan Guo. Our result is the following. There exist such equilibria that are slightly non-isotropic and that decrease from the origin in all directions such that
(i) (electrostatic case) if $\delta B = 0$, it is stable;
(ii) (electromagnetic case) if $\delta B \neq 0$, it is unstable.
Furthermore, if (ii) holds but the spatial period is less than P_0, it still is stable, but if the period is greater than P_0, it is unstable. This result is apparently new, even in a formal sense.

The relativistic Vlasov-Maxwell system is

$$\partial_t f_\pm + \hat{v}_\pm \cdot \nabla_x f_\pm \pm e(E + c^{-1}\hat{v}_\pm \times B) \cdot \nabla_v f_\pm = 0,$$
$$\partial_t E - c\nabla \times B = -j, \qquad \nabla \cdot E = \rho,$$
$$\partial_t B + c\nabla \times E = 0, \qquad \nabla \cdot B = 0.$$

where the phase-space density of electrons (ions) is f_- (f_+), the electromagnetic field is (E, B), the momentum is v, the velocity is $\hat{v}_\pm = v/\langle v\rangle_\pm$ with $\langle v\rangle_\pm = (m_\pm^2 + |v|^2/c^2)^{1/2}$, the energy is $= c^2\langle v\rangle_\pm$, and the density and current are

$$\rho = 4\pi e \int (f_+ - f_-)dv, \quad j = 4\pi e \int (\hat{v}_+ f_+ - \hat{v}_- f_-)dv.$$

Furthermore, c is the speed of light, e is the charge of an ion, $-e$ of an electron, and $m_\pm$ is the mass of an ion (electron). For purposes of exposition, we will choose the various constants $c, e, m_\pm$ to be one, and thereby omit the subscript in $\langle v\rangle$ and $\hat{v}$.

Some types of equilibria that have been studied include the following.

(a) A homogeneous equilibrium $f_\pm = \mu_\pm(v)$ with $E = B = 0$ can be stable or unstable [P]. For instance, if $\mu_\pm$ are functions of $\langle v\rangle$ alone and are strictly decreasing, the equilibrium is electromagnetically stable.

(b) A BGK equilibrium $f_\pm = \mu_\pm(\langle v\rangle \pm \Phi(x))$ with $E = \nabla\Phi, B = 0$ is unstable under various conditions, including periodic, solitary and shock cases [BGK, GS1, GS2].

───────────

*This research is supported in part by NSF grant 97-03695.

[†]Lefschetz Center for Dynamical Systems, Department of Mathematics, Brown University, Providence, RI 02912. Email: wstrauss@math.brown.edu.

(c) A magnetic equilibrium $f_{\pm} = \mu_{\pm}(\langle v \rangle \pm \Phi(x), v_2 \pm \Psi(x))$ with $E = \nabla\Phi, B = \nabla\Psi$ can be either stable or unstable, depending on the shapes of $\mu_{\pm}, \Phi$ and Ψ [G1, G2, GS3].

The magnetic equilibria motivate a deeper study of magnetic perturbations, so we now return to the relatively simple, homogeneous equilibria but permit general electromagnetic perturbations. For simplicity, besides choosing the constants equal to one, we will sometimes take the following special coordinates:

$$\begin{pmatrix} x \\ 0 \\ 0 \end{pmatrix}, \quad \begin{pmatrix} v_1 \\ v_2 \\ v_3 \end{pmatrix}, \quad \begin{pmatrix} E_1 \\ E_2 \\ 0 \end{pmatrix}, \quad \begin{pmatrix} 0 \\ 0 \\ B \end{pmatrix}.$$

We will only consider solutions of period P in x. This is the simplest possible boundary condition. The relativistic Maxwell-system in these coordinates is well-posed [GSc].

2. Formal stability theory. Now we consider the simplest kind of equilibrium

$$f_{\pm} = \mu_{\pm}(v), \quad E = B = 0$$

which is neutral. We linearize and Fourier transform formally with respect to x and t. This procedure leads to a dispersion relation, as follows. An exponential

$$\begin{pmatrix} f_{\pm} \\ E \\ B \end{pmatrix} = e^{i(kx - \omega t)} \begin{pmatrix} f_{\pm}^{\sharp}(v) \\ E^{\sharp} \\ B^{\sharp} \end{pmatrix}$$

is put into the linearized system where k, E and B are real vectors and ω is a complex scalar. In the resulting equations, we eliminate $f_{\pm}^{\sharp}(v)$ and $B^{\sharp} = \frac{k}{\omega} \times E^{\sharp}$ to obtain, after some elementary calculations, the equations

$$(1) \qquad (\omega^2 - k^2)\, k \times E^{\sharp} = -\int (k \times \hat{v}) \left[E^{\sharp} + \frac{(\hat{v} \cdot E^{\sharp})k}{\omega - \hat{v} \cdot k} \right] \cdot \nabla_v \mu \, dv$$

and

$$\omega k \cdot E^{\sharp} = -\int \frac{(\hat{v} \cdot E^{\sharp})k}{\omega - \hat{v} \cdot k} \cdot \nabla_v \mu \, dv.$$

where $\mu = \mu_+ + \mu_-$.

Purely electric (or "longitudinal") perturbations occur when $B^{\sharp} = 0$, that is, k is parallel to $E^{\sharp}$. It is a standard fact that if μ is decreasing in all directions from the origin in v-space, then ω is real. This means that μ is linearly stable.

"Purely magnetic" (or "transverse") perturbations occur when $k \cdot E^{\sharp} = 0, E^{\sharp} \neq 0$. If $\mu = \mu(\langle v \rangle)$ is decreasing, then it is still stable; this is the isotropic case. If $\mu = \mu(\langle v \rangle, v_2)$, say, then a sufficient condition for instability is

$$\left(\frac{2\pi}{P_0}\right)^2 = \int_{R^3} \hat{v}_2 \, \partial_2 \mu \, dv > 0.$$

This is proven via a Penrose type of analysis [P] involving some contour integration, which we omit here.

For example, let $\mu = e^{-\langle v \rangle}(1 - \gamma(v_2))$, where γ is a little bump function with support near 0. Then μ deviates arbitrarily little from a maxwellian but nevertheless is electromagnetically linearly unstable!

3. Nonlinear stability. Let us consider an equilibrium $\mu_{\pm}(\langle v \rangle, v_2)$ as in the example just given.

THEOREM 1. *Assume either $\delta B = 0$ or $P < P_0$. $\forall \epsilon > 0, \exists \delta > 0$ such that if the initial data is within δ of the equilibrium (in a certain norm), then the solution of the full nonlinear problem satisfies*

$$\sup_{0 \leq t < \infty} \|(solution) - (equilibrium)\|_{L^1} < \epsilon.$$

The L^1 norm is taken in all the components. (The critical period P_0 is defined above.)

Proof. Consider initial data of (f, E, B) that are near $(\mu, 0, 0)$. Denote $\partial_x \Psi = B$ with $\int_0^P \Psi \, dx = 0$. Define $Q_{\pm}(\mu_{\pm}, v_2)$ by

$$\partial_1 Q(\mu_{\pm}(\langle v \rangle, v_2), v_2) = -\langle v \rangle,$$

which is legal since $\mu_{\pm}$ are monotone in $\langle v \rangle$. Define

$$J(f, E, B) = \sum_{\pm} \iint \{Q_{\pm}(f_{\pm}, v_2 \pm \Psi) + \langle v \rangle f_{\pm}\} dv dx + \frac{1}{2}\int\{|E|^2 + |B|^2\}dx.$$

J is an invariant with respect to the time evolution. In the following we omit some of the subscripts $\pm$ for the sake of exposition. Now

$$J(f, E, B) - J(\mu, 0, 0) = \sum \iint \{Q(f, v_2 \pm \Psi) - Q(\mu, v_2)$$
$$+ \langle v \rangle(f - \mu)\} dv dx + \frac{1}{2}\int\{|E|^2 + |B|^2\}dx.$$

We require this expression to be positive, and in fact to be bounded below by a norm of the difference. Let us Taylor expand Q around (μ, v_2). All the first-order terms drop out by definition of Q. We get

$$\frac{1}{2}\iint\{\partial_1^2 Q \cdot (f - \mu)^2 + \ldots + \partial_2^2 Q \cdot \Psi^2\} \, dv dx + \frac{1}{2}\int\{|E|^2 + |B|^2\}dx$$

where the dots indicate certain other terms that we can handle. Now

$$
\begin{aligned}
\int \partial_2^2 Q(\mu, v_2) dv &= -\int (\hat{v}_2 \partial_1 \mu + \partial_2 \mu) \cdot \partial_1 \partial_2 Q dv + \int \frac{\partial}{\partial v_2}(\ldots) dv \\
&= -\int \hat{v}_2 \partial_2 \mu \, dv - \int \partial_2 \mu \cdot \partial_1 \partial_2 Q \, dv,
\end{aligned}
$$

where the last term is balanced by other terms, and

$$
\int |B|^2 dx \geq \left(\frac{2\pi}{P}\right)^2 \int \Psi^2 dx
$$

since $\int_0^P \Psi dx = 0$. So two of the terms above combine to give

$$
\frac{1}{2} \left[\left(\frac{2\pi}{P}\right)^2 - \int \hat{v}_2 \partial_2 \mu \, dv \right] \int \Psi^2 dx > 0,
$$

because of the assumption $P < P_0$, where $\mu = \mu_+ + \mu_-$. □

4. Nonlinear instability. By the formal theory given above, we know that if $P > P_0$, then the linearized problem with period P has some unstable point spectrum $e^{\lambda t}$ with $\Re\lambda > 0$. It is by no means a trivial matter to iterate this growth in the nonlinear case because the nonlinear term contains first-order derivatives. However, we can prove the following result. We continue to consider an equilibrium $\mu_\pm(\langle v \rangle, v_2)$ as in the example.

THEOREM 2. *Let $P > P_0$. There exist positive constants ϵ_o, δ_o and a family of solutions f^δ defined for $0 < \delta \leq \delta_o$ such that*

$$
\|f^\delta(0) - \mu\|_{W^{1,1}} < \delta
$$

but

$$
\sup_{0 \leq t \leq C|\log \delta|} \|f^\delta(t) - \mu\|_{L^1} > \epsilon_o.
$$

By $\|\ldots\|_{L^1}$ we mean the sum of the L^1 norms of all five components (f_+, f_-, E_1, E_2, B).

Proof. Write the Vlasov equation in the simplified notation

$$
(\partial_t + \hat{v}_1 \partial_x) f \pm (E + \hat{v} \times B) \cdot \nabla_v f = 0
$$

or

$$
(\partial_t + \hat{v}_1 \partial_x)(f - \mu) \pm (E + \hat{v} \times B) \cdot \nabla_v \mu = \pm(E + \hat{v} \times B) \cdot \nabla_v(f - \mu)
$$

or

$$
(\partial_t + A + K) \begin{pmatrix} f - \mu \\ E \\ B \end{pmatrix} = \mathcal{N} \begin{pmatrix} f - \mu \\ E \\ B \end{pmatrix}
$$

where $A = \hat{v}_1 \partial_x$, K represents the linear term involving $\nabla_v \mu$, and $\mathcal{N}$ represents the nonlinear term on the right side. K is not a compact operator; nevertheless we prove

LEMMA 3. *The operator $Ke^{-tA}K$ is compact in L^1 for all $t > 0$.* The proof uses the estimates of Glassey and Schaeffer [GSc] and the fact that the characteristic speeds of the Maxwell and Vlasov parts are distinct. From the Lemma and the theorem of Vidav [V] follows the

COROLLARY. *The spectrum of $e^{-tL} \equiv e^{-t(A+K)}$ is discrete outside the unit circle.*

Let $e^{\lambda t}$ be an eigenvalue where $|e^{\lambda t}|$ is maximal. For simplicity of exposition below, we assume that λ is real. By the linearized instability in II above, $\lambda > 0$. The corresponding eigenvector, denoted (f^o, E^o, B^o) is integrable. Even though it does not satisfy an elliptic problem, we can bootstrap to show it is smooth.

In order to prove the nonlinear instability, we now solve the full nonlinear problem with

$$f_{\pm}\Big|_{t=0} = \mu_{\pm} + \delta \cdot f_{\pm}^o, \quad E\Big|_{t=0} = \delta \cdot E^o, \quad B\Big|_{t=0} = \delta \cdot B^o$$

for any $0 < \delta \le \delta_o$ (with δ_o to be chosen later). That is, the initial data are close to equilibrium. Now we write the PDE for $f - \mu$ in integral form:

$$f(t) - \mu = e^{-Lt} \delta f^o + \int_0^t e^{-L(t-\tau)} \mathcal{N}(f(\tau) - \mu) d\tau.$$

Taking L^1 norms (of each component), we estimate

$$\|f(t) - \mu\|_{L^1} \ge \delta \|e^{-Lt} f^o\|_{L^1} - \int_0^t e^{\lambda(t-\tau)} \|\mathcal{N}(f(\tau) - \mu)\|_{L^1} d\tau.$$

The last norm (the nonlinear part) is bounded by

$$(\|E\|_{L^\infty} + \|B\|_{L^\infty}) \|\nabla_v(f(\tau) - \mu)\|_{L^1}.$$

The loss of derivatives makes this estimate appear hopeless. Nevertheless, it is possible to estimate the derivatives as follows, using once again the estimates of [GSc].

LEMMA 4 (Main Bootstrap Lemma). *If $\|\langle v \rangle (f(0) - \mu)\|_{W^{1,1}} \le \delta$ and*

$$\|f(t) - \mu\|_{L^1} \le c_1 \delta e^{\lambda t} \quad \text{for } 0 \le t \le T,$$

then there exist c_2, θ depending only on c_1 and λ such that

$$\|E(t)\|_{W^{1,\infty}} + \|B(t)\|_{W^{1,\infty}} + \|\langle v \rangle \nabla_v(f(t) - \mu)\|_{L^1} \le c_2 \delta e^{\lambda t}$$

for $0 \le t \le \min\{T, \frac{1}{\lambda} \log \frac{\theta}{\delta}\}$. Assuming this lemma, we get

$$\|f(t) - \mu\|_{L^1} \ge \delta e^{\lambda t} - c \int_0^t e^{\lambda(t-\tau)} (\delta e^{\lambda \tau})^2 d\tau$$

$$= \delta e^{\lambda t} - c(\delta e^{\lambda \tau})^2 > \epsilon_o$$

by choosing a time $t = t_\delta$ such that $\delta e^{\lambda t} = O(\epsilon_o)$. Since δ is arbitrarily small and ϵ_o is fixed, this proves the instability of the equilibrium μ.   ◻

REFERENCES

[BGK] BERNSTEIN, I., GREENE, J., AND KRUSKAL, M., *Exact nonlinear plasma oscillations*, Phys. Rev. **108**(3): 546–550 (1957).

[G1] GUO, Y., *Stable magnetic equilibria in collisionless plasmas*, Comm. Pure Appl. Math., Vol. **L**, 0891–0933 (1997).

[G2] GUO, Y., *Stable magnetic equilibria in a symmetric plasma*, Commun. Math. Phys., **200**: 211–247 (1999).

[GS1] GUO, Y. AND STRAUSS, W., *Instability of periodic BGK equilibria*, Comm. Pure Appl. Math. Vol. XLVIII, 861–894 (1995).

[GS2] GUO, Y. AND STRAUSS, W., *Unstable BGK solitary waves and collisionless shocks*, Commun. Math. Phys. **195**: 249–265 (1998).

[GS3] GUO, Y. AND STRAUSS, W., *Unstable oscillatory-tail solutions*, SIAM J. Math. Anal., **30**(5): 1076–1114 (1999).

[GSc] GLASSEY, R. AND SCHAEFFER, J., *On the 'one and one-half-dimensional' relativistic system*, Math. Methods Appl. Sci. **13**: 169–179 (1990).

[P] PENROSE, O., *Electrostatic instability of a non-Maxwellian plasma*, Phys. Fluids. **3**: 258–265 (1960).

[V] VIDAV, I., *Spectra of perturbed semigroups with applications to transport theory*, J. Math. Anal. Appl. **30**: 264–279 (1970).

COMBINED LIST OF WORKSHOPS PARTICIPANTS FOR IMA VOLUMES 135: TRANSPORT IN TRANSITION REGIMES AND 136: DISPERSIVE TRANSPORT EQUATIONS AND MULTISCALE MODELS

IMA WORKSHOP: *DISPERSIVE CORRECTIONS TO TRANSPORT EQUATIONS, May 1-5, 2000*

- Javier Armendariz, Institute for Mathematics and its Applications
- Anton Arnold, Fachbereich Mathematik, Geb. 27. 1, Universitat des Saarlandes
- Donald G Aronson, Mathematics Department, University of Minnesota
- Bruce Ayati
- Claude Bardos, University of Paris VII
- Naoufel Ben-Abdallah, University of Toulouse
- Tony Bloch, Department of Mathematics, University of Michigan
- Jared Bronski, Department of Mathematics, University of Illinois Urbana Champaign
- David Cai, Courant Institute of Mathematical Sciences
- John Chadam, Department of Mathematics and Statistics, University of Pittsburgh
- Fred Dulles, Institute for Mathematics and its Applications
- Yalchin Efendiev, Institute for Mathematics and its Applications
- Nicholas Ercolani, Department of Mathematics, University of Arizona
- William R. Frensley, Erik Jonsson School of Engineering and Computer Science, University of Texas-Dallas
- Irene Gamba, Department of Mathematics, University of Texas at Austin
- Carl Gardner, Department of Mathematics, Arizona State University
- Ingenuin Gasser, Department of Mathematics, University of Hamburg
- Patrick Gerard, Department of Mathematics, University of Paris-Sud
- Francois Golse, Departement de Mathematiques, Ecole Normale Superieure
- Takumi Hawa, Institute for Mathematics and its Applications
- Sabina Jeschke, Fachbereich Mathematik, Technische Universitat Berlin

- Yuji Kodama, Department of Mathematics, Ohio State University
- Irina Kogan, School of Mathematics, University of Minnesota
- David Levermore, Department of Mathematics, University of Maryland
- Tong Li, Department of Mathematics, University of Iowa
- Tai-Ping Liu, Department of Mathematics, Stanford University
- Norbert Mauser, Institut für Mathematik Universität Wien
- David McLaughlin, New York University-Courant Institute
- Ken T.R. McLaughlin, Department of Mathematics, University of Arizona
- Willard Miller, Institute for Mathematics and its Applications
- Dave Nicholls, School of Mathematics, University of Minnesota
- Alexei Novikov, Institute for Mathematics and its Applications
- Hans Othmer, School of Mathematics, University of Minnesota
- Peter A Rejto, Department of Mathematics, University of Minnesota
- Christian Ringhofer, Department of Mathematics, Arizona State University
- Fadil Santosa, IMA and Minnesota Center for Industrial Mathematics (MCIM)
- Stephen Shipman, Department of Mathematics, Duke University
- Marshall Slemrod, Department of Mathematics, University of Wisconsin
- Kent Smith, Lucent Technologies, Bell Laboratories
- Henning Struchtrup, Department of Mechanical Engineering, University of Victoria
- Fei-Ran Tian, Department of Mathematics, Ohio State University

IMA WORKSHOP: SIMULATION OF TRANSPORT IN TRANSITION REGIMES, May 22–26, 2000

- Evans Afenya, Department of Mathematics, Elmhurst College
- Ramesh Agarwal, Department of Aerospace Engineering, Wichita State University
- Dinshaw Balsara, National Center for Supercomputing, University of Illinois
- Claude Bardos, University of Paris VII
- Daniel Bentil, Deparment of Mathematics and Statististics, University of Vermont
- Christoph Borgers, Department of Mathematics, Tufts University
- Jose Carrillo, Department of Mathematics/C1200, University of Texas at Austin
- John Chadam, Department of Mathematics and Statistics, University of Pittsburgh
- Pierre Charrier, Departamento de Mathematiques Appliquees, Universite Bordeaux I

- Pierre Degond, CNRS, UMR MIP 5640, Université Paul Sabatier
- Fred Dulles, Institute for Mathematics and its Applications
- Yalchin Efendiev, Institute for Mathematics and its Applications
- Byung Chan Eu, Department of Chemistry, McGill University
- Max Fischetti, Research Division, T.J. Watson Res. Cntr.
- Irene Gamba, Department of Mathematics, University of Texas at Austin
- Dirk Gillespie, Department of Physiology and Biophysics, University of Miami: School of Medicine
- Robert T. Glassey, Department of Mathematics, Indiana University
- Matthias Gobbert, Department of Mathematics and Statistics, University of Maryland, Baltimore County
- David Goldstein, Department of Aerospace Engineering and Engineering Mechanics, University of Texas at Austin
- Thierry Goudon, Laboratoire J.A. Dieudonne, Universite de Nice - Sophia Antipolis
- Clinton Groth, Institute of Aerospace Studies, University of Toronto
- Takumi Hawa, Institute for Mathematics and its Applications
- Jeffrey Hittinger, Department of Aerospace Engineering, University of Michigan
- Reinhard Illner, Department of Mathematics and Statistics, University of Victoria
- Joseph W. Jerome, Department of Mathematics, Northwestern University
- Shi Jin, Department of Mathematics, Georgia Tech
- Ansgar Jungel, Fachbereich Mathematik und Statistik, Universitat Konstanz
- Michael Junk, Fachbereich Mathematik, University of Kaiserslautern
- Junseok Kim, University of Minnesota
- Dimitri Kirill, Institute for Mathematics and Its Applications
- Axel Klar, FB Mathematik und Informatik, TU Darmstadt
- Mohammed Lemou, CNRS, UMR MIP 5640, Université Paul Sabatier
- C. David Levermore, Department of Mathematics, University of Maryland
- Hailiang Liu, Department of Mathematics, UCLA
- Paulo Lugli
- Mitch Luskin, School of Mathematics, University of Minnesota
- Peter Markowich, Institute for Mathematics, University of Vienna
- Nader Masmoudi, Courant Insitute of Mathematical Sciences
- Norbert Mauser, Institut für Mathematik, Universität Wien
- Willard Miller, Institute for Mathematics and its Applications

- Inrina Mitrea, University of Minnesota
- Orazio Muscato, Dipartimento di Matematica, Universita di Catania
- Rho Shin Myong, Department of Aero-Mechanical Engineering, Gyeongsang National University
- Anne Nouri, Department of Mathematical Modeling, INSA-Lyon Scientific and Technical University
- Alexei Novikov, Institute for Mathematics and its Applications
- Lorenzo Pareschi, Department of Mathematics, University of Ferrara
- Benoit Perthame, Departement de Mathematiques et Applications, Ecole Normale Superieure
- Christian Ringhofer, Department of Mathematics, Arizona State University
- Philip Roe, Department of Aerospace Engineering, University of Michigan
- Massimo Rudan, DEIS, Universita di Bologna
- Fadil Santosa, IMA and Minnesota Center for Industrial Mathematics (MCIM)
- Marco Saraniti, Department of Electrical and Computer Engineering, Illinois Institute of Technology
- David Sattinger, Department of Mathematics and Statistics, Utah State University
- Jack Schaeffer, Department of Mathematical Sciences, Carnegie Mellon University
- Christian Schmeiser, Institut für Angewandte und Numerische Mathematik, TU Wien
- Chi-Wang Shu, Division of Applied Mathematics, Brown University
- Marshall Slemrod, Department of Mathematics, University of Wisconsin
- Yoshio Sone, Department of Aeronautic Astronautios, Kyoto University
- Walter A. Strauss, Department of Mathematics, Brown University
- Henning Struchtrup, Department of Mechanical Engineering, University of Victoria
- Holger Teismann, Department of Mathematics, North Dakota State University
- Moulay Tidriri, Department of Mathematics, Iowa State University
- Kun Xu, Department of Mathematics, Hong Kong University of Science and Technology
- Wen-Qing Xu, Department of Mathematics and Statistics, University of Massachusetts Amherst

IMA WORKSHOP: MULTISCALE MODELS FOR
SURFACE EVOLUTION AND REACTING FLOWS
June 5-9, 2000

- Donald G. Aronson, Mathematics Department, University of Minnesota
- Jean-Pierre Boon, Departement de Physique, Universite Libre de Bruxelles
- Len Borucki, Motorola
- Timothy S. Cale, Department of Chemical Engineering, Rensselaer Polytechnic Institute
- Larry Carson, SEMS Technology Center, 3M
- Robert Crone, IBM
- Jacques Dalla Torre, Bell Laboratories, Lucent Technologies
- Fred Dulles, Institute for Mathematics and its Applications
- Yalchin Efendiev, Institute for Mathematics and its Applications
- Avner Friedman, MCIM, University of Minnesota
- Matthias Gobbert, Department of Mathematics and Statistics, University of Maryland, Baltimore County
- Thierry Goudon, Laboratoire J.A. Dieudonne, Universite de Nice - Sophia Antipolis
- Youngae Han, Department of Mathematics, University of Minnesota
- Takumi Hawa, Institute for Mathematics and its Applications
- Markos A. Katsoulakis, Mathematics and Statistics, University of Massachusetts
- Junseok Kim, University of Minnesota
- Yang Jin Kim, Deparatment of Mathematics, University of Minnesota
- John King, Theoretical Mechanics, University of Nottingham
- Matthew Laudon, Motorola
- C. David Levermore, Department of Mathematics, University of Maryland
- Andres F. Sole Martinez, Computer Vision Center (Spain)
- Willard Miller, Institute for Mathematics and its Applications
- David Misemer, 3M
- Peter Mucha, Department of Mathematics, Massachusetts Institute of Technology
- Alexei Novikov, Institute for Mathematics and its Applications
- Peter O'Sullivan, Bell Labs, Lucent Technologies
- Peter Olver, Department of Mathematics, University of Minnesota
- Stanley Osher, Department of Mathematics, UCLA
- David Porter, Department of Astronomy, University of Minnesota
- Christian Ratsch, Department of Mathematics, UCLA

- Fernando Reitich, Department of Mathematics, University of Minnesota
- Christian Ringhofer, Department of Mathematics, Arizona State University
- Alric Rothmayer, Aerospace Engineering and Engineering Mechanics, Iowa State University
- Victor Roytburd, Department of Mathematics, Rensselaer Polytechnic Institute
- Fadil Santosa, IMA and Minnesota Center for Industrial Mathematics (MCIM)
- James Sethian, Department of Mathematics, University of California-Berkeley
- Marshall Slemrod, Department of Mathematics, University of Wisconsin
- Henning Struchtrup Department of Mechanical Engineering University of Victoria
- Paul Tupper, SCCM Program, Stanford University
- Dionisios G Vlachos, Department of Chemical Engineering, University of Massachusetts
- Paul R. Woodward, Department of Astronomy, University of Minnesota
- Darrin York, Department of Chemistry, University of Minnesota

1999–2000	Reactive Flows and Transport Phenomena
2000–2001	Mathematics in Multimedia
2001–2002	Mathematics in the Geosciences
2002–2003	Optimization
2003–2004	Probability and Statistics in Complex Systems: Genomics, Networks, and Financial Engineering
2004–2005	Mathematics of Materials and Macromolecules: Multiple Scales, Disorder, and Singularities
2005-2006	Imaging

IMA SUMMER PROGRAMS

1987	Robotics
1988	Signal Processing
1989	Robust Statistics and Diagnostics
1990	Radar and Sonar (June 18–29)
	New Directions in Time Series Analysis (July 2–27)
1991	Semiconductors
1992	Environmental Studies: Mathematical, Computational, and Statistical Analysis
1993	Modeling, Mesh Generation, and Adaptive Numerical Methods for Partial Differential Equations
1994	Molecular Biology
1995	Large Scale Optimizations with Applications to Inverse Problems, Optimal Control and Design, and Molecular and Structural Optimization
1996	Emerging Applications of Number Theory (July 15–26)
	Theory of Random Sets (August 22–24)
1997	Statistics in the Health Sciences
1998	Coding and Cryptography (July 6–18)
	Mathematical Modeling in Industry (July 22–31)
1999	Codes, Systems, and Graphical Models (August 2–13, 1999)
2000	Mathematical Modeling in Industry: A Workshop for Graduate Students (July 19–28)
2001	Geometric Methods in Inverse Problems and PDE Control (July 16–27)
2002	Special Functions in the Digital Age (July 22–August 2)
2003	Probability and Partial Differential Equations in Modern Applied Mathematics (July 21–August 1)
2004	n-Categories: Foundations and Applications (June 7–18)

IMA "HOT TOPICS" WORKSHOPS

- Challenges and Opportunities in Genomics: Production, Storage, Mining and Use, April 24–27, 1999

- Decision Making Under Uncertainty: Energy and Environmental Models, July 20–24, 1999
- Analysis and Modeling of Optical Devices, September 9–10, 1999
- Decision Making under Uncertainty: Assessment of the Reliability of Mathematical Models, September 16–17, 1999
- Scaling Phenomena in Communication Networks, October 22–24, 1999
- Text Mining, April 17–18, 2000
- Mathematical Challenges in Global Positioning Systems (GPS), August 16–18, 2000
- Modeling and Analysis of Noise in Integrated Circuits and Systems, August 29–30, 2000
- Mathematics of the Internet: E-Auction and Markets, December 3–5, 2000
- Analysis and Modeling of Industrial Jetting Processes, January 10–13, 2001
- Special Workshop: Mathematical Opportunities in Large-Scale Network Dynamics, August 6–7, 2001
- Wireless Networks, August 8–10 2001
- Numerical Relativity, June 24–29, 2002
- Operational Modeling and Biodefense: Problems, Techniques, and Opportunities, September 28, 2002
- Data-driven Control and Optimization, December 4–6, 2002
- Agent Based Modeling and Simulation, November 3–6, 2003

SPRINGER LECTURE NOTES FROM THE IMA:

The Mathematics and Physics of Disordered Media
  Editors: Barry Hughes and Barry Ninham
  (Lecture Notes in Math., Volume 1035, 1983)

Orienting Polymers
  Editor: J.L. Ericksen
  (Lecture Notes in Math., Volume 1063, 1984)

New Perspectives in Thermodynamics
  Editor: James Serrin
  (Springer-Verlag, 1986)

Models of Economic Dynamics
  Editor: Hugo Sonnenschein
  (Lecture Notes in Econ., Volume 264, 1986)

Forthcoming Volumes:

Geometric Methods in Inverse Problems and PDE Control

Mathematical Foundations of Speech and Language Processing

Time Series Analysis and Applications to Geophysical Systems